The Lost Subways of
North America

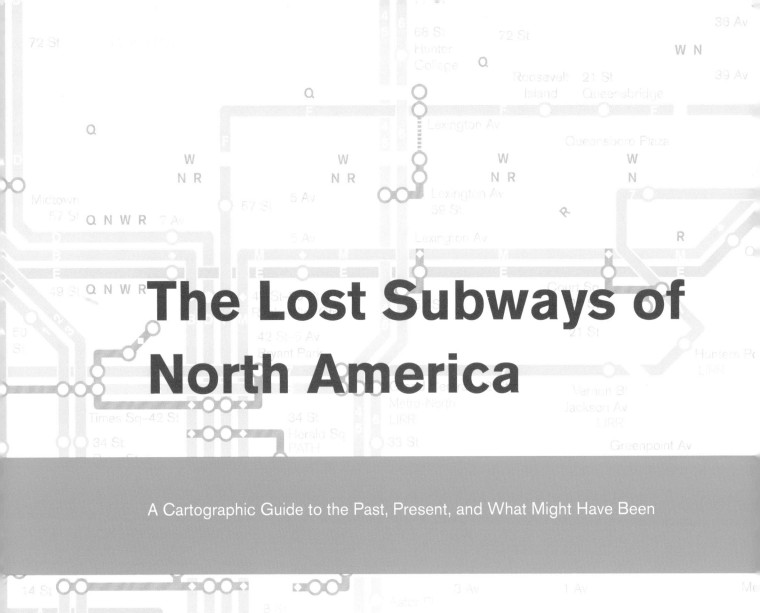

The Lost Subways of North America

A Cartographic Guide to the Past, Present, and What Might Have Been

Written and illustrated by

Jake Berman

The University of Chicago Press
Chicago & London

The University of Chicago Press, Chicago 60637
The University of Chicago Press, Ltd., London
© 2023 by Jake Berman
Published 2023
Printed in Canada

32 31 30 29 28 27 26 25 24   2 3 4 5

ISBN-13: 978-0-226-82979-1 (cloth)
ISBN-13: 978-0-226-82980-7 (e-book)
DOI: https://doi.org/10.7208/chicago/9780226829807.001.0001

Library of Congress Cataloging-in-Publication Data

Names: Berman, Jake, author.
Title: The lost subways of North America : a cartographic guide to the past,
    present, and what might have been / Jake Berman, Jake Berman.
Description: Chicago : The University of Chicago Press, 2023. | Includes bibliographi-
    cal references and index.
Identifiers: LCCN 2023002893 | ISBN 9780226829791 (cloth) | ISBN 9780226829807
    (ebook)
Subjects: LCSH: Subways—United States. | Urban transportation—United States.
Classification: LCC HE308 .B33 2023 | DDC 388.40973—dc23/eng/20230526
LC record available at https://lccn.loc.gov/2023002893

♾ This paper meets the requirements of ANSI/NISO z39.48-1992 (Permanence of Paper).

Nobody's gonna drive this lousy freeway when they can take the Red Car for a nickel.

EDDIE VALIANT, *Who Framed Roger Rabbit*

Contents

Introduction *1*

A Brief Primer on Transit and Urban Development *3*

1 **Atlanta**
The City Too Busy to Hate *7*

2 **Boston**
Urban Institutions, Megaprojects, and City Revival *17*

3 **Chicago**
The Loop Elevated, Beloved Steel Eyesore *29*

4 **Cincinnati**
A Short History of a Never-Used Subway *43*

5 **Cleveland**
Transit and the Perils of Waterfront Redevelopment *49*

6 **Dallas**
They Don't Build Them Like They Used To *59*

7 **Detroit**
The City-Suburban Rift and the Most Useless *67*
Transit System in the World

8 **Houston**
The City of Organic Growth *79*

9 **Los Angeles**
72 Suburbs in Search of a City *87*

10 **Miami**
Overpromise, Underdeliver *103*

11 **Minneapolis–St. Paul**
The Mob Takeover of Twin City Rapid Transit *115*

12 **Montreal**
The Metro as Showcase Megaproject *123*

13 New Orleans
How a Big City Grew into a Small Town   *133*

14 New York City
The Tortured History of the Second Avenue Subway   *143*

15 Philadelphia
How Not to Run a Railroad   *157*

16 Pittsburgh
How to Make Buses Work   *169*

17 Richmond
The First Streetcar System   *177*

18 Rochester
The Only City to Open a Subway, Then Close It   *181*

19 San Francisco
The View from Geary Street   *185*

20 Seattle
Consensus through Exhaustion   *199*

21 Toronto
Subway Line as Political Football   *209*

22 Vancouver
An Exceptional Elevated   *221*

23 Washington, DC
The Freeway Revolt and the Creation of Metro   *231*

Conclusion   *241*

Acknowledgments   *243*
Notes   *245*
Further Reading   *264*
List of Archives Used   *266*
Index   *267*

INTRODUCTION

At 1:40 a.m., Sunday, June 19, 1955, the last regularly scheduled train of the Pacific Electric Railway left the Italianate, marble-floored Subway Terminal, 417 South Hill Street, Downtown Los Angeles. The Pacific Railway Society and the Electric Railroaders Association of Southern California held quiet ceremonies later that day, and the occasion came and went with little fanfare. The *Los Angeles Times* didn't bother to send a reporter. The last train carried a banner that read "To Oblivion."

The Pacific Electric had built the mile-long tunnel and its downtown terminal in the 1920s, at the then-enormous cost of $4 million, and it was believed that the subway would eventually form the basis of a regionwide rapid transit system. At the time, the Pacific Electric stood astride Southern California like a colossus, with over a thousand miles of electric rail service. And transportation wasn't the only thing the Pacific Electric had its tentacles in—the Pacific Electric was also the largest real estate developer in Southern California. Towns kowtowed before the company to attract a station. One town, Pacific City, had even renamed itself Huntington Beach, after the Pacific Electric's founder.

But by 1955, the once-mighty Pacific Electric was broke, and Los Angeles was going full speed ahead with the construction of its freeway system. The Pacific Electric had run out of cheap land to develop, the company hadn't turned a profit in over 10 years, its trains were unreliable at best, and its infrastructure was decrepit. Making matters worse, the Pacific Electric's Red Cars would get stuck in traffic behind the hordes of new motorists who jammed LA's roads during the prosperous 1950s. The last Red Cars ran in 1961.

I originally became aware of this history because of Los Angeles's infamous gridlock. A little over a decade ago, I was living in LA and was caught in an interminable traffic jam on the 101 Freeway. It was a hot summer day. The air conditioning in my car wasn't working particularly well. I had been stopped behind some guy in a Jeep with too many bumper stickers for half an hour. Bored, frustrated, and questioning my decision to leave New York for Los Angeles, my mind began to wander. I asked aloud, to the empty car, "Why doesn't LA have good public transit?"

Unable to figure out a good answer while stuck in traffic, I stopped by the LA Central Library a little while later. There, I stumbled on an ancient map of the Red Car system, showing a spiderweb of electric railway lines extending all across Southern California. In a corner of the map, a long-dead cartographer proudly printed, in all caps, "LARGEST ELECTRIC RAILWAY SYSTEM IN THE WORLD."

The largest electric railway system in the world? In *Los Angeles*?

I had assumed that cars in Los Angeles were just a fact of life, like beaches, palm trees, and tacos. But that wasn't the case at all. Gridlock was a choice that the people of Los Angeles had made.

Los Angeles was far from the only city to radically reshape itself for cars and freeways. Scenes like this repeated themselves across North America, as cities turned against public transportation and embraced the car after World War II. But not every city is the same, and the results in otherwise-similar cities were often dramatically different. Los Angeles would become the poster child for freeways, suburbs, and lousy traffic thanks in part to its experience with the Pacific Electric. In contrast, rival San Francisco opened a municipal streetcar company to challenge its privately owned streetcar monopoly. The city-owned Municipal Railway ultimately outcompeted the privately owned Market Street Railway and bought it out. San Francisco's leaders were thus more receptive to public transit expansion during the freeway-mad 1950s and 1960s. Coupled with a grassroots revolt against urban freeways, the region built the Bay Area Rapid Transit subway system instead.

This book sheds some light on that history through maps of past, present, and never-built transit systems. I examine four types of cities: American cities built in the pre-automobile era, like New York; American cities that came of age in the automobile era, like Houston; Canadian cities, which mostly decided not to build downtown freeways and followed a decidedly different trajectory than American cities; and smaller transit systems of special historical interest, like the abandoned Rochester Subway. I haven't included Mexico, because Mexico's urban history is so unlike that of the United States and Canada. With each city, I've focused on a few key factors that influenced metropolitan transport and land use. But that's not to say that these are the only factors, or that those factors aren't present elsewhere.

A few notes about the maps themselves: I've drawn the maps in this book using period-influenced design and typography, but I have freely used the mapping conventions of modern public transit agencies. Thus, if there is the potential for clarity and geographical accuracy to conflict, I have erred on the side of clarity. I've also used full color, which would have been prohibitively expensive for most of the 19th and 20th centuries. Where sources conflict or I've had to reconstruct station locations from contemporary transit-planning practices, I've noted it. If my original sources don't provide definitive names, I've assigned route designations and labeled stations. Period-correct place names have also been used. For example, the map of the 1962 subway proposal for Washington, DC, has stations on "Jeff Davis Highway," even though the road has since been renamed. All of the maps are also online at www.lostsubways.com.

I hope that you, the reader, feel the same sense of discovery leafing through this as I did when I first stumbled onto that map of the Pacific Electric in a hot Los Angeles summer.

A BRIEF PRIMER ON TRANSIT AND URBAN DEVELOPMENT

Mass Transit Technology

Good mass transit should be fast, frequent, and reliable, and it should go where people want to go. This doesn't require a specific technology, but it does require choosing the right tool for the job. To distinguish these technology types, I've employed a consistent terminology, using appropriate regionalisms. For example, Philadelphia uses the word *trolley*, while New Orleans uses the word *streetcar*.

Subway, *elevated*, and *metro* all describe the same thing: an electric railway system that is fully separated from other traffic, like the New York Subway, the Chicago Elevated, and the Montreal Metro. They are the highest-capacity form of mass transit, but the most expensive to build. *Subway* and *elevated* are general descriptors. Large portions of New York's subway are elevated above city streets, and the Chicago Elevated has underground tunnels. Primitive elevated systems in the late 1800s were pulled by steam locomotives, but all were converted to electric power before World War I.

Light rail is an electric railway system that has dedicated lanes or trackways. Stations average about every half mile in the urban center and about every mile outside city cores. Light rail systems still have some intersections with cross traf-

fic. This makes light rail slower than subways, but cheaper to build. Some light rail systems are brand new, like Seattle's. Others, like Pittsburgh's, are upgrades of legacy trolley systems.

Interurbans, the predecessors of light rail, provided similar service deep into suburban and rural areas. Unlike light rail, interurbans often lacked dedicated lanes within city centers. The interurban is extinct, except for a handful of lines in Chicago and Philadelphia. The most famous interurban system is Los Angeles's old Red Car system, which was four times the size of today's London Underground.

Streetcars and *trolleys* are electric railway systems that provide local transportation. They run in normal traffic, stopping every few blocks. Their golden age was in the 1920s, before the automobile and motorbus were mature technologies. Most streetcar networks closed in the mid-20th century, but surviving systems still run in New Orleans and Toronto. A few small streetcar networks also opened in the late 20th and early 21st centuries to promote downtown development. A related technology is the electric *trolleybus*, an electric bus powered by streetcar-style overhead wire. Common in Europe, trolleybuses are rare in North America.

The *horsecar* and *cable car* were the predecessors of the streetcar. Horsecars, dragged by horses or mules, are extinct. Cable cars are pulled by a continuously running mechanical cable. They survive only in San Francisco.

Regional rail and *commuter rail* primarily serve workers going to and from downtown. Service is frequent during rush hours and spotty at other times. Busy regional rail systems can be upgraded to metro-like standards, as is currently underway in Montreal and Toronto.

Monorails are electric rail systems that run on one rail. Monorails were in vogue in the 1950s and 1960s, but they offer few advantages over traditional metro or light rail systems. Today, the busiest North American monorail is at Disney World.

People movers are low-capacity, fully automated electric rail lines. A few people movers are used as urban transport, like the Miami Metromover. But they are most commonly found at airports and tourist attractions, like the Atlanta airport's Plane Train, or the Hogwarts Express at Universal Studios Florida. People movers are the lineal descendant of the *pod car*, a sort of horizontal elevator.

Busways, also known as *bus rapid transit*, are a hybrid between light rail and the humble city bus. A fully built busway system has dedicated lanes, ticket machines, dedicated platforms, and priority at traffic signals. When done well, like the G Line in Los Angeles, busways provide nearly as good service as light rail at a fraction of the cost.

Freeways, *expressways*, and *turnpikes* are the high-speed roadways that were built after World War II. They make up the backbone of the North American transportation network today.

Land Use Regulation and Real Estate Development

Land use policy and transit policy can't be decoupled. High-capacity transit works best when it serves the places where lots of people live and work. This, in turn, is heavily influenced by local *land use laws*. Land use laws define, at a granular level, which buildings are legal to build and live in, which businesses can operate, the amenities required, and so on. These laws subtly influence every aspect of daily life.

Before the 20th century, land use laws tended to define the size and scale of buildings, and set out health and safety requirements. But early land use laws generally did not restrict what could be carried out where. It was entirely normal for businesses, single-family homes, boardinghouses, and warehouses to coexist on the same block.[1] This all changed in the first decades of the 20th century. In 1916, New York was the first major city to establish a comprehensive *zoning law*. This zoning law didn't just limit building size. It also limited what activities were legal on particular pieces of land.

The most common type of North American zoning law strictly separates residential, commercial, and industrial *uses* of land. This means, for example, that one can't build a house in a commercial zone or open a store in a residential zone. After New York introduced zoning, these laws spread quickly throughout North America. In the United States, the Supreme Court held that zoning was constitutional in 1926.[2] Between the world wars, zoning laws guided urban growth, but generally did not limit it. For example, in 1935, Los Angeles's apartment zones could theoretically accommodate 22 million inhabitants, in a city with 1.2 million residents.[3] In this period, the intensity of development largely matched the value of the land.

In New York, one can see the results of interwar zoning policy by following the Brooklyn-bound A train from downtown Manhattan. The Chambers Street station serves the World Trade Center and other skyscrapers. Four miles outbound at Nostrand Avenue, the neighborhood is mostly four- to six-story apartment buildings and rowhouses. Four miles further at Euclid Avenue, the apartment buildings and rowhouses are two to four stories tall. At Lefferts Boulevard, the end of the line, the housing is mostly single-family homes with off-street parking.

After World War II, zoning laws gradually grew more restrictive. Most modern zoning laws provide exacting specifications for what can and can't be done on particular pieces of land. Zoning also regulates lot sizes, the dimensions of buildings, off-street parking requirements, and the kinds of activities allowed. To illustrate, Levittown, New York, half an hour east of Lefferts Boulevard, was built after World War II. Levittown looks much the same as it did in the mid-20th century. The building stock is mostly 70-year-old, two-story tract homes with yards and low-slung strip malls. Levittown is prosperous. Its median household income in 2020 was $121,260, 70 percent higher than the national average. Despite this wealth, population growth has been stagnant since 1990 because the local zoning law bans denser development.[4]

But it's hard to make broad generalizations on this subject. The difficulty of developing real estate varies wildly, even within a metropolitan area. For example, in Emeryville, California, a suburb of San Francisco, zoning regulations are permissive. It's straightforward to get approval for new buildings. The city of Lafayette, 10 miles east of Emeryville, is quite the opposite. Since 2011, Lafayette's residents have blocked housing construction on a vacant lot near the city's subway station. The political fight has been brutal, making headlines as far away as New York.[5] If there's ever a way to illustrate the axiom that "all politics is local politics," it's in land use.

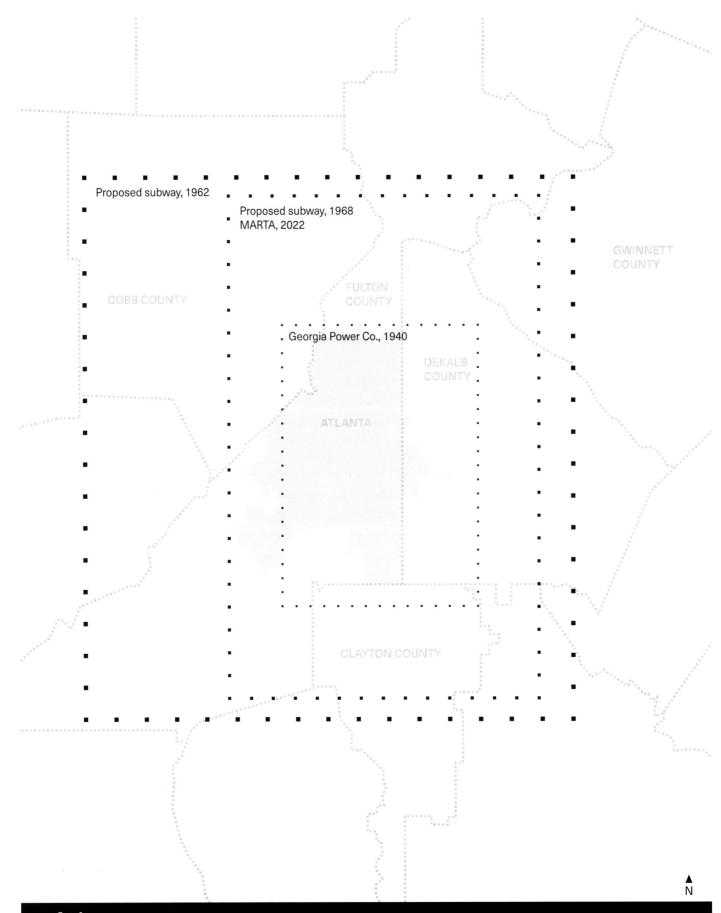

Proposed subway, 1962

Proposed subway, 1968
MARTA, 2022

Georgia Power Co., 1940

GWINNETT
COUNTY

COBB COUNTY

FULTON
COUNTY

DEKALB
COUNTY

ATLANTA

CLAYTON COUNTY

N

Atlanta

5 miles/8 km

1

ATLANTA

The City Too Busy to Hate

1.1 Map of greater Atlanta.

There's a joke I heard when I was last in Georgia: "In Macon, they ask you where you go to church. In Savannah, they ask you what you'd like to drink. And in Atlanta, they ask you what your business is." Pithy, to be sure, but it's also representative of the image Atlanta (fig. 1.1) projects to the world. For decades, Atlanta sold itself as the most Northern of Southern cities. It proudly branded itself "the city too busy to hate," open to modernity in a very un-Southern way. The truth is more complicated. Race has played a major role in Atlanta transit policy, especially during the mid-20th century, when the Metropolitan Atlanta Rapid Transit Authority (MARTA) was planning to build a subway.[*]

Atlanta has always been an atypical Southern city. In most of the Old South, the economy depended on slave agriculture. Entrenched plantation aristocrats ruled. But not in Atlanta. The city's economy depended on the railway in the 19th century. As such, Atlanta has always had a strong Northern influence and a flair for self-promotion. As W. E. B. Du Bois wrote in 1906, Atlanta is "South of the North, yet North of the South."[1]

Atlanta grew rapidly in the late 19th and early 20th centuries. By 1940, it was the second-largest city in the South. It had a large streetcar system run by Georgia Power, the local electric utility. (Typical for the period,

---

[*] Atlanta is hardly the only city where the tenor of race relations played a major role in transport policy. For other examples, see chapter 7 on Detroit and chapter 23 on Washington, DC.

Georgia Power's streetcar system was racially segregated.) After World War II, large numbers of people moved to the Sun Belt, and Atlanta took full advantage. But Atlanta's business and political elite approached postwar growth differently than other Southern cities. In the early 1960s, Atlanta's elite were pushing for both expressways and a modern subway system. Voters approved the subway in 1971 after major political difficulties. MARTA's first rail line opened in 1979, but the system ended up much smaller and less comprehensive than the original proposals. This is because of the interplay between race, transport, and money.

Gate City of the New South

Atlanta's origin is a terminus, with apologies to Oscar Wilde. In 1836, the State of Georgia approved the construction of the Western and Atlantic Railroad. That railroad ran from central Georgia to the port town of Chattanooga on the Tennessee River. The town that grew up around the final rail station was named Terminus, Latin for "a boundary, limit, or end." Other railways soon converged on the little town. By the eve of the Civil War in 1860, that town, now renamed Atlanta, was the railway hub of the South. When Atlanta fell to federal troops under General William Tecumseh Sherman in 1864, it sealed the fate of the rebellion.

Atlanta's culture of boosterism and its appreciation of the nouveau riche appeared early in its history. Atlanta's first news magnate, Henry Grady, of the *Atlanta Constitution*, promoted the city as the "Gate City of the New South" in the decades after the Civil War. Grady sold Atlanta as a place

for industry, commerce, and the white middle class, as contrasted with the "Old South" of slavery, treason, and oligarchy. Atlanta openly courted Northern money and Northern transplants. The results were growth and prosperity.[2] By the 1880 census, Atlanta had passed Savannah as the largest city in Georgia. The electric streetcar was introduced in the 1890s, and the streetcar suburb soon became the primary form of urban expansion. By 1940, Atlanta was the South's second-largest city, behind only New Orleans.

Georgia Power's streetcar network covered nearly all of what is considered the urban core of greater Atlanta today (fig. 1.2). Georgia Power was Atlanta's electric utility, then as now. This combination of businesses was common in the early 20th century, though to a 21st-century observer the combination of transit and electricity might seem strange. Transit and electricity were linked because streetcar infrastructure was also used to connect neighborhoods to the power grid. The Wheeler-Rayburn Act of 1935 disrupted this business model. As an antitrust measure, Wheeler-Rayburn forced most electric utilities to divest from their transit operations.[3] The last trains ran in 1949. Trolleybuses replaced most of the streetcars.

"We Are Simply Used to Winning"

After World War II, cheap residential air conditioning turned Atlanta into a boomtown. Like Houston and Dallas, Atlanta was ideally positioned to take advantage of postwar auto-oriented suburbanization, and it made plans for a large expressway system. But in the 1950s and 1960s, At-

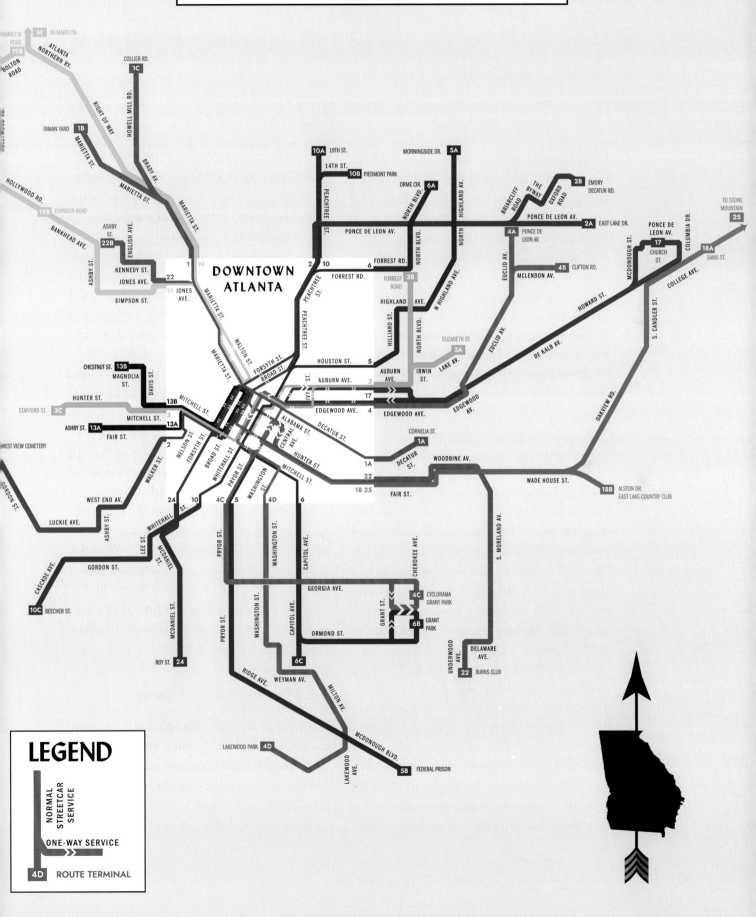

TRANSPORTATION MAP
of Atlanta
SHOWING STREETCAR LINES OF THE GEORGIA POWER CO.
REVISED FOR 1940

LEGEND

NORMAL STREETCAR SERVICE

ONE-WAY SERVICE

4D ROUTE TERMINAL

lanta's regional planning commissions realized that expressways alone wouldn't be enough to handle anticipated growth.[4] These commissions believed that a subway was a necessary complement to the under-construction expressway network.

The subway proposals capitalized on Atlanta's image as the most racially progressive city in the former Confederacy. Atlanta was, in its boosters' words, "the city too busy to hate." At the time, the city was in a state of racial détente. An alliance between the white business elite and middle-class blacks dominated city politics. Broadly speaking, the white business class agreed to back gradual desegregation, and in return, black voters backed the elites' economic development plans.[5] MARTA was one such plan.

Early MARTA plans were meant to cover all five central counties of metropolitan Atlanta: Fulton, DeKalb, Clayton, Gwinnett, and Cobb. (Atlanta proper straddles Fulton and DeKalb.) In the early 1960s, Atlanta's white political and business class lined up behind the subway project, just as it had with all of Atlanta's other civic improvements (fig. 1.3). With the plans in hand, and after three hard-fought elections, the state legislature created MARTA in 1965.

By 1965, the local political dynamic in Atlanta had changed dramatically from just 10 years before. The civil rights movement was in full swing, and the era of legal segregation had just ended. Atlanta's buses were desegregated in 1959, after a two-year court fight. White ridership dropped by double digits.[6] With the end of legal segregation, whites left central Atlanta neighborhoods en masse. Large numbers of these whites moved to suburbs opened to settlement by the new expressways. Expressways were routed through black neighborhoods like Sweet Auburn, immediately east of downtown, and Summerhill, south of downtown, so suburban, majority-white auto commuters could reach downtown Atlanta. Atlanta was not unique in this regard. Black Bottom in Detroit was demolished to build I-375; in the San Francisco Bay Area, the elevated Cypress Freeway cut West Oakland in two; and Chicago's South Side was divided by the Dan Ryan Expressway.

Affluent, white, and reactionary Cobb County, northwest of downtown Atlanta, opted out of MARTA immediately.[7] Cobb County's decision came as no surprise, as Cobb's favorite son was virulent segregationist Lester Maddox. Maddox got his big break in politics by openly defying the Civil Rights Act, refusing to serve blacks at his diner near Georgia Tech in 1964. His defiance made newspapers from Washington State to Florida. Like his contemporary George Wallace in Alabama, Maddox rode a pro-segregation platform all the way to the governor's mansion.[8]

These changing political winds should have given pause to Atlanta's downtown establishment. But the city's business and political elites were used to pliant voters who would approve ambitious civic improvements on speculation. After all, when it came time to attract major league sports to Atlanta in 1964, Mayor Ivan Allen proposed to "build a stadium on land we don't own with money we don't have for a team we don't have."[9] By 1967, the stadium was finished and Major League Baseball, the National Football League, and the North American Soccer League had teams playing there.

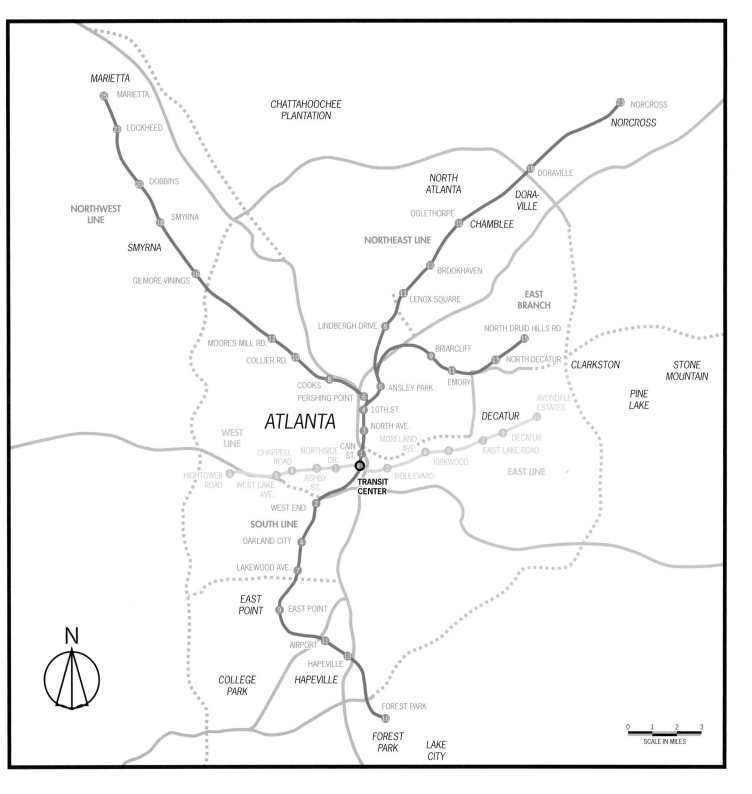

PROPOSED RAPID TRANSIT SYSTEM

Metropolitan Atlanta Study Commission
December 1962

② MINUTES FROM TRANSIT CENTER

━━●━━ EAST-WEST LINE

━━◆━━ NORTH-SOUTH LINE AND BRANCHES

━━━ EXISTING EXPRESSWAY

••••• PROPOSED EXPRESSWAY

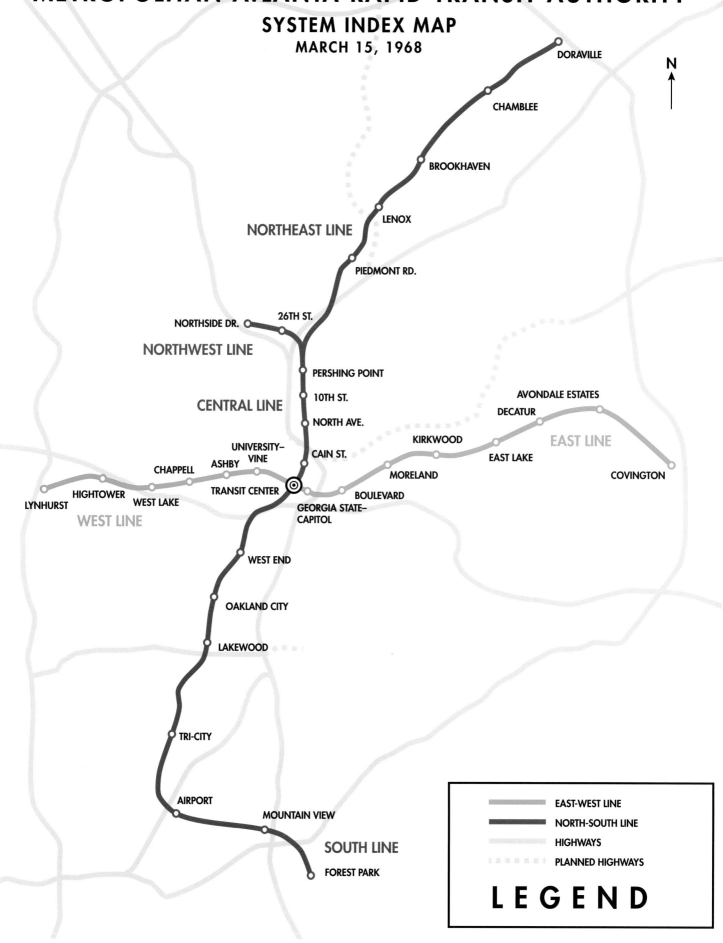

METROPOLITAN ATLANTA RAPID TRANSIT AUTHORITY
SYSTEM INDEX MAP
MARCH 15, 1968

N

DORAVILLE

CHAMBLEE

BROOKHAVEN

NORTHEAST LINE

LENOX

PIEDMONT RD.

NORTHSIDE DR.   26TH ST.

NORTHWEST LINE

PERSHING POINT

10TH ST.

CENTRAL LINE

NORTH AVE.

AVONDALE ESTATES

DECATUR

CAIN ST.

UNIVERSITY–
VINE

KIRKWOOD

EAST LINE

ASHBY

MORELAND

EAST LAKE

COVINGTON

CHAPPELL

HIGHTOWER

WEST LAKE

TRANSIT CENTER

BOULEVARD

LYNHURST

GEORGIA STATE–
CAPITOL

WEST LINE

WEST END

OAKLAND CITY

LAKEWOOD

TRI-CITY

AIRPORT

MOUNTAIN VIEW

SOUTH LINE

FOREST PARK

LEGEND

▬▬▬	EAST-WEST LINE
▬▬▬	NORTH-SOUTH LINE
▬▬▬	HIGHWAYS
┈┈┈	PLANNED HIGHWAYS

1.4 Proposed system rejected in the first
     MARTA referendum, 1968.

In 1968, Mayor Allen and his downtown allies charged forward with a referendum to build a MARTA subway in the four remaining counties, Fulton, DeKalb, Clayton, and Gwinnett (fig. 1.4). It should come as no surprise that MARTA's backers expected to lose badly in the mostly white suburbs and to make up the difference with black and liberal white votes in the urban core. (Suburban racists have joked for decades that MARTA stands for "Moving Africans Rapidly through Atlanta.")[10] As expected, suburban whites opposed MARTA. But urban blacks opposed it as well. Black community leaders had warned MARTA backers, to no avail, that black voices had been excluded from the planning process, that MARTA needed to provide better rail service to Atlanta's black neighborhoods, and that bus improvements needed to be part of any referendum. The 1968 referendum went down in flames, losing by 10 points. In the aftermath, one of the directors of Atlanta's Chamber of Commerce privately wrote to MARTA's chair: "None of us should be ashamed of losing. Our only problem is we are simply used to winning."[11]

Back to the Drawing Board

After the loss, the MARTA board and its allies conducted a postmortem. The 1968 referendum had failed because of three reasons, in their estimation: not enough input from Atlanta's black community; not enough immediate improvements to the existing transit system; and worries about financing, as the 1968 proposal was slated to be financed with property taxes.[12]

To deal with the first issue, pro-MARTA forces conducted aggressive outreach to voters all over greater Atlanta, but in particular to black voters. The Committee for Sensible Rapid Transit, which organized the pro-MARTA campaign, was deliberately set up with black and white cochairs. MARTA mounted a professional public relations campaign, holding public forums in all of Atlanta's neighborhoods. A third of these forums took place in black districts.

To respond to the demand for immediate improvements, MARTA proposed to take over and modernize the privately owned Atlanta Transit System. Atlanta Transit System had bought the transit system from Georgia Power, replacing the old trolleybuses with motor buses. Black voters' desires for immediate bus improvements were addressed directly. MARTA promised more frequent bus service, more bus shelters, better coverage in minority neighborhoods, better transit information services, lower fares, and new air-conditioned buses. (Atlanta averages 90 degrees Fahrenheit and 74 percent humidity in July, so air conditioning was a major improvement.)

The importance of the efforts to reduce fares should not be understated, either. Atlanta Transit System's one-way bus fare was $0.60. MARTA's supporters promised to lower the fare to $0.15. In inflation-adjusted 2022 dollars, that's a drop from $4.20 to $1.03.

Last, the financing plan was changed. Rather than use property taxes to finance the system, which suburban officials opposed, the 1971 plan called for increasing the sales tax.[13] Lester Maddox, by that time serving as lieutenant governor, fought

13

it to the end. Maddox himself hamstrung the system by blocking the state government from subsidizing MARTA's day-to-day transit operations. This policy has remained solidly in place for half a century.[14]

A Referendum on Race

When election day came in 1971, the pro-MARTA forces got the system approved, but only by a razor-thin margin and only in the urban counties. In DeKalb County, the referendum won by 3,358 votes out of 70,000 cast. In Fulton County, the margin of victory was 471 votes, with 106,000 cast. The two suburban counties, Clayton and Gwinnett, voted against the referendum by a 3–1 margin. The Clayton and Gwinnett sections were canceled. The greatly reduced rail system, operating in only DeKalb and Fulton, opened in 1979 (fig. 1.5).

It's hard not to see the role of race in the two-decade-long fight to establish the MARTA system. As an *Atlanta Magazine* writer put it in a retrospective: "Votes against MARTA by residents of Cobb, Clayton, and Gwinnett weren't votes about transportation. They were referendums on race. Specifically, they were believed to be about keeping the races apart."[15] Segregationists like Lester Maddox fought an urban rapid transit system tooth and nail, while Georgia threw billions of dollars into new expressways. Between 1960 and 1970, the city of Atlanta lost 20 percent of its white population, and the black population grew to be a majority of the city. The suburban counties absorbed this white flight.

These decisions made in the 1960s and 1970s still have ramifications half a century later. Atlanta is the second most segregated metropolitan area in North America by some measures.[16] For the first 50 years of its existence, MARTA didn't get a cent of support from the State of Georgia.[17] When the state appropriated money for MARTA for the first time in 2021, it was considered newsworthy.[18] (The $6 million state outlay was 0.4 percent of MARTA's $1.3 billion budget that year.)[19] While Clayton County joined the MARTA district in 2014, there is no money to extend the subway there now. Cobb and Gwinnett have remained intransigent.

Thanks to the white voters of greater Atlanta trying to keep blacks out of the suburbs in the mid-20th century, contemporary Atlanta has some of the continent's worst traffic congestion and smog. There's no good way to attack the problem because most of the metropolis's physical infrastructure is built to serve only the car. To paraphrase William Faulkner: "The past isn't dead and buried. In fact, it isn't even past."

1

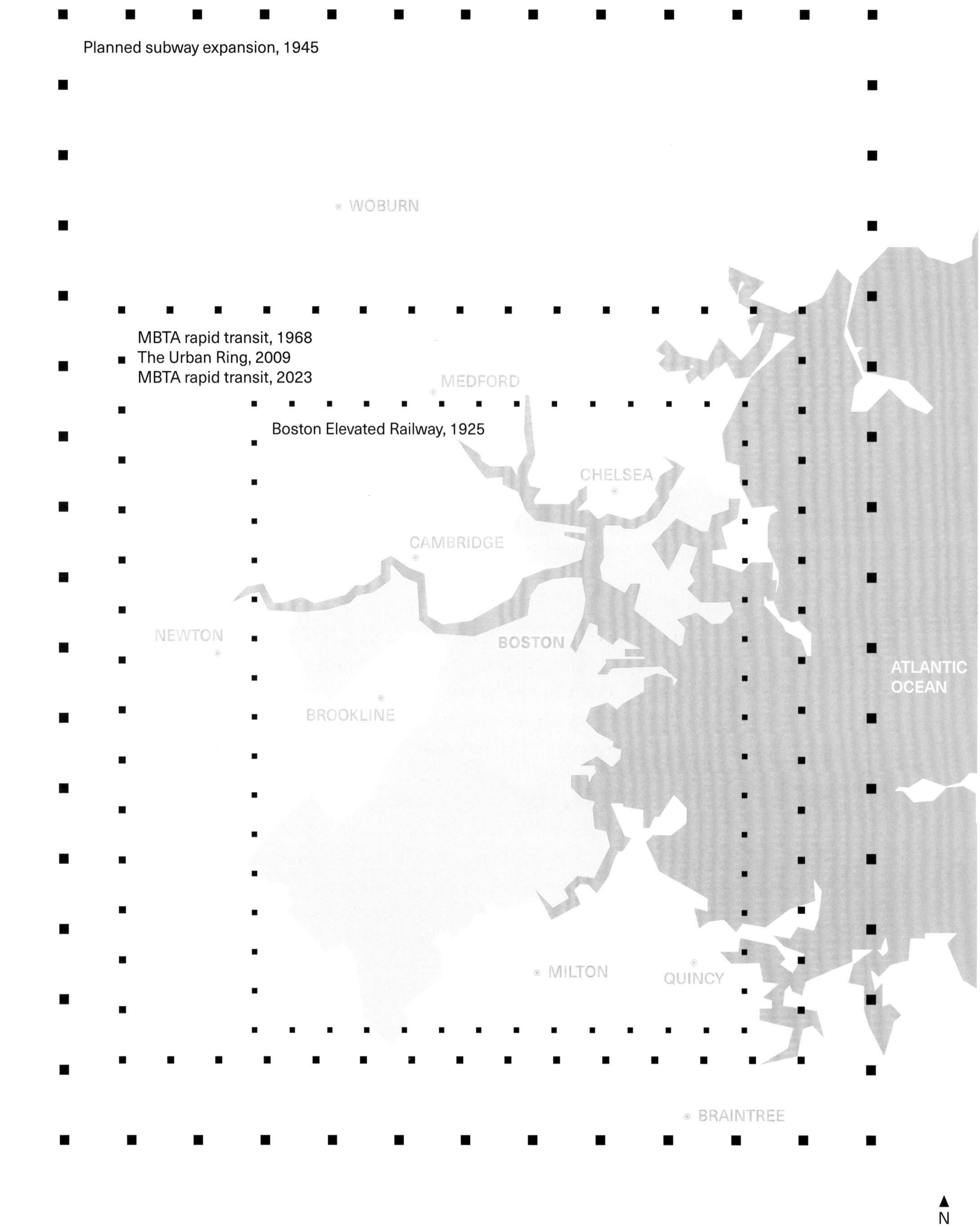

Planned subway expansion, 1945

WOBURN

MBTA rapid transit, 1968
The Urban Ring, 2009
MBTA rapid transit, 2023

Boston Elevated Railway, 1925

MEDFORD

CHELSEA

CAMBRIDGE

BOSTON

NEWTON

ATLANTIC
OCEAN

BROOKLINE

MILTON

QUINCY

BRAINTREE

N

Boston

2 miles/3.2 km

2

BOSTON

Urban Institutions, Megaprojects, and City Revival

2.1 Map of greater Boston.

In the 19th century, Boston was second only to New York as the financial and shipping capital of North America. It was the economic hub of New England. This prosperity led to large-scale infrastructure investments of all kinds, including the first underground electric railway in North America. Much of this wealth was also invested into the metropolis's educational, medical, and scientific institutions, creating the continent's largest concentration of institutions of higher learning. That period of prosperity was beginning to end by the 1920s. Never a major industrial center, Boston (fig. 2.1) entered 30 years of stagnation, followed by 30 years of decline. Over those six decades, Boston's leaders were willing to try almost anything to stanch the bleeding. This included large-scale investment in the region's transit infrastructure. But nothing seemed to work.

Then a very strange thing happened: Boston became hot stuff again in the 1980s. The national economy shifted toward financial services, research, and high technology. This played to the strengths of Boston's legacy institutions. Boston had no shortage of banks, insurers, hospitals, and universities. Its well-educated workforce was perfectly placed to take advantage. Greater Boston came back roaring, and it has never looked back. Boston is once again a flourishing metropolis in the early 21st century. But even as the region has used its legacy institutions to return to prosperity, Boston's state and local governments have been unable to deliver new public infrastructure to match. Much-needed transport projects have regularly turned

into boondoggles, and the procurement and construction bureaucracies are in dire need of reform.

Yankee Ingenuity

Plentiful water power and Yankee ingenuity made
New England into one of the world's first industrialized regions. New England pioneered the mass
production of watches, textiles, apparel, and shoes,
and Boston was its hub. But Boston's role in this
economic system was not to produce goods. The
factories themselves tended to be in satellite cities
like Lowell, Waltham, and New Bedford. Rather,
Boston provided the financial, legal, logistical, and
transport services for New England's factories.

Boston's inner-ring suburbs developed around
horsecar and railway lines in the late 19th century,
and around streetcar and subway lines in the early
20th century. The first underground rapid transit
in North America, which now carries the Massachusetts Bay Transportation Authority (MBTA)
Green Line, opened in 1897. Three decades of extensive subway and elevated expansion followed.
By 1925, widespread tunnel construction through
downtown Boston had removed streetcars from
downtown Boston's streets entirely. The rapid transit system had assumed a form recognizable to Bos-
tonians today, with a quartet of downtown stations
making up the system's core: Park Street; Scollay
(now Government Center); Winter-Summer/
Washington (now Downtown Crossing); and Dev-
onshire/Milk-State (now State) (fig. 2.2). This network was unified in the early 20th century under
a privately owned monopoly, the Boston Elevated
Railway.

At the 1920 census, Boston proper had a population of 748,000. Between 1920 and 1950, growth
was basically stagnant. The population increased at
a rate of only 0.2 percent per year, far below the
national average.[1] During World War II, a blue-
ribbon commission led by Senator Arthur Coolidge
suggested that the key to reviving Boston was infrastructure (fig. 2.3).[2] There would be new freeways
to relieve traffic congestion on Boston's infamously
tangled streets. Large-scale mass transit extensions
would bring commuters downtown. Aging streetcars and unsightly elevated lines would be replaced
with modern buses and subways. In particular, the
elevateds were widely viewed as causing blight,
no matter how useful they were as transport. The
Commonwealth of Massachusetts would take over
the Boston Elevated Railway system, giving control
to the predecessor of the MBTA. Out with the old,
in with the new, was the thinking.

"What the Hell Is That?"

Boston's wholesale urban renovation went full
speed ahead after the war. A third of Boston's historic core was torn down. The working-class residents of the racially integrated West End were
evicted. Their neighborhood was demolished and
replaced with tower blocks. Between 1950 and
1970, the West End's population dropped by 85
percent. The same happened in the historic but
decaying Scollay Square district, where another
20,000 working-class residents were removed to
build government office buildings.[3]

The focal point of the Scollay Square redevelopment was a new city hall, built from blocky con-

2

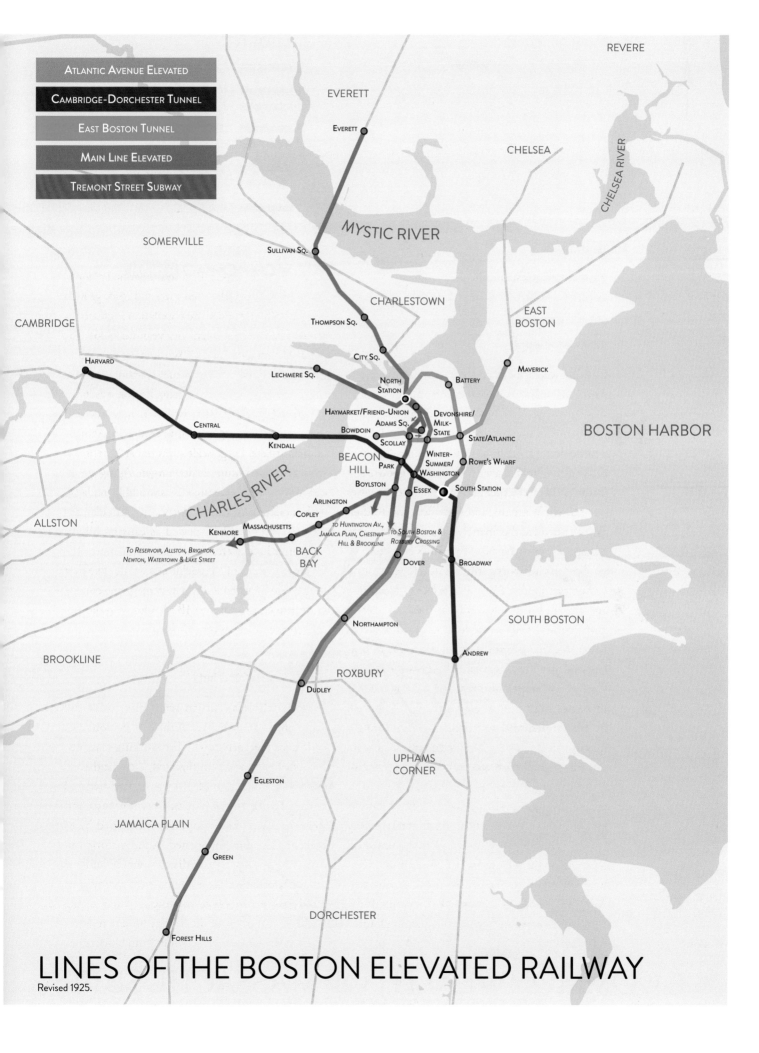

LINES OF THE BOSTON ELEVATED RAILWAY

Revised 1925.

COMMONWEALTH OF MASSACHUSETTS

PRESENT RAPID TRANSIT SYSTEM AND PROPOSED SUBURBAN EXTENSIONS

APRIL, 1945

2.3 Coolidge Commission transit
   expansion proposal, 1945.

crete. Its brutalist design pleased few, except the architectural community and the politicians who had commissioned it in the first place. In "Proudest Achievements of American Architecture in the Nation's First 200 Years," the American Institute of Architects ranked the new city hall eighth best, two places ahead of the Brooklyn Bridge.[4] Robert A. M. Stern, dean of Yale's architecture school, remarked that the building allowed Boston to be "seen as the great new urban experimental center, where new work could go side by side with Faneuil Hall."[5] But public opinion tended to agree with the dignitary who remarked at the unveiling of the building's architectural plans, "What the hell is that?"[6] Even today, Boston City Hall is popularly known as "the world's ugliest building."[7]

The same brute-force redevelopment happened with the transportation network. The commonwealth of Massachusetts rammed a two-decked aboveground freeway called the Central Artery through downtown Boston. The commonwealth also extended the Massachusetts Turnpike to downtown by building a trench directly through densely populated Newton and Brighton. The old urban elevated lines were dismantled. The elevateds were replaced with modern subway lines, mostly as proposed by the Coolidge Commission (fig. 2.4). Suburban subway extensions were built with large garages so commuters could park and ride.

Boston tired early of this postwar urban renewal. The downtown extension of the Massachusetts Turnpike caused a local political firestorm. In response, Governor Francis Sargent put a moratorium on all new freeway construction in central

Boston. Similarly, transformative changes to the metropolis's transit network were shelved in favor of incremental improvements. A pod-car system encircling downtown Boston, adding a circumferential line to the existing hub-and-spoke subway system, never even left the drawing board.[8] The same fate befell a mid-1970s plan to convert most of the MBTA's commuter rail lines to carry subway trains.

All these alterations to greater Boston's urban fabric couldn't save the metropolis from the changes in the greater economy. The region continued to lose ground. Between 1950 and 1980, greater Boston's population growth rate was half the national average. Deindustrialization, the population exodus to the Sun Belt, and the invention of air conditioning all took their toll. The textile, leather, and clothing industries, which had been the backbone of Massachusetts's economy for over a century, decamped for the South. By March 1976, Massachusetts's unemployment rate was 38 percent higher than the national average. At the 1980 census, Boston proper had lost a quarter of its population from 60 years earlier—a percentage drop worse than Detroit.

Boston's problems were especially complex in the mid-1970s, because the city was subject to a wide-ranging school desegregation program that further destabilized the urban core.[9] The program bused students from white-majority neighborhoods to black-majority schools, and vice versa. Busing happened all over the United States, but the reaction was unusually strong in Boston. At newly integrated schools in white districts, protesters pelted school buses with rocks, and white par-

21

massachusetts bay transportation authority

rapid transit service
effective july 1967

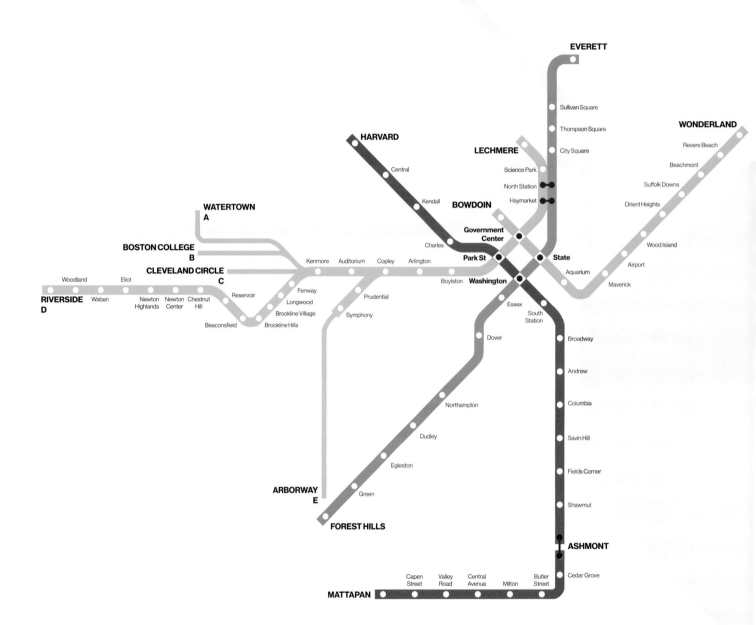

red line subway

orange line subway

green line streetcar

blue line subway

2.4 MBTA subways and light rail, 1967. The old
Washington Street Elevated carried the Orange
Line from Dover station to Forest Hills station.

ents refused to send their children to class. Rioting
was widespread. Forty riots took place between
1974 and 1976 alone.[10] Governor Sargent had to de-
ploy the National Guard. One city official involved
in implementing the busing plan later remarked,
"There wasn't a lot of education to be had during
that period of time. It was basically just trying to
keep kids from being killed."[11] Thousands of urban white parents decided to send their children
to private schools or moved to the suburbs instead,
accelerating the city's decline.

The Massachusetts Miracle

A savvy observer in 1980 could be forgiven for
thinking that Boston would continue to deterio-
rate like cold Northern cities. In the 1970s, unemployment was persistently above the national average. But in the 1980s, computers, biotechnology,
health care, and finance began to take on a bigger
role in the national economy. Greater Boston was
well placed to capitalize on this shift because of
the quantity and quality of its higher education,
medical, and research institutions. Additionally,
due to its mercantile past, Boston had a strong,
well-developed financial services sector. By late
1987, unemployment was 45 percent below the
national average. This turnaround was called the
Massachusetts Miracle.

With the return of prosperity, greater Boston
could afford to think big again about transportation. In the 1980s, the commonwealth proposed
closing the aboveground Central Artery. The Artery would be replaced with a new freeway tunnel
and a new rail tunnel linking North Station and

South Station, Boston's two commuter rail terminals. The rail tunnel would allow commuter trains
to run from one end of greater Boston to the other,
like Philadelphia's Center City Commuter Connection, instead of having all trains terminate at
dead ends downtown. This megaproject was uni-
versally known as the Big Dig.[12] In addition, the
1909-vintage Washington Street Elevated, which
carried the Orange Line south of downtown, was
slated for demolition. The elevated would be replaced with a new branch of the Green Line light
rail on Washington Street. The Orange Line would
be shifted into a trench half a mile to the west.

The Spirit Is Willing, but the Bureaucracy
Is Weak

Ironically, greater Boston's need for more transport
infrastructure came just as its institutional capacity
to build it radically decreased. The Big Dig cost
over five times its original estimate, and construction alone took a quarter century.[13] The freeway
tunnel cost overruns made it financially impossible
to build the rail tunnel. The rail tunnel was canceled.[14] The Washington Street Elevated was torn
down, and the relocated Orange Line opened at
reasonable cost in 1987. However, the Washington
Street Green Line extension became mired in political infighting and cost inflation. Unable to deliver on the promise of a light rail line, the MBTA
suggested a busway instead.

But the MBTA's busway alternative wasn't a
true busway. It was just a rebranding of an existing
bus line.[15] A well-designed busway has dedicated
bus lanes, ticket machines at stations, boarding at

23

2.5 MBTA rapid transit, with
proposed Urban Ring busway and
Silver Line bus tunnel, 2009.

all bus doors, priority at traffic lights, and stops about every half mile. The MBTA's proposal, called the Silver Line, had none of these things. The neighborhood's state representatives dubbed it "the Silver Lie."[16] A decade of political warfare resulted. In the end, the slow, underwhelming Silver Line opened in 2001, 14 years after the closure of the Washington Street Elevated. Later in the decade, a project to build a busway ring around central Boston fell through entirely. So did the bus tunnel needed to run the Silver Line quickly through downtown Boston (fig. 2.5).

In the early 2020s, costs are still out of control. For example, when the MBTA extended the Green Line north of Lechmere in 2022, it cost twice as much per mile as comparable projects did a generation ago (fig. 2.6).[17] Some of this is due to the interminable bureaucracy involved in funding such projects. But it's also because mass transit projects are a useful way to deal out political pork. The new Green Line stations were overdesigned, large, and expensive, suited more to a subway line than a simple light rail line. Most of Boston's aboveground light rail stations are simple, including only a platform, tracks, a shelter, and a ticket machine. At the end of the day, the MBTA made the necessary cuts, but only after $700 million had been burned in the first iteration of the design.[18]

The core of the issue is that the MBTA lacks the administrative expertise and institutional knowledge to get things done cheaply and quickly in the 21st century.[19] (New York's Metropolitan Transportation Authority has similar problems but at an even larger scale.) This is in part because the MBTA is penny-wise and pound-foolish. For instance, the MBTA's in-house bureaucracy is short-staffed by international standards. This makes the MBTA overly dependent on outside project management consultants who have no vested interest in keeping budgets low. Exacerbating the problem, MBTA in-house planners' salaries are tens of thousands of dollars less than comparable wages in the private sector. Brain drain is endemic, as staff tend to leave for private industry after a few years.[20]

These issues are fixable. Reforming the MBTA bureaucracy should not be an impossible task for a wealthy, successful metropolis. In decades past, Boston knew how to deliver major public infrastructure quickly and at reasonable cost. Recreating this capability in the 21st century requires greater Boston to invest in developing functional institutions of government. Boston should be no stranger to these types of investments. After all, Boston's revitalization happened because previous generations poured enormous resources into creating lasting institutions of research, education, health care, and finance.

MBTA RAPID TRANSIT

SERVICE AFTER CONSTRUCTION OF URBAN RING PHASE 2, SILVER LINE PHASE 3 AND GREEN LINE EXTENSION AS PLANNED, JANUARY 2009

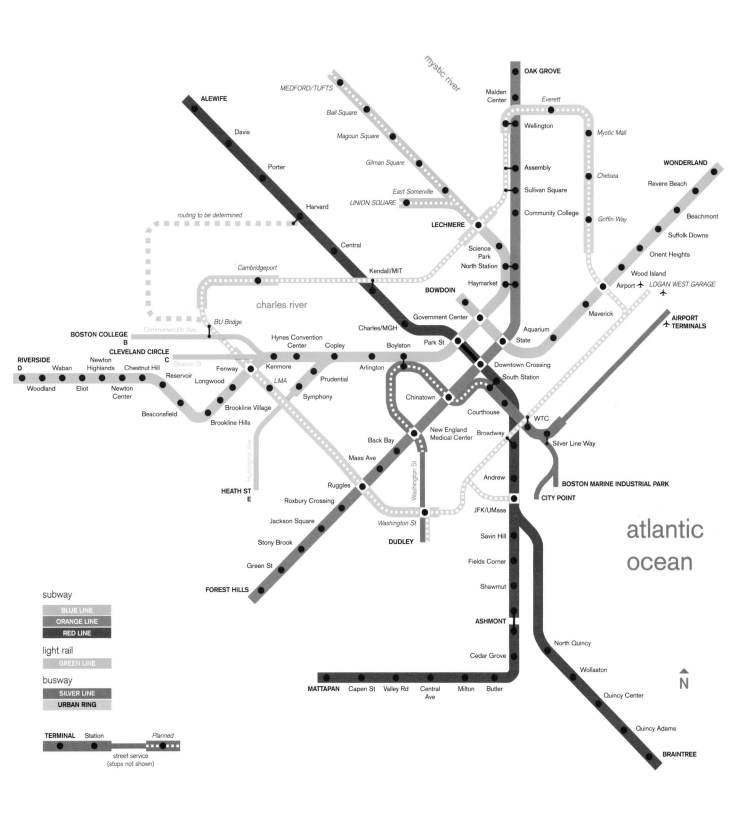

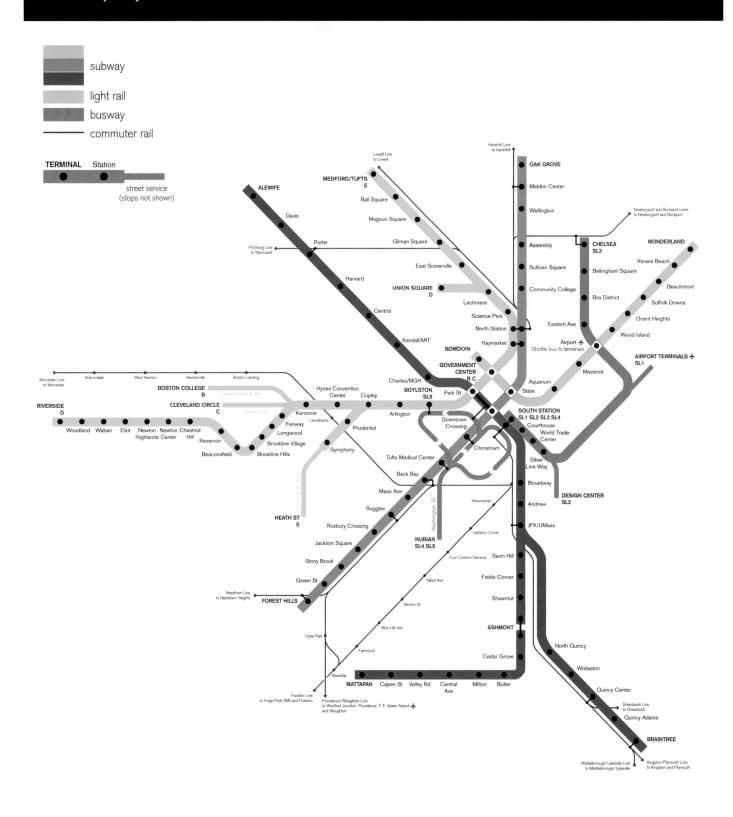

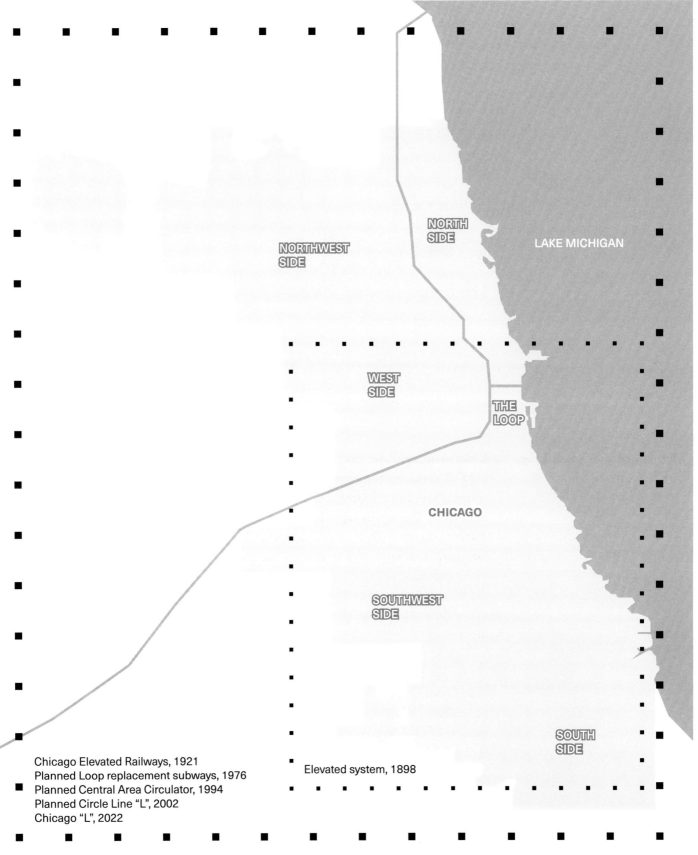

NORTHWEST
SIDE

NORTH
SIDE

LAKE MICHIGAN

WEST
SIDE

THE
LOOP

CHICAGO

SOUTHWEST
SIDE

SOUTH
SIDE

Chicago Elevated Railways, 1921
Planned Loop replacement subways, 1976
Planned Central Area Circulator, 1994
Planned Circle Line "L", 2002
Chicago "L", 2022

Elevated system, 1898

N

Chicago

2 miles/3.2 km

CHICAGO

The Loop Elevated, Beloved Steel Eyesore

3.1 Map of greater Chicago.

Chicago occupies the most important strategic place in the interior of North America, at the narrowest point between the watersheds of the Mississippi River and the Great Lakes. The gap between watersheds is only eight miles. The rapidly expanding United States established its first permanent outpost in what is now downtown Chicago in 1803. The city was incorporated in 1833. By 1860, Chicago had become the ninth-largest city in the United States, and even the destruction of the Great Chicago Fire did little to slow its expansion. For most of the 19th century and the first half of the 20th, Chicago (fig. 3.1) was at the forefront of American urban innovation.

One of those innovations was rapid transit. In many ways, the story of Chicago can be told through the story of the Loop. The Loop is a 1.8-mile, two-track elevated that runs over Lake Street, Wabash Avenue, Van Buren Street, and Wells Street in downtown Chicago. The Loop makes up the core of the Chicago Elevated, or simply the L. (In this chapter, I use the Loop to refer only to the rapid transit tracks, not the namesake central business district.) Built in the 1890s with a quintessentially Chicago mix of sharp elbows and shameless corruption, the Loop has anchored the downtown through good times and bad.

The Loop is an anachronism. For most of the 20th century, Chicago real estate interests, politicians, and architects tried to tear it down and replace it with subways.[1] In their view, the old steel structure was an eyesore, an antiquated relic of the early industrial era. None of these plans came to fruition, because the alternatives couldn't distribute passengers through central Chicago as effectively. Not for a reasonable price, anyway.

By the 1980s, it was clear that the Loop was there to stay. Chicago adjusted to the new reality accordingly. Instead of demolishing the Loop, Chicago's leaders proposed supplementing it instead. With the city's postindustrial rebirth in the 1990s and 2000s, the transit infrastructure helped Chicago reinvent itself as a global city and the capital of the Midwest. But this transformation did not happen evenly. Chicago's downtown and much of the North Side face the challenges of affluence, like housing affordability and traffic congestion. But the South and West Sides face the challenges of poverty, like violence, disinvestment, and poor-quality schools. The city is still heavily segregated by race.

Early 21st-century Chicago is one part thriving world city and one part shrinking Rust Belt city. The state of the L and its downtown Loop reflects that.

Bribery, Blackmail, and Front Companies

Chicago was the 19th century's greatest boomtown. In six decades, Chicago had grown from a sleepy village of 200 to the second-largest city in North America. The Great Chicago Fire of 1871, which left a third of the city's residents homeless, scarcely slowed its expansion. In 1870, Chicago had a population of 298,000. By 1890, the population had increased to 1.1 million. To serve this explosive

growth, Chicago had tried to build elevated lines, but no one company managed to make everything come together.

The first line to break the logjam was the South Side Elevated, a steam-powered elevated that opened its initial three-and-a-half-mile segment from Congress Street south to 39th Street in 1892. In May 1893, the line was quickly extended to Jackson Park, another five miles away, to serve the crowds attending the World's Columbian Exposition—the first world's fair to be held in the Midwest. The contemporary Lake Street Elevated and Metropolitan West Side Elevated would open shortly thereafter.

This nascent network had two major problems. First, the elevated system in 1894–1895 was composed of totally disconnected lines. The South Side, Lake Street, and Metropolitan Elevateds all had separate terminals at the edges of Chicago's business district. Commuters had to walk or take a streetcar from one terminal to the other to transfer. Under Illinois law at the time, a majority of adjacent property owners had to consent to the construction of an elevated. Downtown landowners were loath to potentially reduce their property values by allowing noisy, dirty, and shadow-casting elevated lines to be built above the streets.[2]

Second, the new elevated lines were entering service just as the national economy was in the middle of a full-scale meltdown. The crash started when rope monopoly National Cordage Company and the giant Philadelphia & Reading Railroad became insolvent in early 1893.[3] A chain reaction ensued, and a string of surprise bankruptcies hit every sector of the economy. Across the country,

575 banks failed.[4] Unemployment skyrocketed from 4 percent in 1892 to 18 percent in 1894.[5] With the economic chaos, the elevated companies were desperately short of money.

Into this vacuum stepped the amoral streetcar magnate Charles Tyson Yerkes (rhymes with *turkeys*), a man who once remarked, "The secret of success in my business, is to buy old junk, fix it up a little, and unload it upon other fellows."[6] Yerkes quietly bought out the cash-strapped Lake Street Elevated in July 1894. He proceeded to get property owners' consent for the downtown elevated Loop using bribery, blackmail, and a network of front companies. Notably, the Van Buren Street leg of the Loop came to be because Yerkes got approval from the property owners on the western, industrial half of Van Buren. He then built the elevated over the eastern, commercial half of Van Buren, where property owners had held fast.[7]

By 1897, the Loop was completed. Yerkes had a virtual monopoly over downtown rapid transit. The other elevated companies were left with no choice but to pay Yerkes for the privilege of using his Loop.[8] Commuters demanded direct service to downtown Chicago, and the more distant terminals of Yerkes's competitors weren't going to cut the mustard (fig. 3.2).

Ultimately, though, Yerkes was dependent on the goodwill of Chicago's citizens and their elected officials. The franchise agreements to run his empire were time limited and subject to renewal by the City Council. Shameless and ruthless, he sought to complete his conquest of Chicago transit by securing changes to the law that would give him 50-year franchise agreements. Yerkes miscalculat-

ed. His attempts to strongarm the City Council and the Illinois General Assembly backfired. Despite a rumored $500,000 ($18 million in 2022 dollars) in bribes for the state legislature alone, popular backlash was so strong that the politicians refused to renew Yerkes's franchises altogether.[9] He accepted defeat in 1899. Yerkes departed Chicago, sold his transit empire, and moved to London. On arrival, he proceeded to play hardball with the English, using the same tactics he learned in Chicago to buy up failing companies that would become part of the London Underground.

The construction of the Loop elevated tracks in the 1890s made it possible for Chicago's downtown to develop into a dense business center dominated by some of the tallest buildings on the planet. But by the 1920s, the shortcomings of this arrangement were brutally apparent. Every single elevated train in greater Chicago ran through the Loop, so the Loop acted as a choke point (fig. 3.3).[10] Loop-related delays cascaded through the system, causing frequent commuter meltdowns.

Taking stock of the transit situation in the 1920s, Chicago was a very different place than it is today. The South Side was a major immigrant destination. European immigrants in particular flooded into the city, to work in the great Union Stock Yards that made Chicago "hog butcher for the world." Chicago was also a place of hope for blacks coming from the South as part of the Great Migration. After World War II, ridership losses, expressway construction, and white flight took their toll. Tellingly, of the five South Side elevated branches that existed in the 1920s, three were demolished after World War II.

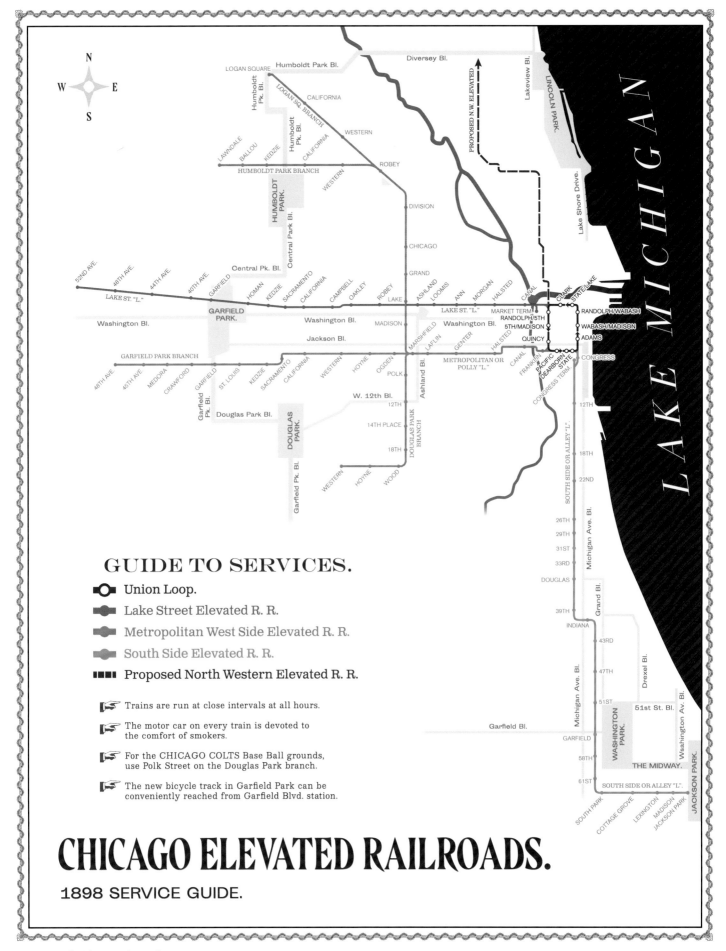

CHICAGO ELEVATED RAILROADS.

1898 SERVICE GUIDE.

GUIDE TO SERVICES.

○ Union Loop.
● Lake Street Elevated R. R.
● Metropolitan West Side Elevated R. R.
● South Side Elevated R. R.
▪▪▪▪ Proposed North Western Elevated R. R.

☞ Trains are run at close intervals at all hours.

☞ The motor car on every train is devoted to the comfort of smokers.

☞ For the CHICAGO COLTS Base Ball grounds, use Polk Street on the Douglas Park branch.

☞ The new bicycle track in Garfield Park can be conveniently reached from Garfield Blvd. station.

3.2 Early Chicago elevateds, 1898.

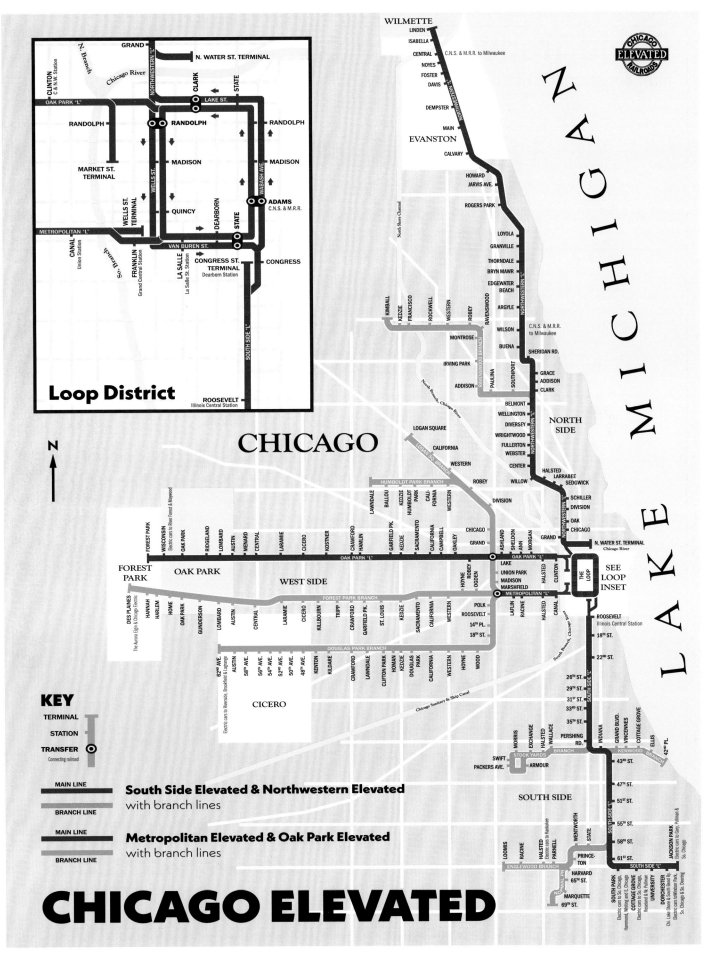

3.3 Chicago Elevated Railways system, 1921.

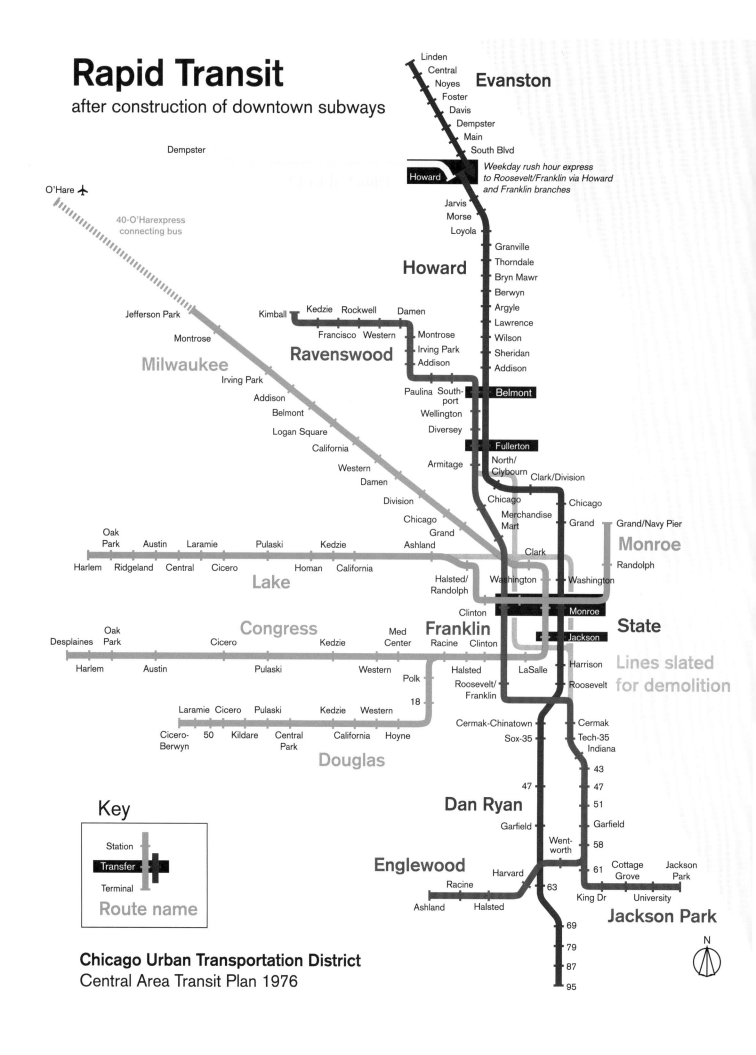

3.4 Proposed Loop replacement subways, 1976.

From Eyesore to Landmark

The decades after World War II were a time of retrenchment for Chicago. Greater Chicago expanded in population and in size, but the city proper began to suffer from urban decay, especially on the industrial South and West Sides. The Chicago Transit Authority (CTA), which took over the L system from the remaining private operators in 1947, responded by paring down the transit network. The CTA eliminated elevated branch lines, replaced its entire streetcar fleet with buses, and rerouted much of the downtown-bound L service into newly built downtown subways. Like most other American cities, Chicago also decided to demolish huge sections of the city core to build expressways. Unlike most other American cities, Chicago used the new expressways as an opportunity to tear down old elevated branch lines and replace them with brand-new ones. The creaky Metropolitan and Garfield Park lines were replaced with a newly built line in the median of the Eisenhower Expressway. Similarly, most of the Northwest Elevated was moved into the Kennedy Expressway.

A part of this modernization effort was Chicago's last serious attempt to demolish the obsolete Loop and replace it with subways (fig. 3.4). In 1976, the federal government allocated money to design and build the subways.[11] Around the same time, the federal government provided money for the companion Crosstown Expressway, which was slated to run circumferentially around the city.

But as resistance and price tags kept rising, Mayor Jane Byrne and Governor James "Big Jim"

Thompson made a deal in 1979 to kill both the Crosstown Expressway and the subway project. Their agreement freed up over $2 billion ($8.2 billion, inflation-adjusted) for the city and state to use elsewhere.[12] Much of it went to deferred maintenance, but the money also funded L extensions to O'Hare and Midway airports.

If You Can't Beat 'Em, Join 'Em

The repairs and extensions of the 1980s and early 1990s were desperately needed, given the decaying state of the L system. However, they didn't solve the downtown capacity issues that had plagued the Loop since its construction in the late 19th century. All plans in recent memory have accepted this reality. Proposals for new lines are meant to supplement rather than supplant the Loop.

In the 1990s, Mayor Richard M. Daley threw his support behind a downtown light rail system called the Circulator and lined up the money for the project (fig. 3.5). The federal, state and city governments each assumed one-third of the cost.[13] The Circulator project came to grief due to political changes in Washington, DC. In the 1994 midterm elections, the Republican Party took over Congress, promising to reduce federal spending. Virginia Congressman Frank Wolf, the new House Transportation Subcommittee chairman, zeroed out federal funding for the Circulator. Daley went to Washington expecting to talk Wolf into a deal. But when Wolf and Daley met, Wolf flatly refused to commit to any future federal funding to the Circulator.[14] Daley went back to Chicago empty-handed.

35

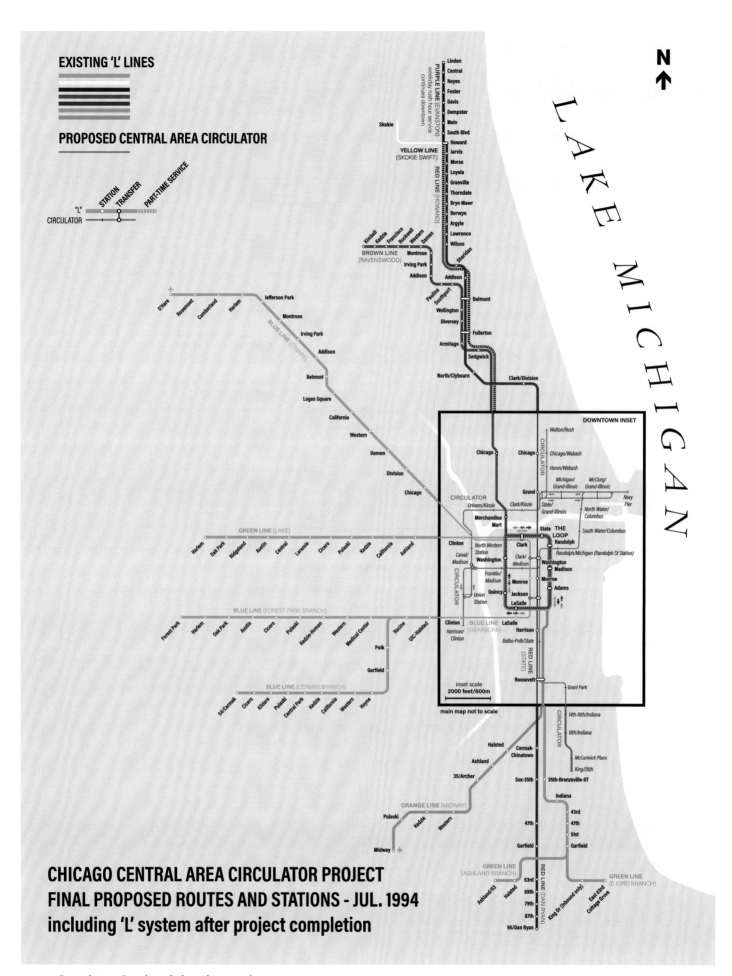

3.5 Central Area Circulator light rail proposal, 1994.

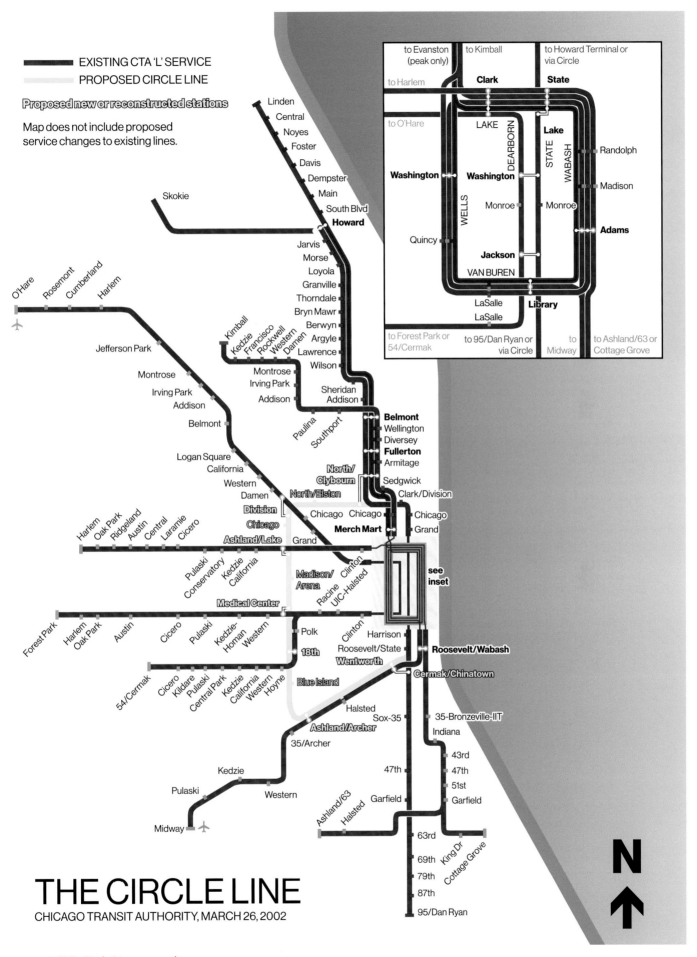

THE CIRCLE LINE

CHICAGO TRANSIT AUTHORITY, MARCH 26, 2002

3.6 CTA Circle Line proposal, 2002.

In 2002, the CTA introduced plans to create a Circle Line by using a mix of existing and new L tracks to run a full circle around the outer downtown, providing much-needed crosstown service and improving service to the 560-acre hospital complex known as the Illinois Medical District (fig. 3.6).[15] The CTA studied the Circle Line for seven years. The delay ended up fatal for the project. In 2002, greater Chicago was still suffering the aftereffects of the dot-com recession, but the economy was on the road to recovery. Metropolitan unemployment was about 7 percent. But by the time planning studies were complete in 2009, the stock market had crashed, and the economy was in freefall. Unemployment hit 11 percent that year. With the CTA already maxed out financially, the Circle Line was quietly dropped from planning consideration.[16]

Half Manhattan, Half Detroit

In the early 2020s, the elevated Loop is still at capacity, as it has been for a century. The L has expanded minimally since the airport extensions reached O'Hare in 1984 and Midway in 1993. Downtown Chicago hasn't gotten any new L lines since the Blue Line's downtown subway was finished in 1951. All the same, L ridership increased by 70 percent in the pre-coronavirus decades. Chicago reinvented itself as a modern postindustrial city with a strong service economy.[17] This growth has not been evenly distributed.

Broadly speaking, 21st-century Chicago is one part Manhattan-style prosperity, and one part Detroit-style disinvestment. If you walk around downtown or Near North Side communities like River North and Goose Island, you'll see flourishing neighborhoods, with construction cranes everywhere. Parking is hard to come by, traffic is a pain, and housing is expensive—albeit still far cheaper than, say, New York or Los Angeles. This version of Chicago is largely populated by whites, with significant minorities of Asians and Hispanics. That's not the case if you go to places like Englewood on the South Side. Walking around Englewood, you could be forgiven for thinking that Chicago's population is mostly black, impoverished, and still suffering from the effects of deindustrialization.

This split personality comes out in the CTA's management of the L (fig. 3.7). When the Brown Line on the white, wealthy North Side was closed for renovation in the early 2000s, the CTA determined that all the line's stations had to be kept open. Every station platform had to be expanded at great cost.[18] This was a questionable idea from a transit-planning perspective. In particular, the Ravenswood branch of the Brown Line, between Belmont and the Kimball terminus, has a station every four-tenths of a mile. This is too close together for efficient service. For comparison, Vancouver's Millennium Line elevated opened in 2002 and typically has stations spaced every half mile to mile. The B Line subway in Los Angeles, completed in 2000, stops every one to two miles. The CTA had no similar compunction about closing stations on the South Side. When the CTA closed the entire Green Line for renovation in the mid-1990s, it permanently closed five South Side stations and respaced the stops for faster service.[19]

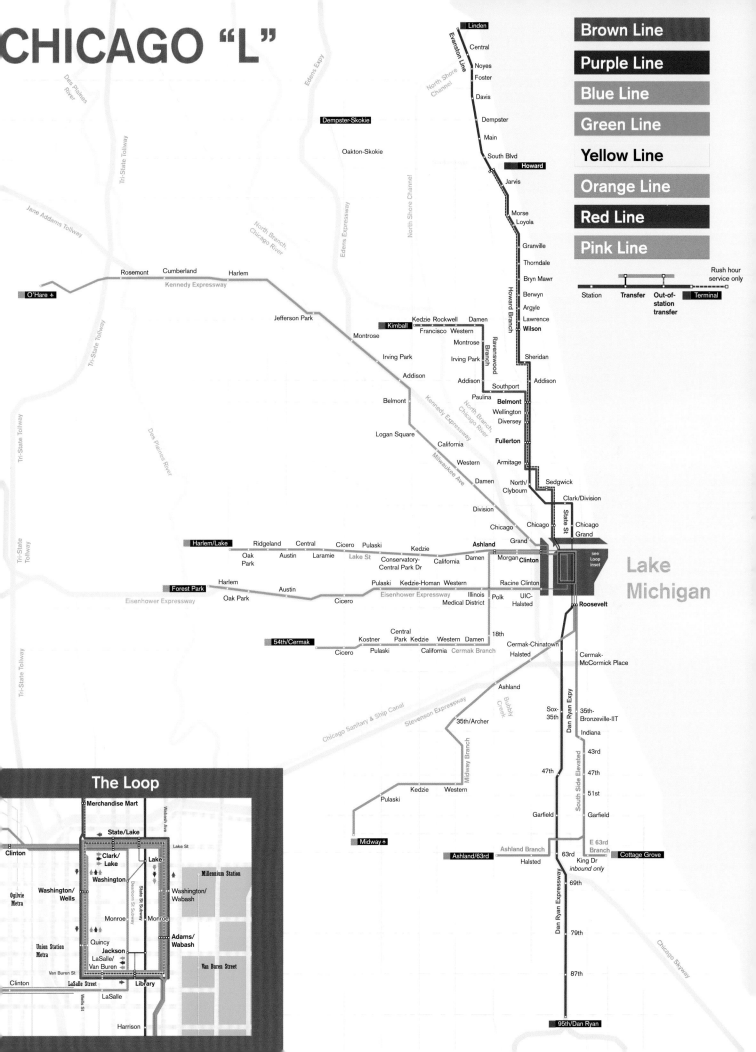

As for the Loop tracks themselves, they are in no danger of removal today. In more recent decades, the CTA has continuously invested in the Loop, closing redundant stations and opening new ones. Having survived a final attempt at demolition, the Loop has become part of Chicago's cultural identity, as much as the cable cars of San Francisco or the streetcars of New Orleans. Establishing shots depicting Chicago on screen now include the Loop as surely as they include the Willis Tower, Navy Pier, and the Art Institute. A century and a half later, the Loop abides.

③

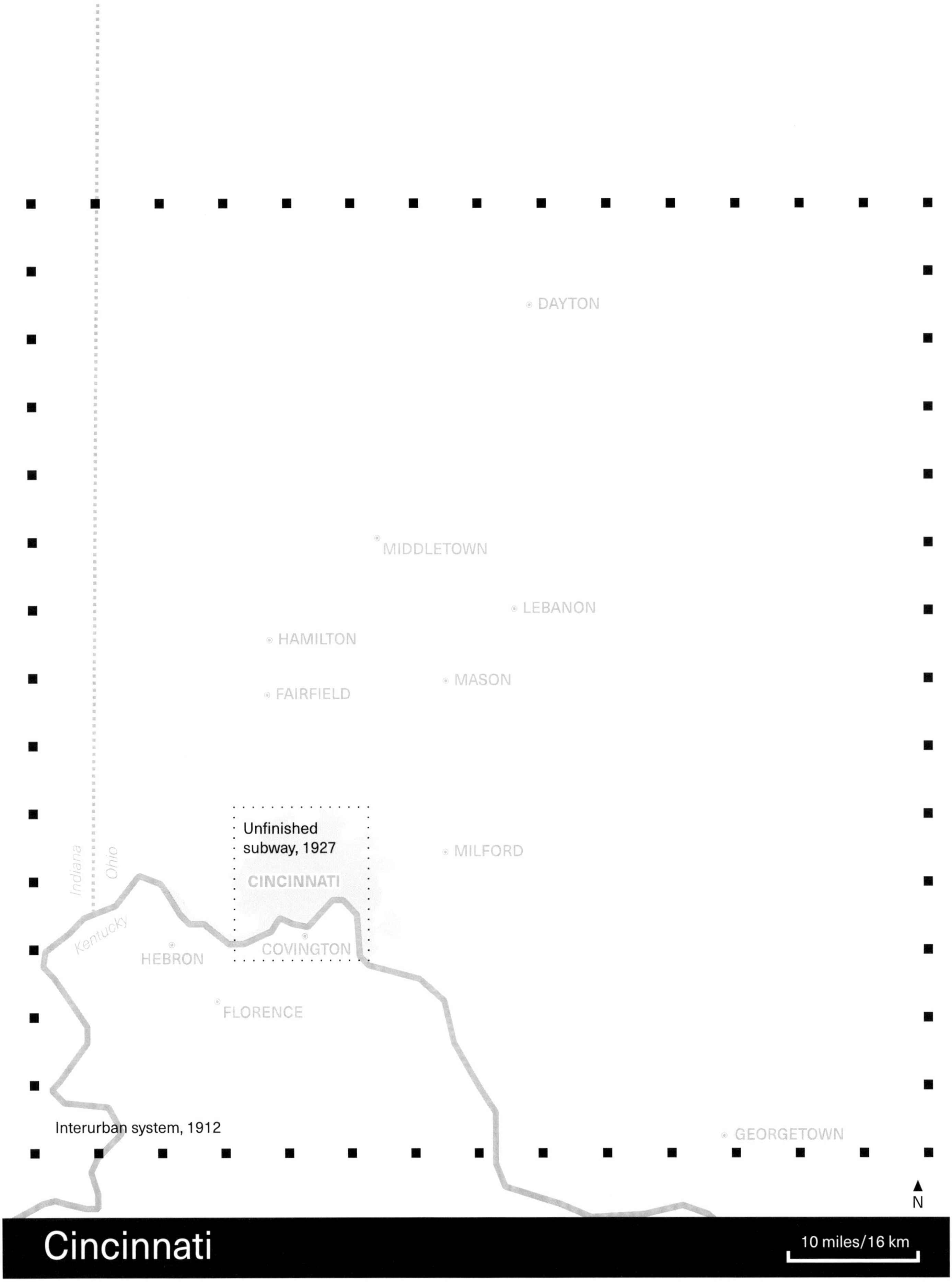

DAYTON

MIDDLETOWN

LEBANON

HAMILTON

MASON

FAIRFIELD

Indiana

Ohio

Unfinished
subway, 1927

CINCINNATI

MILFORD

Kentucky

HEBRON

COVINGTON

FLORENCE

Interurban system, 1912

GEORGETOWN

N

Cincinnati

10 miles/16 km

CINCINNATI

A Short History of a Never-Used Subway

4.1 Map of greater Cincinnati.

Cincinnati, uniquely among North American cities, began to dig a subway, stopped digging midway through the project, and never used the unfinished tunnels for their intended purpose. A century later, no trains have ever run.

The idea to build a subway in a midsized city like Cincinnati (fig. 4.1) was due to the traffic congestion of the late interurban era in the 1910s. Like most Midwestern cities, Cincinnati had a complex network of interurbans. Cincinnati's interurbans went as far as Dayton, 53 miles to the north (fig. 4.2). The interurbans were the anchors of greater Cincinnati's transport network. In a time when paved roads were uncommon and motor vehicles even less so, the interurbans were surrounding towns' lifelines, transporting both freight and passengers to the bustling, prosperous city of Cincinnati.

Cincinnati's interurbans had no dedicated tracks to reach downtown and no centralized terminal facility. This led to gridlock. Further complicating the situation, Cincinnati's municipal government had decreed that the local streetcar network be built to nonstandard dimensions. The overwhelming majority of North American railways have a track gauge (i.e., the distance between the rails) set at 4 feet, 8.5 inches. Cincinnati required rails to be placed 5 feet, 2.5 inches apart. This decision was originally made to prevent steam trains from running on city streets, and it created a major dilemma for Cincinnati's interurban companies.[1]

An interurban company that built its railway to the usual North American track gauge could buy its equipment off the rack. Existing steam rail-

INCINNATI AND VICINITY
ELECTRIC INTERURBAN RAILWAYS, 1912

INDIANA

OHIO

KENTUCKY

DAYTON

MIAMISBURG

FRANKLIN

MIDDLETOWN

TRENTON

HAMILTON

LINDENWALD

Mill Creek Route

Cincinnati & Dayton Route

HICKENLOOPER

SYMMES

PLEASANT RUN

SPRINGDALE

NEW BURLINGTON

GLENDALE

WYOMING

MT. HEALTHY

HARTWELL

CARTHAGE

ST. BERNARD

COLLEGE HILL

SPRING GROVE

CINCINNATI
BUSINESS DISTRICT

Harrison
Branch

HARRISON

WHITE WATER PARK

VALLEY JCT.

CLEVES

NORTH BEND

SEKATIAN

ADDYSTON

FERN BANK

HOME CITY

ELIZABETHTOWN, O.

L'BURG JCT., IND.

HOMESTEAD

LAWRENCEBURG

AURORA

Lawrenceburg
Branch

DELHI

ST. JOSEPH'S

TROUTMAN'S

ANDERSON'S FERRY

WATER
WORKS

CALIFORNIA JCT.

CALIFORNIA

CONEY ISLAND

SWEET WINE

BRACHMANS

FRUIT HILL

8 MILE

9 MILE

10 MILE

NEW PALESTINE

BLAIRVILLE

NEW RICHMOND

Cincinnati & Eastern Line

Rapid Line

LEBANON

S. LEBANON

MASON

KINGS MILLS

MILTOMSON

BRECON

HAZELWOOD

WINSLOW PARK

BLUE ASH

TERRA ALTA

ROSSMOYNE

DEER PARK

SILVERTON

KENNEDY HGTS.

PLEASANT
RIDGE

CYPRESS

NORWOOD

MADISONVILLE

MADEIRA

MOLBERRY

TERRACE PK.

INDIAN HILL

MILFORD

PLAIN-
VILLE

MT.
WASH.

CEDAR PT.

CARREL

JCT.
WASH.

FRUIT HILL

FORESTVILLE

TOBASCO

WITHAMS-
VILLE

AMELIA

CLOUGH PIKE

CHERRY GROVE

MT. SUMMIT

MT. CARMEL

SUMMERSIDE

GLEN ESTE

HIGHLAND PK.

BRAZIERS

JUDDS

HAMLET

MT. HOLLY

BANTAM

HAMLET

WILTSEE

HULINGTON

S. BANTAM

SWINGS

BETHEL

BETHEL

WALKER'S MILLS

HAMERSVILLE

N. FEESBURG

GILLETTS

SUNSHINE

RUSSELLVILLE

GEORGETOWN

BATAVIA

Batavia Branch Line

Suburban Line

Georgetown
Interurban

Kroger Line

BLANCHESTER

WOODSVILLE

EDENTON

GOSHEN
STATION

NEWTONSVILLE

OWENSVILLE
(BOSTON)

MONTEREY

MARATHON

STONELICK

PERINTOWN

MOLBERRY

DODSONVILLE

HILLSBORO

Swing Line

FAYETTEVILLE

STATION | TRANSFER STATION | STREETCAR CONNECTION | SERVICE ON LOCAL STREETS

Suburban, Cincinnati & Eastern, and Rapid Lines (I.R.&T.)

Mill Creek and Cincinnati & Dayton Routes (O. E. Ry.)

Georgetown Interurban (C.G. & P.)

Kroger Line (C. M. & L.)

Lawrenceburg and Harrison Routes (C.L. & A.)

Swing Line (C. & C.)

SERVICE GUIDE

road lines could be converted for interurban use cheaply and quickly. But entering the Cincinnati city center was not possible without prohibitively expensive construction, so passengers and freight would have to transfer at the urban periphery. In figure 4.2, the Cincinnati, Lawrenceburg, and Aurora (in orange), the Ohio Electric Railway's Cincinnati & Dayton route (in red), the Swing Line (in blue), and the Georgetown Interurban (in pink) all chose this approach.

The rest of Cincinnati's interurbans built their lines to the Cincinnati track gauge. This allowed them to serve downtown directly, via the local streetcar tracks. But this approach was also unsatisfactory. Existing steam railroad lines required costly construction to convert. The interurban companies were forced to buy pricier nonstandard equipment. Interurbans had to creep downtown behind local streetcars, a process that took 45 minutes to an hour each way. This was not a tenable situation.

The simplest way to solve the problem was a downtown subway with rails built to the North American standard track gauge. A subway could solve multiple problems at once. It would remove interurban trains from city streets, shuttle suburban commuters directly downtown, and reduce traffic congestion. With the support of George "Boss" Cox's political machine, the project went forward.[2] At the time, Cincinnati conveniently had a derelict canal running through its downtown that was suitable for subway conversion. (Similar canal-to-subway conversions also occurred in Newark, New Jersey and Rochester, New York.) This reuse of the old canal bed meant that costs

were expected to be much lower than in Boston and New York, where more extensive tunneling was needed.[3]

In 1916, the City Council approved bonds to build the subway loop to the tune of $6 million ($163 million in 2022 dollars). The electorate ratified the bond issue in April 1917. Then fate intervened two weeks later. The United States entered World War I, and construction was put on hold. By the time subway construction started on January 28, 1920, the price tag had doubled, and nearly all the interurbans were in deep financial trouble. Only three days after subway construction began, the Swing Line interurban was sold for scrap.[4] The next year, the Ohio Electric Railway went bankrupt. The Interurban Railway & Terminal Company collapsed in 1922. The rest of the interurbans vanished one by one over the next 15 years.

All the same, subway construction continued for most of the 1920s, until money ran out in 1928, with seven stations built and the loop half completed. But by that time, the problem the subway was designed to solve no longer existed, as the interurban companies were dead or dying. Boss Cox's machine was out of office. Without the Cox machine's support, the subway had become a political orphan. The new mayor, reformer Murray Seasongood, was loath to throw good money after bad. Seasongood deliberately allowed the mandate of the Rapid Transit Commission, which was building the line, to expire on January 1, 1929.[5] When the Great Depression hit later that year, it killed the subway project for good (fig. 4.3).

The aboveground portions of the subway were demolished after World War II to build Interstate

45

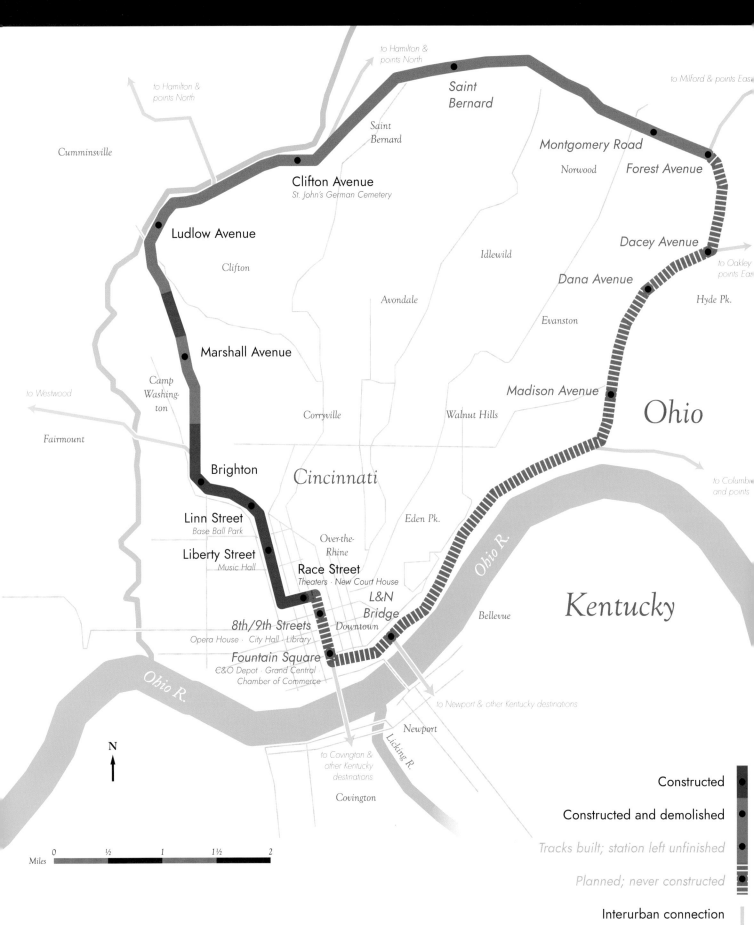

CINCINNATI SUBWAY.

to Hamilton &
points North

to Milford & points Eas

Saint
Bernard

Montgomery Road

Norwood

Forest Avenue

to Hamilton &
points North

Saint
Bernard

Cumminsville

Clifton Avenue
St. John's German Cemetery

Dacey Avenue

Ludlow Avenue

Idlewild

Dana Avenue

Clifton

to Oakley
points Eas

Avondale

Evanston

Hyde Pk.

Marshall Avenue

Madison Avenue

Ohio

to Westwood

Camp
Washing-
ton

Corryville

Walnut Hills

Fairmount

Cincinnati

Brighton

Eden Pk.

Ohio R.

to Columbe
and points

Linn Street
Base Ball Park

Over-the-
Rhine

Liberty Street
Music Hall

Race Street
Theaters · New Court House

Kentucky

Ohio R.

L&N
Bridge

Bellevue

8th/9th Streets
Opera House · City Hall · Library

Downtown

Fountain Square
C&O Depot · Grand Central
Chamber of Commerce

Licking R.

to Newport & other Kentucky destinations

Newport

N

to Covington &
other Kentucky
destinations

Covington

Constructed

Constructed and demolished

Tracks built; station left unfinished

Planned; never constructed

Miles   0   ½   1   1½   2

Interurban connection

4.3 Cincinnati's unfinished subway, 1929.

75. The underground sections are still intact beneath the city's Central Parkway. The tunnels carry water mains and fiberoptic cables today. Various proposals have been raised over the past 90 years to use the tunnels as a part of a light rail network, but none has passed muster with the electorate.

The closest Cincinnati ever got to using the tunnels for transit was in 2002. That year, the Southeast Ohio Regional Transit Authority (SORTA) put a sales tax measure on the ballot to build a five-line regional light rail network called Metro-Moves. MetroMoves was competently engineered, forward thinking, and totally unrealistic in Cincinnati's political environment. Simply put, SORTA had misread the electorate's appetite for new taxes. Part of the blame can be laid at the feet of a race riot the year before in the Over-the-Rhine district, which at the time was largely poor and black. But part of the blame can also be laid at the feet of pro football.[6]

Eight years earlier, voters had approved a sales tax hike to build a football stadium for the local NFL team, the Cincinnati Bengals. This stadium deal was a slow-motion disaster. About $455 million later ($888 million in 2022 dollars), the stadium was 55 percent over budget. Attendance was no better than before, the Bengals kept losing, and the cost of the bonds put the county under major financial stress. Cincinnati became a national laughingstock. The *Wall Street Journal* later called the stadium deal "one of the worst professional sports deals ever struck."[7] With the memory of the stadium debacle still fresh, MetroMoves failed by a 2–1 margin.[8]

Since then, there has been little serious effort to use the tunnels for their original intended purpose. As a practical matter, it's hard for Cincinnati to justify the large investment necessary to build out a useful light rail network. When the Cincinnati Subway began construction in 1920, the city's population was about 400,000. Most workers commuted on foot or by transit, and the automobile was a relative rarity. A century later, the city's population has dropped to 310,000, the vast majority of the population has access to a car, and metropolitan population growth has been slow. A single three-and-a-half-mile downtown streetcar line opened in 2016, but ridership has not met initial projections.[9] Further extensions seem unlikely. Tellingly, when SORTA asked the voters to raise taxes in 2020, the money was for improved bus service, not to build light rail.[10]

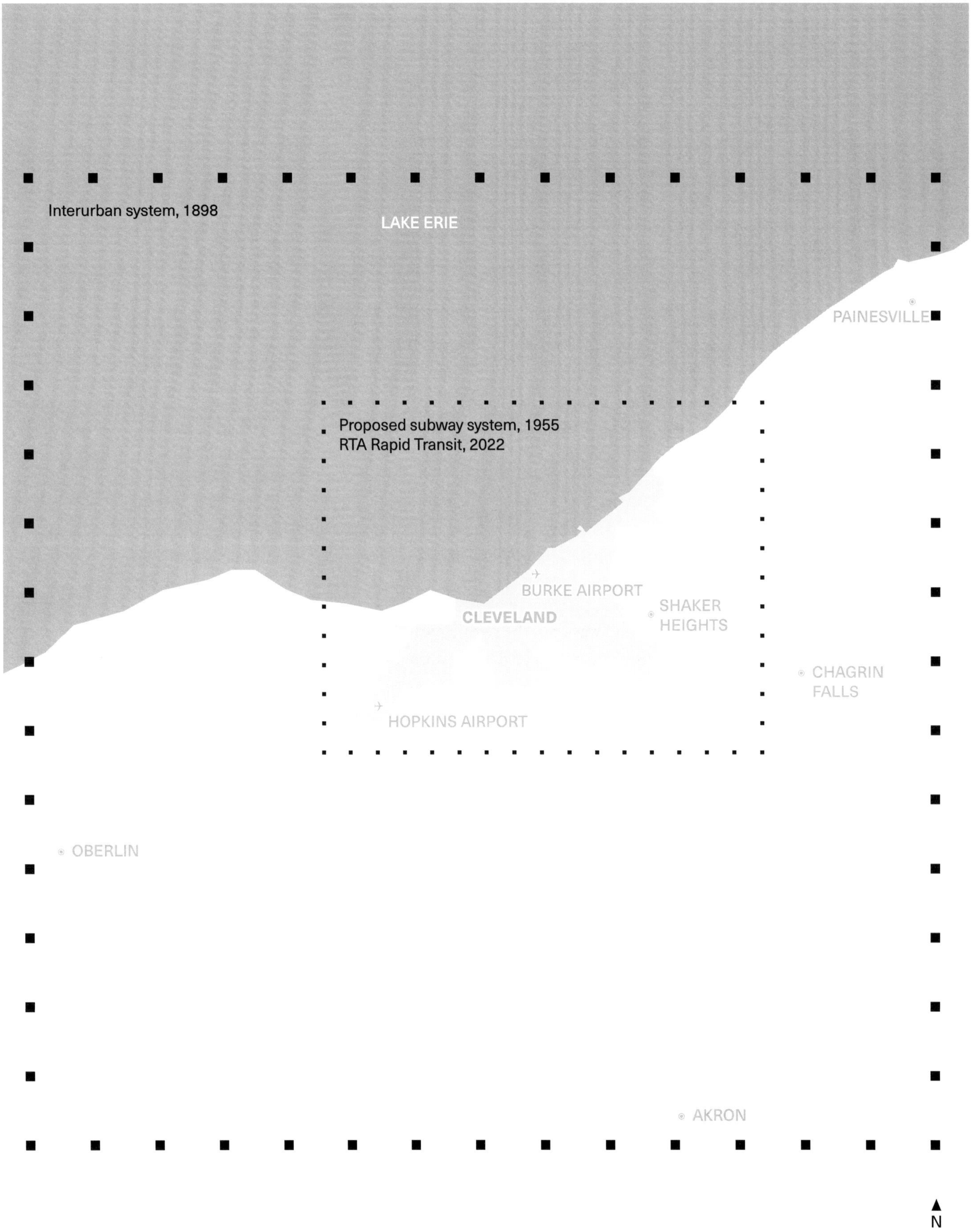

Interurban system, 1898

LAKE ERIE

PAINESVILLE

Proposed subway system, 1955
RTA Rapid Transit, 2022

BURKE AIRPORT

SHAKER
HEIGHTS

CLEVELAND

CHAGRIN
FALLS

HOPKINS AIRPORT

OBERLIN

AKRON

N

Cleveland

5 miles/8 km

5

CLEVELAND

Transit and the Perils of Waterfront
Redevelopment

5.1 Map of greater Cleveland.

Cleveland was once the sixth-largest city in the United States. It entered a prolonged period of deterioration during the second half of the 20th century, like most Northern manufacturing cities. To stave off the forces of urban decay, Cleveland (fig. 5.1) tried megaprojects, among other things. One of those megaprojects was to redevelop waterfront industrial land.

Urban waterfront redevelopment was a common theme in the late 20th and early 21st centuries. This was due to macro-level forces that ended the need for most cities to have traditional industrial waterfronts. Far fewer dockworkers and warehouse workers were needed after the shipping container was introduced. At the same time, urban heavy industries requiring large amounts of semiskilled labor moved overseas or introduced automation. These trends left immense amounts of land ripe for redevelopment. This fact was not lost on planners and politicians. Thus, New York turned Manhattan's Hudson River waterfront into a linear park. Boston's Seaport is now full of skyscrapers and a new courthouse. San Francisco redeveloped its waterfront, the Embarcadero, by building a stadium, relocating its science museum there, and building a light rail line.

Cleveland did the same as San Francisco, at least on paper. Cleveland already had light rail lines that could be extended along the waterfront. These lines were the sole survivors of Cleveland's once-extensive streetcar and interurban systems. But while San Francisco's Embarcadero redevelopment is celebrated by critics and the public alike, Cleveland's waterfront transit expansion has done little to affect the greater urban fabric or promote nearby growth. The Embarcadero is greater than the sum of its parts. Cleveland's waterfront is quite the opposite.

The three-mile-long Embarcadero is busy at all hours. It is one of San Francisco's major draws for both locals and tourists. A continuous string of parks, museums, tourist attractions, offices, and apartments runs along San Francisco Bay, anchored by Fisherman's Wharf at the north end and San Francisco's main commuter rail station at the south end.[1] The F-Market streetcar and N-Judah light rail run in the median of the grand boulevard that replaced the old Embarcadero Freeway. In 2019, the F carried 19,700 passengers per day, and the N carried 43,000.[2]

In contrast, the Cleveland waterfront is cut off from the rest of the city by the Shoreway freeway. The waterfront itself is mostly sterile despite pockets of success. The large-scale infrastructure investment has done little to promote redevelopment nearby. Before it was shut down due to coronavirus, Cleveland's light rail line, the Waterfront Line, carried so few passengers that the press dubbed it "the Ghost Train."[3] In 2016, 20 years after its opening, the Waterfront Line carried only 400 passengers per day, even with trains running every 15 minutes.[4] Service was suspended indefinitely when the coronavirus pandemic hit in 2020. Since then, one of the Waterfront Line bridges has been deemed unsafe. Repairs are ongoing, but service isn't expected to resume until 2024.[5]

Something went wrong in Cleveland, and it's worth examining just how it did. The following sections examine each of the seven Waterfront Line stations in turn, starting at the Tower City station

downtown and finishing about two miles away at South Harbor station, the end of the line.

Tower City

The Waterfront Line starts at Tower City station in downtown Cleveland. Tower City is the major transfer point between Cleveland's one subway line and two light rail lines. It is the only station located in downtown proper. During the streetcar era, all major interurban lines and most local streetcars terminated at nearby Public Square (fig. 5.2). The modern Green and Blue Lines, which opened in the first half of the 20th century, are the two survivors of this period. Both continued to run because they used a dedicated trench to access the underground Tower City terminal on Public Square. Another reason the lines survived is that they served tony Shaker Heights, at the time the most expensive municipality in greater Cleveland.[6]

Tower City station is poorly situated, because it's at the extreme southwest corner of downtown. Cleveland's planners have tried multiple times to solve this problem and better connect Tower City to the commercial core. In the mid-1950s, Cleveland's city council approved a two-line subway with a combination of existing railway lines and a new underground loop through downtown (fig. 5.3). The suburban subway branches (today's Red Line) were built. But the all-important downtown underground loop was gone, after Cuyahoga County engineer Albert Porter nixed the loop in favor of freeways.[7]

The next major push to fix this problem came in the 1970s, when the federal government was

5

LEGEND.

LORAIN & CLEVELAND Ry.
CLEVELAND & CHAGRIN FALLS ELECTRIC R. R.
AKRON, BEDFORD & CLEVELAND R. R.
CLEVELAND, PAINESVILLE & EASTERN R. R.
CLEVELAND BEREA ELRIA & OBERLIN R. R.

HUB   MAJOR   MINOR
      STATION STATION

LAKE ERIE

PAINESVILLE
HEISLEY
MENTOR
W. MENTOR
REYNOLDS
WILLOUGHBY
NOBLE
NOTTINGHAM
WICKLIFFE
COLLINWOOD
COITS
EUCLID
COLLAMER
GLENVILLE
FOREST HILL
CASE
AVE.
CLEVELAND
PUBLIC SQ.
FAIRMOUNT
AVON BEACH PARK   AVON
SHEFFIELD
ROCKY RIVER
PARDEE
W. DOVER
N. DOVER
WARRENSVILLE
LORAIN
DOVER BAY
W. CLEVELAND
ROCKY RIVER BRIDGE   ROCKPORT
DOVER
KAMMS
CHAGRIN FALLS
NEWBURG
WHITE HOUSE CROSSING
S. DOVER
N. OLMSTED
PURITAS SPRINGS
COUNTY LINE
DEWARS
N. RIDGEVILLE
BEREA
BEDFORD
TOWN LINE
ELYRIA
NORTHFIELD
LITTLE YORK
ERLIN
BOSTON LEDGES
SEASONS
Rocky R. West Branch
Rocky R. East Branch
Cuyahoga R.
Chagrin R.
Cuyahoga R.
SILVER LAKE
CUYAHOGA FALLS
AKRON

INTERURBAN ELECTRIC RAILWAYS
OF
CLEVELAND, OHIO
1898 Revision.

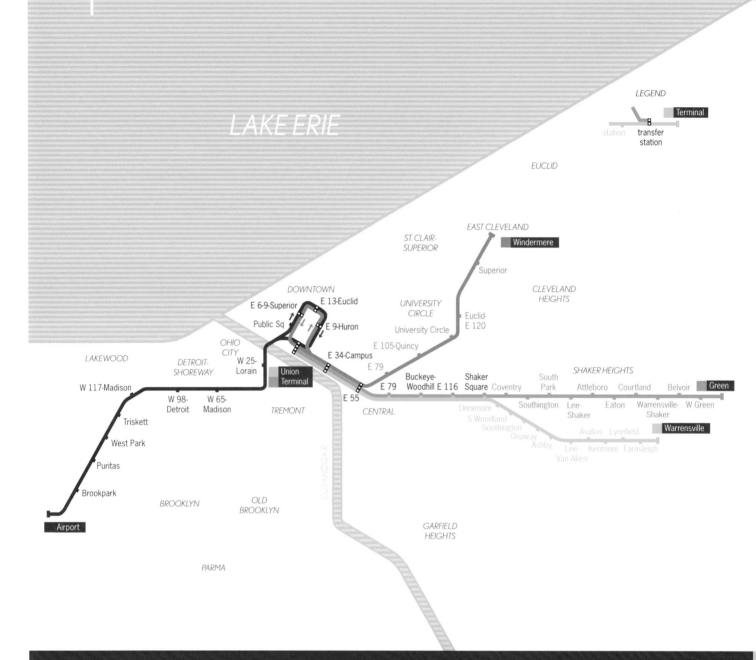

LEGEND

station transfer station Terminal

N

LAKE ERIE

EUCLID

ST. CLAIR-SUPERIOR

EAST CLEVELAND

Windermere

Superior

CLEVELAND HEIGHTS

DOWNTOWN

E 6-9-Superior   E 13-Euclid

Public Sq   E 9-Huron

UNIVERSITY CIRCLE

University Circle

Euclid-E 120

E 105-Quincy

OHIO CITY

W 25-Lorain

E 34-Campus

E 79

SHAKER HEIGHTS

LAKEWOOD

DETROIT-SHOREWAY

Union Terminal

E 79

Buckeye-Woodhill E 116

Shaker Square

South Park

Attleboro   Courtland   Belvoir   Green

W 117-Madison

E 55

Coventry

Southington   Lee-Shaker   Eaton   Warrensville-Shaker   W Green

W 98-Detroit   W 65-Madison

TREMONT

CENTRAL

Drexmore

S Woodland   Southington

Avalon   Lynnfield   Warrensville

Triskett

Onaway   Ashby   Lee-Van Aken   Kenmore   Farnsleigh

West Park

Puritas

Brookpark

BROOKLYN   OLD BROOKLYN

GARFIELD HEIGHTS

Airport

CUYAHOGA

PARMA

CLEVELAND TRANSIT SYSTEM
and
SHAKER HEIGHTS RAPID TRANSIT

As planned with downtown subway loop, 1955

CTS Subway	Shaker Heights Rapid Transit
Airport Line	Shaker Line
Windermere Line	Van Aken Line

5.3 Cleveland's 1955 subway proposal. Tower City
is called by its old name, Union Terminal.

heavily funding investments in downtown people movers. The City of Cleveland won a $41 million federal grant ($280 million, adjusted for inflation) to design an automated downtown people mover.[8] But the project had two glaring problems, obvious even to casual observers. First, the proposed loop ran in only one direction, so it would often be faster to walk. Second, automated trains weren't quite ready for prime time in the 1970s. Three weeks after it opened in 1972, San Francisco's pioneering Bay Area Rapid Transit system suffered a high-profile crash that made national news.

Eventually, political opposition gained momentum. Opponents questioned whether the people mover could be delivered on budget and without hurting local bus service. Forces against the people mover found an ally in mayor Dennis Kucinich. Kucinich canceled the people mover in 1977 and returned the money to the federal government.[9] The federal government sent the money to Miami, Detroit, and Jacksonville, where the results mostly proved Kucinich right. The Miami Metromover has become an integral part of downtown Miami's transit, albeit at a high construction cost.[10] But the people movers in Detroit and Jacksonville ended up being white elephants.[11] Since then, the only two additions to Cleveland's rapid transit network have been the HealthLine busway in 2008 and, of course, the Waterfront Line, which opened in 1996 (fig. 5.4). Cleveland's transit authority does have two other bus lines that it markets as busways, but they are not effective, lacking prepaid tickets, dedicated lanes, and specialized stations.

Settlers Landing and Flats East Bank

The first two stations after Tower City are good examples of ex-industrial waterfront redevelopment. The stations, Settlers Landing and Flats East Bank, are situated in the Flats district, on the banks of the Cuyahoga River. In the 1960s, the Flats were lined with heavy industry. So much industrial waste was dumped into the Cuyahoga that the river notoriously caught fire in 1969.[12]

Fifty years later, the Cuyahoga River has been cleaned up. Kayakers on the river are a common sight. The Flats are the terminus of an 87-mile linear bike trail that runs through the heart of Ohio. New construction has filled the Flats with parks, shops, housing, and corporate offices.[13] Companies like elite consulting firm McKinsey, global accounting firm Ernst & Young, and sports network ESPN have relocated there. In summary, the Flats are healthy, lively and active. This success has not been replicated at the other Waterfront Line stations.

West Third

Leaving Flats East Bank station, the Waterfront Line takes a graceful right turn up and over the freight tracks of the Norfolk Southern Railway before arriving at the West Third Street station. The area surrounding West Third creates a discontinuity between the vibrant activity of the Flats and the lakefront attractions further down the line. The station lies beneath the Shoreway overpass, so the station is cut off from both the Flats and the rest

RTA RAPID TRANSIT

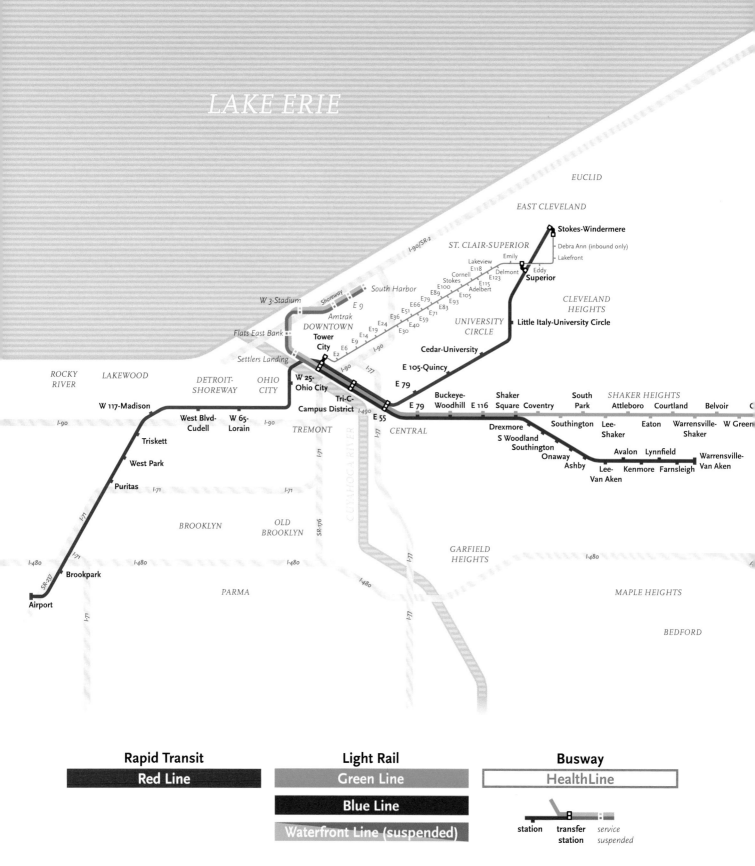

LAKE ERIE

EUCLID

EAST CLEVELAND

Stokes-Windermere

Debra Ann (inbound only)

ST. CLAIR-SUPERIOR

Lakefront

Lakeview

Emily

E118

Delmont

Eddy

Cornell

E123

Superior

Stokes

E115

E100

Adelbert

CLEVELAND

South Harbor

E89

E105

HEIGHTS

W 3-Stadium

Shoreway

E79

E93

E 9

E66

E83

Amtrak

E51

E71

UNIVERSITY

Little Italy-University Circle

DOWNTOWN

E36

E59

CIRCLE

Flats East Bank

E24

E40

Tower

E19

E30

Cedar-University

City

E14

Settlers Landing

E9

E2

I-90

E 105-Quincy

E6

I-90

I-77

E 79

ROCKY

LAKEWOOD

DETROIT-

OHIO

RIVER

SHOREWAY

CITY

W 25-

Ohio City

E 79

Buckeye-

Shaker

South

SHAKER HEIGHTS

W 117-Madison

Tri-C-

Woodhill

E 116

Square

Coventry

Park

Attleboro

Courtland

Belvoir

C

Campus District

I-490

West Blvd-

W 65-

I-90

E 55

TREMONT

CENTRAL

E 79

Cudell

Lorain

Drexmore

Southington

Lee-

Eaton

Warrensville-

W Green

Triskett

S Woodland

Shaker

Shaker

I-71

I-77

Southington

Avalon

Lynnfield

West Park

Onaway

Warrensville-

Ashby

Lee-

Kenmore

Farnsleigh

Van Aken

Puritas

I-71

Van Aken

I-71

I-71

BROOKLYN

OLD

GARFIELD

BROOKLYN

HEIGHTS

I-480

I-480

I-480

I-480

I-480

I-480

SR-237

I-71

Brookpark

MAPLE HEIGHTS

PARMA

I-77

Airport

BEDFORD

Rapid Transit	Light Rail	Busway
Red Line	**Green Line**	**HealthLine**
	Blue Line	
	Waterfront Line (suspended)	station / transfer station / *service suspended*

5.4 Cleveland RTA Rapid Transit, 2022.

of downtown. The Shoreway's traffic, noise, and pollution makes the station area an unpleasant place to be. There is little nearby except for acres of surface parking and Browns Stadium (officially FirstEnergy Stadium), home of the namesake NFL team. Unless the Browns are playing a home game, there is little reason to use West Third station.

There are two major reasons Cleveland's decision to build a dedicated NFL stadium on Lake Erie was unconducive to the development of a thriving waterfront. First, football-specific stadiums don't see enough use to permit a surrounding entertainment district to develop. Excluding playoff games, NFL stadiums host only 10 home games a year, and other events are rare.[14] In the first 17 years of its existence, Browns Stadium hosted only 33 ticketed non-football events—fewer than two a year.[15] Few musicians, religious revivals, rodeos, and so on can fill a 70,000-seat stadium, especially in a midsized metropolis like Cleveland. In contrast, the National Hockey League stadium in Columbus, Ohio, two hours to the south, has a capacity of 20,000 and averages 4.16 non-hockey events per month.[16] Second, although the parking lots around Browns Stadium are useful for gameday tailgating, they create a dead zone around the stadium when the Browns aren't playing.[17] The surface parking lots are designed to handle gameday crowds, which is overkill the other 355 days of the year. These two factors mean that the stadium and its adjoining station contribute little to Cleveland's day-to-day urban fabric.

In contrast to Cleveland, the City of San Francisco demolished the old Embarcadero Freeway and replaced it with a surface boulevard to make the area walkable. San Francisco built a baseball stadium, today known as Oracle Park, and encouraged real estate development near the stations. Baseball stadiums like Oracle Park tend to do a better job of promoting neighborhood growth than football stadiums do, mostly because Major League Baseball teams play more games. Each MLB team plays 81 home games per year, eight times as many home games as the NFL. Oracle Park has comparatively little on-site parking, so San Francisco's transit authorities run extra trains for fans to get to the game. All these factors make for a pleasant built environment, like other baseball-anchored neighborhoods such as Wrigleyville, Chicago, and Fenway, Boston. Visitors are encouraged to take transit and enjoy the neighborhood before games, compared to Cleveland's culture of driving in and tailgating in a parking lot.* In a sense, this creates a self-reinforcing loop. Cleveland favors suburban drivers with its infrastructure, so more people drive. San Francisco favors pedestrians and transit, so more people walk or take the train.

Amtrak, East Ninth, and South Harbor

A thousand feet to the east of West Third is the Waterfront Line's stop at Cleveland's Amtrak station. In theory, this is a useful connection. In practice, Amtrak's service hours mean that there was little reason to build a station here. The *Lake Shore Lim-*

---

* Ironically, Browns Stadium sits on the same site as the unloved Cleveland Municipal Stadium, which was home to both the Browns and Major League Baseball's Cleveland Indians.

55

ited, the only Amtrak service that stops in Cleveland, arrives in the wee hours of the morning. But the pre-pandemic Waterfront Line stopped running at 10:20 p.m.

Continuing a thousand feet further east, the next Waterfront Line station is East Ninth Street. East Ninth theoretically serves the northern part of downtown Cleveland and waterfront attractions like the Rock & Roll Hall of Fame, the Great Lakes Science Center, and Voinovich Bicentennial Park. But East Ninth station is located so awkwardly that it's just as convenient to walk from Tower City or drive. East Ninth is built below an overpass, squeezed between the freight tracks and the Shoreway. To reach Cleveland City Hall or other downtown buildings, passengers must cross a long, windswept bridge over the freight tracks. Making matters worse, the Waterfront Line's circuitous route means it's just as fast for City Hall–bound passengers to get off at Tower City and walk over. From Tower City to City Hall, it's 14 minutes on foot. From East Ninth, City Hall is a seven-minute walk, but the Waterfront Line takes seven minutes to go from Tower City to East Ninth. A similar problem applies if riders want to visit Cleveland's waterfront attractions. To reach Voinovich Bicentennial Park, for instance, Waterfront Line riders have to walk a third of a mile and cross the Shoreway. Drivers can park across the street from the park for $4, less than the round-trip Waterfront Line fare.

Further east is the Waterfront Line's terminus, South Harbor station. The only thing accessible on foot from South Harbor is a 15-acre city parking lot bracketed by the freight tracks to the south and the Shoreway to the north. The station wouldn't have much utility even if appropriate pedestrian connections existed. Beyond the freight tracks, there is a nondescript section of downtown Cleveland with a mix of light industrial facilities and government offices. On the far side of the Shoreway is Burke Lakefront Airport.

Burke is a crumbling 450-acre facility that no commercial airline uses. It loses up to $2 million per year.[18] In theory, Burke is meant to serve as a relief airport for overcrowding at Cleveland's main airport, Hopkins International Airport. In practice, Hopkins has enough capacity for the foreseeable future, so Burke receives only cargo, charter, and private planes. Passenger loads at Hopkins peaked in the year 2000. Since 2014, an entire Hopkins concourse has lain empty, for lack of demand.[19]

The Devil Is in the Details

There have been many opportunities for Cleveland to make the Waterfront Line useful and to create a continuous corridor of active urban waterfront. All the ingredients are there: new development, museums, parks, and so on. But these improvements exist in isolation. The Shoreway cuts downtown off from the waterfront. The stadium and its parking lots separate the healthy Flats from the waterfront attractions. Jet-engine noise from Burke Airport further reduces the attractiveness of Cleveland's waterfront to new development. The solution is relatively straightforward: downgrade the Shoreway to a surface-level boulevard, develop the parking lots, and close Burke. These three things combined would create a continuous two-mile-long transit-connected greenway, with copious amounts of vacant land newly available for apartments, offices, parks, and shops.

These are major changes for a city, but they aren't unprecedented. SoFi Stadium near Los Angeles, which opened in 2020 and is home to the NFL's LA Rams and LA Chargers, is at the center of the privately funded Hollywood Park commercial and residential complex. Hollywood Park was designed to be a year-round destination, not just a football destination. A people mover between SoFi Stadium and the LA Metro is in the works. Elsewhere in greater Los Angeles, local authorities have received final approval to close Santa Monica Airport. Like Burke, Santa Monica Airport sits on extremely valuable land and is no longer necessary as an airstrip. Final shutdown is set for 2028.

To Cleveland's credit, big ideas like closing the airport are now part of the local public dialogue. It appears at least possible that these types of improvements are in the cards. Jimmy Haslam, owner of the Cleveland Browns, has proposed that the Shoreway be downgraded and the surplus land be redeveloped.[20] The city government has begun to study the closure of Burke Airport.[21]

Cities, like individuals, follow trends. But blindly following trends without understanding why a particular trend works is liable to cause expensive mistakes. The Cleveland waterfront revitalization project and its accompanying infrastructure was one of those mistakes. Before its suspension due to coronavirus, Cleveland's Waterfront Line was useful on special occasions, like attending a Browns game, visiting the science museum, and going to the Rock & Roll Hall of Fame. But the area as a whole is a collection of missed opportunities that doesn't act in concert with the new growth in the Flats. Cleveland should have followed the model of San Francisco's Embarcadero redevelopment, and unified its parks, stadium, and attractions to create a popular, pleasant everyday neighborhood.

DENTON

MCKINNEY

FRISCO

DART, 2022

PLANO

FARMERS
BRANCH

RICHARDSON

GARLAND

SOUTHLAKE

Dallas Railway Co.,
1919

DALLAS

FORT WORTH

ARLINGTON

MIDLOTHIAN

WAXAHACHIE

CLEBURNE

N

Dallas

5 miles/8 km

6

DALLAS

They Don't Build Them Like They Used To

6.1 Map of Dallas and surroundings.

In the first half of the 20th century, it was axiomatic that mass transit infrastructure be built where riders were, and that transit systems aim to attract as many riders as possible. In urban areas, this meant putting transit lines through the most densely populated neighborhoods. In suburban areas, transit extensions went hand in hand with real estate development. That is, when a new line opened, the transit operator would make sure that new homes and businesses were built nearby. Today, that's not the norm. In cities that grew up in the automobile age, it's distressingly common for these lessons to have been forgotten.

Thus, in city cores, busy corridors with lots of potential riders get passed over. Long suburban rail extensions get built instead. These suburban stations are often surrounded by multiple-acre parking lots rather than places for people to live and work. Few take the train in locales like this. Dallas (fig. 6.1), and the Dallas Area Rapid Transit (DART) light rail system, is a good case study to analyze how the flexible land use practices of the past acted in concert with transit, compared with today's more rigid rules. To look at how things were built in the old days, the first stop is the Bishop Arts District in Dallas proper. Bishop Arts is a typical 1920s streetcar suburb. Bishop Arts' development pattern will be contrasted with the modern DART Mockingbird and Parker Road stations to illustrate the weaknesses of modern transit routing and land use practices, and why old-fashioned neighborhoods like Bishop Arts don't get built today.

6.2 Dallas Railway streetcars and connecting interurban service, 1919. Bishop Arts is located on the Junius-Tyler line southwest of downtown, and Southern Methodist University is at the terminal of the SMU branch north of downtown.

Bishop Arts

The Bishop Arts District is located in Dallas's Oak Cliff section. Bishop Arts was built as a shopping district around the old Dallas Railway's Junius-Tyler streetcar line.[1] It is Dallas's largest surviving early 20th century commercial district. (On the Dallas streetcar map in figure 6.2, Bishop Arts is at the southwestern end of the Junius-Tyler line.) Originally a white neighborhood, whites left and minorities moved there in large numbers after the end of segregation; the neighborhood struggled economically. In the 1970s and 1980s, artists looking for cheap studio space colonized the place, as happens in many cities. In the early 2020s, it has become a flourishing center of art and culture. Bishop Arts is beginning to see pressures associated with gentrification, like increasing real estate prices and displacement of existing residents. That said, gentrification is proceeding much more slowly than in coastal cities like San Francisco. In the early 2000s, over three-quarters of Bishop Arts residents were Hispanic. At the 2020 census, the area was still about 70 percent Hispanic. (Over the same period, San Francisco's Mission District dropped from 50 percent to 34 percent Hispanic.)

Land use in Bishop Arts is extremely efficient. Residents and shoppers were expected to arrive by train, so commercial buildings were commonly built up to the property line. Nearly every square foot of street frontage was used, so pedestrians could get as quickly as possible from station to storefront. Few concessions were made to drivers. The parking lots that exist in Bishop Arts today were formerly occupied by housing and businesses.

The residential portions are similarly compact, with a mix of single-family homes, duplexes, and small two- to three-story apartment buildings.

Bishop Arts is quaint, small-towny, and comfortable. Bishop Avenue, which forms the heart of the district, is a traditional main street. Dozens of businesses are concentrated within a small area, with residential buildings on the side streets. Few buildings are taller than three stories. But all the same, Bishop Arts is densely populated, with 7,086 residents per square mile at the 2020 census.[2] This is roughly the same population density as the city of Baltimore, and nearly double the density of the rest of the city of Dallas.

At the time Bishop Arts was built, there was a natural incentive for this type of development. The old Dallas Railway was a business, and it was in the company's financial interest to encourage as many people as possible to live and work near the stations. With the loose land regulation of the time, this, in turn, meant that businesses tended to cluster near transport hubs.[*] In the case of Bishop Arts, it's why the neighborhood's shopping district is built around the old streetcar station. While neighborhood commercial districts developed around individual lines, major transit hubs became focal points for the metropolis. It's not an accident that Neiman Marcus, Dallas's grand old central department store, sits at Main and Ervay, within a five-minute walk of nearly every streetcar

---

[*] This type of organic development still happens near Houston's light rail stations, due to Houston's unusually loose land use regulations and its lack of traditional zoning (see chapter 8).

dart light rail

and connecting streetcar and commuter rail service

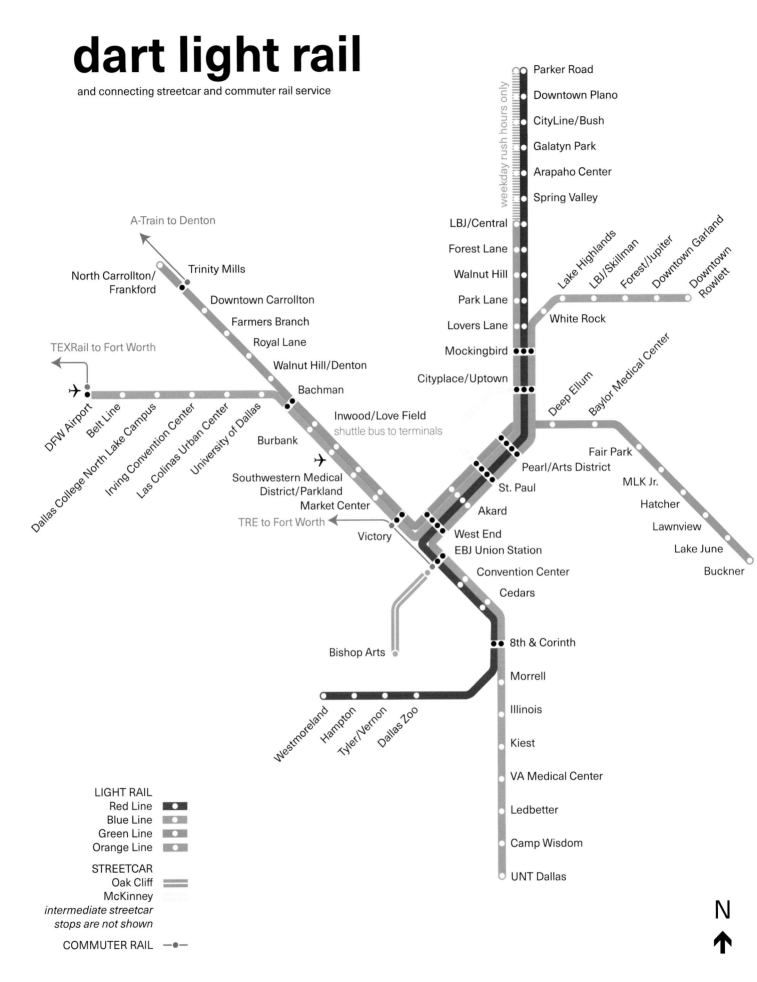

A-Train to Denton

TEXRail to Fort Worth

TRE to Fort Worth

weekday rush hours only

Parker Road
Downtown Plano
CityLine/Bush
Galatyn Park
Arapaho Center
Spring Valley
LBJ/Central
Forest Lane
Walnut Hill
Park Lane
Lovers Lane
Mockingbird
Cityplace/Uptown

Lake Highlands
LBJ/Skillman
Forest/Jupiter
Downtown Garland
Downtown Rowlett
White Rock

Deep Ellum
Baylor Medical Center

North Carrollton/
Frankford
Trinity Mills
Downtown Carrollton
Farmers Branch
Royal Lane
Walnut Hill/Denton
Bachman
Inwood/Love Field
shuttle bus to terminals

DFW Airport
Belt Line
Dallas College North Lake Campus
Irving Convention Center
Las Colinas Urban Center
University of Dallas
Burbank
Southwestern Medical
District/Parkland
Market Center
Victory

Bishop Arts

Westmoreland
Hampton
Tyler/Vernon
Dallas Zoo

Pearl/Arts District
St. Paul
Akard
West End
EBJ Union Station
Convention Center
Cedars

Fair Park
MLK Jr.
Hatcher
Lawnview
Lake June
Buckner

8th & Corinth
Morrell
Illinois
Kiest
VA Medical Center
Ledbetter
Camp Wisdom
UNT Dallas

LIGHT RAIL
Red Line
Blue Line
Green Line
Orange Line

STREETCAR
Oak Cliff
McKinney
*intermediate streetcar
stops are not shown*

COMMUTER RAIL

N

6.3 Dallas Area Rapid Transit, 2022. Bishop Arts
     (in lavender) is the terminal of the Oak Cliff streetcar.
     Mockingbird station is northeast of downtown.
     Parker Road is the northern Red Line terminal.

and interurban route that passed through downtown Dallas. (In figure 6.2, this is where the Ervay line joins the Main and Swiss lines.)

Ironically, Bishop Arts was bypassed when DART was built in the 1990s and 2000s. Instead, Bishop Arts is served by the slow Oak Cliff streetcar and the No. 9 bus. Both of these lines run in normal traffic. Figure 6.3 depicts the modern DART system; Bishop Arts is at the terminal of the Oak Cliff streetcar line.

Missed Connections

The next stop after Bishop Arts is the 1997-vintage DART Mockingbird station on the Orange and Red Lines, four miles north of downtown Dallas. Mockingbird station shows how post–World War II light rail serves existing urban areas poorly. To give credit where credit is due, the City of Dallas has allowed apartments and a shopping mall to be built at Mockingbird. But the station itself is in the wrong place.

The big attraction and cluster of jobs near Mockingbird station is Southern Methodist University (SMU). The university was originally built directly on a streetcar line, and the Dallas Railway's Highland Park–Munger line had its terminal at SMU's front gate (see fig. 6.2). The modern station is not so conveniently situated. To get from the modern Mockingbird station to SMU's main campus, it's a full mile on foot and requires crossing over a freeway. This is not exactly a pleasant walk in the Texas heat and humidity. All this extra distance makes it much more likely that students, faculty, and staff will drive rather than put up with

the hassle. DART does run a shuttle bus from the campus to the station, but that's no substitute for putting the station at the campus.

Mockingbird isn't the only DART station built too far from major destinations. Northwest of downtown Dallas, the Southwest Medical District station is at the very edge of the namesake hospital district. Had DART built the station at the district's center, all major hospitals would be within a 10-minute walk. This did not happen. Instead, the station is 1.4 miles away from the University of Texas's William P. Clements University Hospital—nearly a half hour on foot.

Oceans of Asphalt

Last on this tour of greater Dallas is the Parker Road station, terminal of the DART Red Line. Parker Road is located in suburban Plano, 20 miles to the north of downtown Dallas. Plano was rural until the 1970s, when it became part of the Dallas–Fort Worth metroplex's expanding suburban ring. Major corporations like JCPenney and Frito-Lay moved their headquarters to office parks in Plano, supercharging the area's growth. Between 1960 and 1980, Plano's population grew by 3,300 percent. With its newfound wealth came traffic congestion, and Plano joined the DART district to provide a high-quality transit option. In 2002, Parker Road station opened for service.

Parker Road is an example of what not to do near a transit station. As of 2022, Parker Road is surrounded by 40 acres of parking lots, two pawn shops, two auto shops, a 24-hour bowling alley called the Plano Super Bowl, and a solitary apartment building under construction that was approved in 2021. That's it. There are plenty of places to put your car near the station, and you can go bowling, but it's awfully hard to live, shop, get groceries, go out to dinner, work, or worship there. There's not much of a neighborhood near the station. The transit infrastructure is decidedly underused.

In the 21st century, it's almost unheard of for a quirky, human-scale, and surprisingly dense area like Bishop Arts to grow and thrive near a rail station like Parker Road. But it isn't because of technical limitations. It's because it's illegal.

Zoning Out

Plano's zoning laws ban this type of traditional neighborhood development through limitations on land use, building-size restrictions, and minimum parking requirements. Plano's use restrictions allow commercial, industrial, and governmental uses on the land around Parker Road station. But it's not legal to build homes, like it was in the old days.[3] The solitary apartment building under construction near Parker Road had to get a special variance from the Plano City Council. This variance took years to wind its way through the bureaucracy.[4] The nearest place where it's legal to build housing without special dispensation is three-quarters of a mile away on foot. There's just no convenient way to take the train without driving or transferring from a bus.

The second issue is the building-size law. Plano's zoning bans the types of buildings built in the early 20th century, because older buildings are just too big. In Bishop Arts, buildings are between one and three stories, and sit on lots that are about 5,000 square feet, typical for the era. In the 1920s, it was entirely normal to build structures of up to 10,000 square feet on such lots.[5] That's not allowed in the 2020s. Plano's building-size law caps square footage at the size of the lot, at a 1:1 ratio.[6] Thus, the largest building that's legal to build on a 5,000-square-foot lot is a 5,000-square-foot building. Plano's law also limits where a building can be built on the lot. As mentioned before, storefronts of past eras were built up to the property line on the street side to make it convenient for pedestrians and transit riders. Plano makes that illegal as well. All buildings must be set back at least 50 feet from the property line.[7]

Third, Plano's minimum parking law requires suburban-style construction. Under the law, every new building has to have one parking lot space for every 200 square feet of retail space.[8] Office buildings and houses have similar mandatory parking minimums. Each parking space requires about 400 square feet—150 square feet for the parking spaces themselves and 250 square feet so cars can reach the street.[9] Thus, to build a 10,000-square-foot retail building, Plano requires 20,000 square feet of parking lot. The only way to do this economically is to build a strip mall or a suburban-style subdivision.

Dallas just doesn't build things like they used to. Some of this can be attributed to changing tastes and the onward march of time. But it's also because greater Dallas's local governments missed major destinations when building new transit. These local governments compounded their mistake by making it illegal for lively, traditional neighborhoods like Bishop Arts to grow up near transit infrastructure. This is why suburban light rail stations like Parker Road are surrounded by acres of asphalt rather than offices, homes, and shops. It's a shame, really, because stations like Mockingbird and Parker Road are missed opportunities to do things the old-fashioned way.

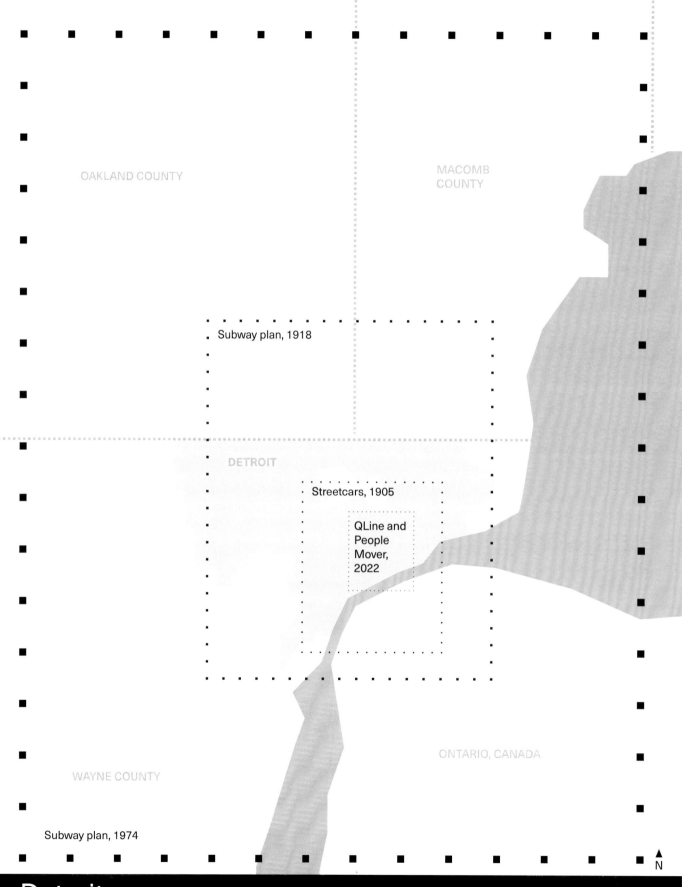

OAKLAND COUNTY

MACOMB
COUNTY

Subway plan, 1918

DETROIT

Streetcars, 1905

QLine and
People
Mover,
2022

ONTARIO, CANADA

WAYNE COUNTY

Subway plan, 1974

N

Detroit

5 miles/8 km

DETROIT

The City-Suburban Rift and the Most Useless Transit System in the World

7.1 Map of Metro Detroit.

After decades of decay that made the city synonymous with urban blight, Detroit has finally turned a corner. In the late 2000s and early 2010s, the city was destitute. There was talk of selling off the Detroit Institute of Arts' collection—one of the largest in the world—to satisfy creditors.[1] Architectural wonders like the 18-story, marble-walled Michigan Central Station were left to rot. For decades, the station's only regular visitors were urban explorers, thieves, and vandals. Chrysler's 2011 Super Bowl commercial asked rhetorically, "What does a town that's been to hell and back know about the finer things in life?"[2] By 2013, Detroit (fig. 7.1) owed $18.5 billion to creditors that it could not pay. The city declared bankruptcy. It was the largest municipal bankruptcy in North American history.

The bankruptcy forced the region to realize that major policy changes were needed to preserve the metropolis's long-run viability.[3] Metro Detroit was up to the challenge. (Within Michigan, "Metro Detroit" is the usual term for the Detroit metropolitan area.) A decade later, Detroit's city center has been revitalized. A forest of new towers has popped up downtown.[*] Regional cooperation saved the Detroit Institute of Arts' collection from creditors, and put the Institute on sound financial footing. Ford bought Michigan Central Station and embarked on a $740 million renovation.[4]

---

[*] In concentrating development downtown, not every neighborhood could be salvaged. The math is cold and ruthless. Detroit was built for 2 million inhabitants, and it cannot maintain that much public infrastructure with a 2020 population of 639,000.

A major factor in this resurgence is the thawing of relations between Detroit and its suburbs. The majority-black city and the majority-white suburbs were at each other's throats for decades. Naked racism was common. For example, long-time county executive L. Brooks Patterson, of suburban Oakland County, remarked in 2014: "What we're going to do is turn Detroit into an Indian reservation, where we herd all the Indians into the city, build a fence around it, and then throw in the blankets and corn."[5] Metro Detroit's transit system reflected the toxicity of this relationship. Although 67 percent of city residents work outside the city, and 75 percent of workers in the city live in the suburbs, most suburban buses weren't allowed to stop in the city as late as the mid-2010s.[6]

The city-suburb split is also why Detroit is home to the least useful piece of mass transit infrastructure in North America. The transit system in question is the Detroit People Mover, a three-mile aerial loop that goes in only one direction, serves a single square mile of downtown Detroit, and has so many stations that it's often faster to walk. The People Mover is an orphan. The People Mover was originally designed to be the downtown circulator for a regional, federally funded subway in the 1970s. The subway suffered a death by a thousand cuts because the city and the suburbs couldn't make nice.

There's plenty of blame to go around for the subway plan's collapse. Detroit's mayor Coleman Young, legendary for wielding profanity like a scalpel, feuded theatrically with the suburbs. The suburbs fired back, and suburban politicians stubbornly refused to commit local funds. In the end, the feds got cold feet. Thus, the only portion of the subway plan actually built was the People Mover, which opened in the mid-1980s.

Post-bankruptcy, transit improvements have mirrored improvements elsewhere in the region. The city and suburbs have agreed to run truly regional bus lines for the first time in Metro Detroit's history. A regional transit authority was finally set up. Things are looking up for the first time in half a century. But it still bears examining how Metro Detroit got into this situation in the first place. This is the story of Detroit's relationship with its suburbs.

An Abusive Monopoly

Detroit's transit system wasn't always subject to brutal city-suburb infighting. In the early 20th century, Detroit had a unified, modern transit system that stretched to faraway cities like Ann Arbor and Lake Orion, which are still at the urban periphery 125 years later (fig. 7.2). The catch was, the transit system was run by an abusive monopoly, the Detroit United Railway (DUR). To navigate Metro Detroit, there was no other practical alternative to the DUR. The car was a toy for rich people, and the bus hadn't been invented yet. To avoid the DUR, the options were to walk or ride a horse. (The DUR's relationship with Detroit mirrors Los Angeles's relationship with the Pacific Electric Railway, and San Francisco's relationship with the Market Street Railway.)

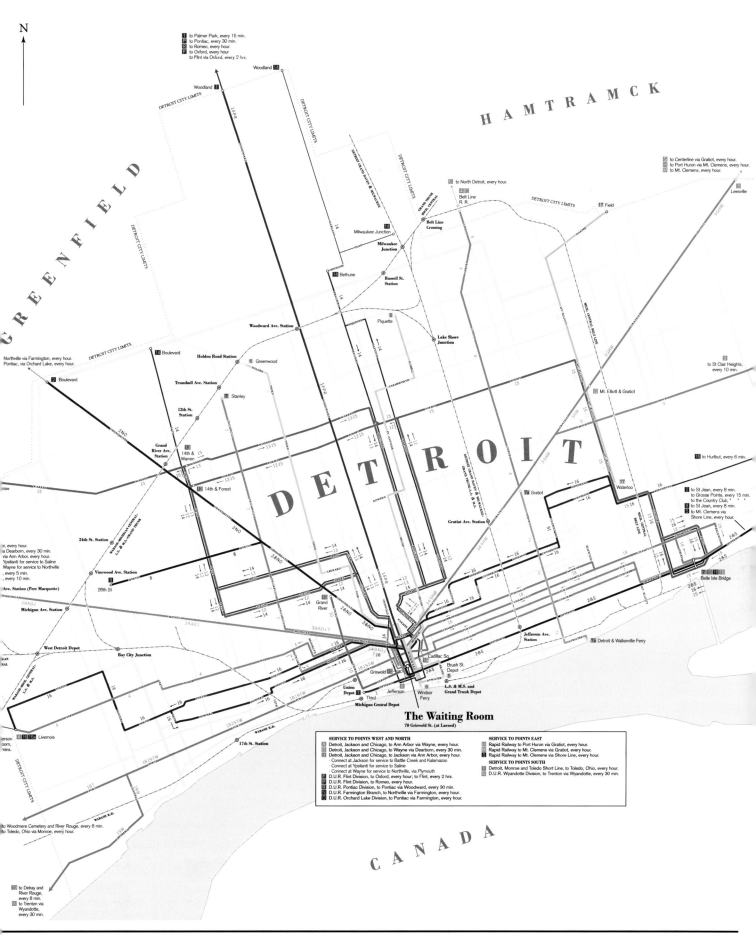

DETROIT

Street car and suburban railway service.
Service updated for 1905.

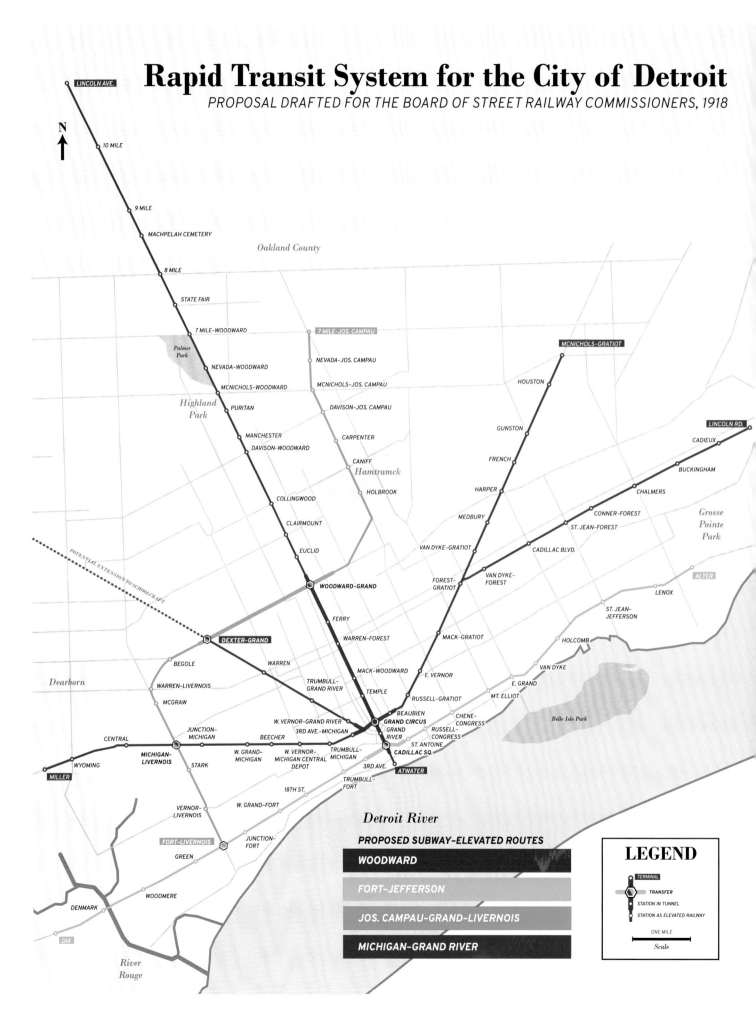

Rapid Transit System for the City of Detroit

PROPOSAL DRAFTED FOR THE BOARD OF STREET RAILWAY COMMISSIONERS, 1918

N

LINCOLN AVE.

10 MILE

9 MILE

MACHPELAH CEMETERY

Oakland County

8 MILE

STATE FAIR

7 MILE-WOODWARD

7 MILE-JOS. CAMPAU

MCNICHOLS-GRATIOT

Palmer Park

NEVADA-WOODWARD

NEVADA-JOS. CAMPAU

HOUSTON

LINCOLN RD.

MCNICHOLS-WOODWARD

MCNICHOLS-JOS. CAMPAU

CADIEUX

Highland Park

PURITAN

DAVISON-JOS. CAMPAU

GUNSTON

BUCKINGHAM

MANCHESTER

CARPENTER

FRENCH

CHALMERS

DAVISON-WOODWARD

CANIFF

Hamtramck

HARPER

CONNER-FOREST

Grosse Pointe Park

COLLINGWOOD

HOLBROOK

MEDBURY

ST. JEAN-FOREST

CLAIRMOUNT

CADILLAC BLVD.

ALTER

EUCLID

VAN DYKE-GRATIOT

LENOX

WOODWARD-GRAND

FOREST-GRATIOT

VAN DYKE-FOREST

ST. JEAN-JEFFERSON

POTENTIAL EXTENSION TO SCHOOLCRAFT

FERRY

HOLCOMB

WARREN-FOREST

MACK-GRATIOT

VAN DYKE

DEXTER-GRAND

E. GRAND

Dearborn

BEGOLE

WARREN

MACK-WOODWARD

E. VERNOR

Belle Isle Park

WARREN-LIVERNOIS

TRUMBULL-GRAND RIVER

TEMPLE

MT. ELLIOT

MCGRAW

RUSSELL-GRATIOT

BEAUBIEN

CHENE-CONGRESS

W. VERNOR-GRAND RIVER

GRAND CIRCUS

JUNCTION-MICHIGAN

3RD AVE.-MICHIGAN

GRAND RIVER

RUSSELL-CONGRESS

CENTRAL

BEECHER

ST. ANTOINE

MICHIGAN-LIVERNOIS

W. GRAND-MICHIGAN

W. VERNOR-MICHIGAN CENTRAL DEPOT

TRUMBULL-MICHIGAN

CADILLAC SQ.

STARK

WYOMING

3RD AVE.

MILLER

ATWATER

TRUMBULL-FORT

18TH ST.

VERNOR-LIVERNOIS

W. GRAND-FORT

Detroit River

FORT-LIVERNOIS

JUNCTION-FORT

GREEN

PROPOSED SUBWAY-ELEVATED ROUTES

WOODWARD

FORT-JEFFERSON

DENMARK

WOODMERE

JOS. CAMPAU-GRAND-LIVERNOIS

DIX

MICHIGAN-GRAND RIVER

River Rouge

LEGEND

TERMINAL

TRANSFER

STATION IN TUNNEL

STATION AS ELEVATED RAILWAY

ONE MILE

Scale

7.3 Subway proposal, released 1918. Mayor James
     Couzens vetoed the proposal in 1920.

The DUR was so hated that Detroit mayor James Couzens, a former Ford executive, vetoed the construction of a subway system in 1920 because the funding plan required the city to enter into a joint venture with the DUR (fig. 7.3).[7] Couzens (rhymes with *dozens*) preferred to take control of the DUR's streetcars first. Once the buyout was complete, Detroit would build a subway. Couzens got his wish. In 1922, the city took control of the DUR's streetcars.[8] Subway planning began shortly thereafter, but the stock market crash of 1929 put the kibosh on these plans.

In the early 21st century, the idea of Detroit annexing a suburb is unthinkable. But in the early 20th century, it was the norm. At the time, the city gradually absorbed nearby streetcar suburbs and brought them into the fold. This form of expansion continued well into the 1920s.[9] These neighborhoods, with rows of 1,000-square-foot bungalows, abound to this day within the Detroit city limits. There was no city-suburban divide at the time, simply put. But as the mass-produced streetcar suburbs built before World War II gave way to the mass-produced freeway suburbs after the war, the fates of the city and the suburbs began to diverge.

Redlining and Freeways

There were two key factors that led the city and the suburbs to split up. The first was the Federal Housing Administration's (FHA) mortgage policies. The FHA refused to subsidize mortgages in black neighborhoods while subsidizing brand-new, whites-only suburban subdivisions. This process is called "redlining."[10] FHA policies represented a formalization of preexisting, unofficial rules used by local banks and realtors. In Detroit, as in the rest of the country, redlining meant that blacks were crowded into older city neighborhoods. Banks would not invest in businesses and mortgages within the redlined neighborhoods. Blacks were barred from living elsewhere, sometimes by law and sometimes by social convention. Black families who attempted to move to white-only neighborhoods were frozen out by realtors. If they persisted in trying to move, those families were often threatened with physical harm. Whites who could afford to leave the inner city left.[11]

The second factor was the Federal-Aid Highway Act of 1956, which supplied large-scale federal subsidies to build freeways. Detroit, automobile capital of the world, used this funding to build a half dozen freeways through the city core. Detroit's freeway system made it easy for white, suburban commuters to live outside the city and work in the city. Freeways serving downtown were routed through black neighborhoods, leaving those areas impoverished and polluted, and compounding the effects of redlining.[12] But redlining and freeways happened nearly everywhere in the United States. What made Detroit's situation particularly toxic was the interplay of redlining, freeways, and local race relations.

Exodus

Detroit's prosperity had attracted newcomers for decades, but worker shortages still persisted. To staff the factories, automakers recruited both black and white Southerners, as well as poor Appalachian whites. This process went into overdrive when the United States entered World War II. Migrants poured into Detroit by the trainload. Detroit's housing shortage and race relations, never great to begin with, only got worse. The city was riven by wartime race riots in 1942 and 1943.

After the end of the war, white flight for the suburbs began in earnest, as residents and jobs left the city. At its peak in 1950, Detroit proper had 1.9 million residents, 84 percent white and 16 percent black. By 1970, the city population had shrunk to 1.5 million, 55 percent white and 43 percent black. Over the same period, Metro Detroit's population grew by 40 percent, from 3.2 million to 4.5 million. By and large, the suburbs were whiter than a ghost in a snowstorm. At the 1970 census, Dearborn, population 104,199, had 13 black residents; Livonia, population 110,199, had 41; Troy, population 39,419, had 1.

The urban core was beginning to decay. To keep people and jobs within the city, many transit proposals were mooted. There were plans to build a monorail, multiple schemes involving express buses on freeways, and a few different subway plans. All fell through for one reason or another.[13] The proposal that got the most traction was the 1970s subway plan under the auspices of the now-defunct Southeastern Michigan Transportation Authority (SEMTA).[14] The fight over SEMTA's subway (fig. 7.4) showcased Metro Detroit's racial and ethnic fault lines.[15] In the end, the only thing to come out of the 1970s subway proposal was the sad, useless People Mover.

Stop the Robberies, Enjoy Safe Streets

The Michigan Legislature formed SEMTA on July 10, 1967.[16] The agency was initially established to unify the region's fragmented, financially unstable transit operators under one umbrella, including a half dozen bus operators and two commuter train services.[17] SEMTA's other major goal was to build a full-scale regional subway system, following the model of Bay Area Rapid Transit in greater San Francisco. The federal government offered to pay the bulk of the subway cost, but only if the region could agree on how to provide a local contribution.[18] This did not happen. Metro Detroit was rapidly fragmenting and an agreement would be impossible to achieve.

Two weeks after SEMTA's formation, Detroit exploded into riots again. The riots kicked off because the Detroit police raided a black speakeasy in the early hours of the morning on July 23, 1967, and arrested everyone inside.[19] A crowd gathered, and violence broke out. The overmatched police retreated, leaving the city to descend into chaos. Arson, looting, and sniper fire stretched across the city, day and night. The authorities declared a curfew and called in reinforcements from the county sheriff, the state police and the National Guard. Even so, the situation took only a few hours to spin out of control. Two days later, on July 25, President Lyndon Johnson deployed 4,700 paratroopers from the elite 82nd and 101st Airborne Divisions, plus 8,000 more National Guardsmen. It took three days for federal troops to restore order. Forty-three people died.

N

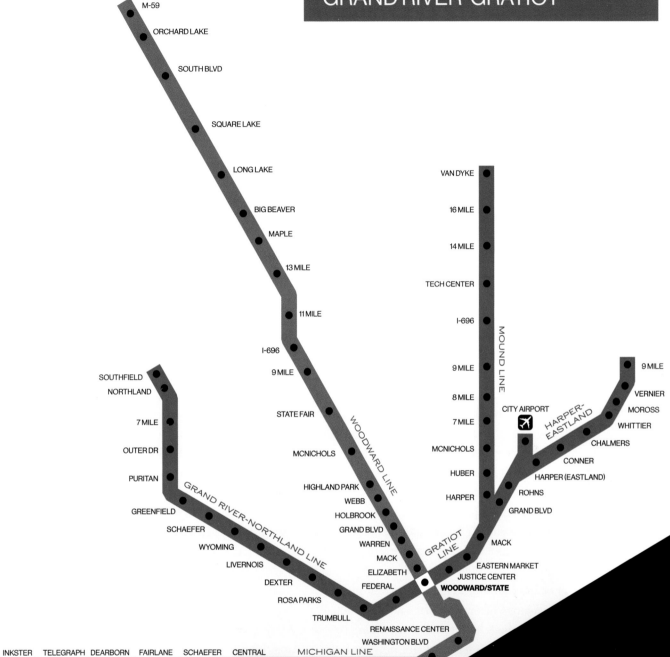

M-59

ORCHARD LAKE

SOUTH BLVD

SQUARE LAKE

LONG LAKE

BIG BEAVER

MAPLE

13 MILE

11 MILE

I-696

9 MILE

STATE FAIR

MCNICHOLS

HIGHLAND PARK

WEBB

HOLBROOK

GRAND BLVD

WARREN

MACK

ELIZABETH

FEDERAL

WOODWARD LINE

VAN DYKE

16 MILE

14 MILE

TECH CENTER

I-696

9 MILE

8 MILE

7 MILE

MCNICHOLS

HUBER

HARPER

MOUND LINE

CITY AIRPORT

HARPER-EASTLAND

9 MILE

VERNIER

MOROSS

WHITTIER

CHALMERS

CONNER

HARPER (EASTLAND)

ROHNS

GRAND BLVD

MACK

EASTERN MARKET

JUSTICE CENTER

WOODWARD/STATE

GRATIOT LINE

SOUTHFIELD

NORTHLAND

7 MILE

OUTER DR

PURITAN

GREENFIELD

SCHAEFER

WYOMING

LIVERNOIS

DEXTER

ROSA PARKS

TRUMBULL

GRAND RIVER-NORTHLAND LINE

RENAISSANCE CENTER

WASHINGTON BLVD

INKSTER   TELEGRAPH   DEARBORN   FAIRLANE   SCHAEFER   CENTRAL

MICHIGAN LINE

AIRPORT LINE

VAN BORN

I-94

METRO AIRPORT

GRAND BLVD   RIVERVIEW

LIVERNOIS

SPRINGWELLS

SCHAEFER

OUTER DRIVE

SOUTHFIELD

EMMONS BLVD

NORTHLINE ROAD

PENNSYLVANIA

FORT LINE

semta

proposed metro detroit rapid transit network, march 1974

A backlash was inevitable. In the 1969 election, the law-and-order candidate, former Wayne County sheriff Roman Gribbs, was elected mayor. As mayor, Gribbs promptly brought the hammer down, creating a secret, elite undercover police unit called STRESS (Stop the Robberies, Enjoy Safe Streets) to reduce street crime. STRESS officers wandered around what were considered dangerous areas posing as old ladies, alcoholics, priests, and other easy targets. When an alleged mugger appeared, supporting units pounced. STRESS operations routinely ended in violence. Between 1971 and the unit's disbandment in 1974, STRESS officers shot and killed 22 civilians, the vast majority black men.[20] STRESS was popular among whites and detested by blacks.[21] Crime rates kept rising.

STRESS's existence and the high crime rate became the key issues in the 1973 mayoral election. By then, Detroit's population was evenly split between whites and blacks, and thoroughly polarized on racial lines.[22] Gribbs was not running for re-election. The two candidates to succeed him could not be more opposite. One was the foul-mouthed state senator Coleman Young, who was emphatically vocal about his black identity. Young vowed to abolish STRESS if elected. The other was white police commissioner John Nichols, part of the brain trust that had created STRESS. Young got 92 percent of the black vote, while Nichols got 91 percent of the white vote, giving Young a narrow win.[23]

Hit Eight Mile

January 2, 1974 was a frigid day in Detroit, with a high of 16 degrees Fahrenheit. It was inauguration day for Coleman Young. In his inaugural address, Young said: "I issue open warnings now to all dope pushers, to all rip-off artists, to all muggers: it's time to leave Detroit; hit Eight Mile [the dividing line between city and suburbs]. And I don't give a damn if they are black or white, or if they wear Superfly suits or blue uniforms with silver badges. Hit the road."[24] Blacks understood this speech as a promise to fight crime and to reform a police department known for its excesses. Whites thought Young was encouraging Detroit's riffraff to prowl the suburbs north of Eight Mile Road.[25] At one swift stroke, Young endeared himself to blacks in the city and infuriated whites in the suburbs. Shortly thereafter, he fired Nichols and abolished STRESS.[26] Nichols left the city, like so many other whites already had. Nichols eventually was elected sheriff in suburban Oakland County, on the other side of Eight Mile Road.

As mayor, Young courted big business and federal dollars, and attempted to arrest the urban decay caused by white flight and deindustrialization. He never abandoned his aggressive style. As one *Detroit Free Press* columnist put it: "He didn't mince words, he sometimes adopted an angry tone, and he often swore. For many whites in the 1970s, Young was the first elected leader to confront them with the race issue, and many resented being lectured by a militant black man."[27]

This dynamic came into play when President Gerald Ford offered Metro Detroit $600 million ($3.4 billion in 2022 dollars) to build a SEMTA subway in 1976. In retrospect, it was a foregone conclusion that the majority-black city and the majority-white suburbs wouldn't be able to reach an accord. All 30 local newspapers in Oakland County published a supplemental "Public Service Supplement" opposing "Coleman Young's subway."[28] L. Brooks Patterson, then Oakland County prosecutor, was against it. Simply put, the suburbs were not swayed. Without suburban buy-in, the federal government wouldn't fund the subway.[29] Revised versions of SEMTA's rail plan in 1979 and 1984 also fell flat. Suburban support for regional transit was so dismal that Metro Detroit's two remaining commuter rail services closed without fanfare in 1983 and 1984.

Nonetheless, the People Mover project kept chugging along, with the backing of Mayor Young, and Detroit's congressional delegation secured federal funding. While the People Mover was under construction in the 1980s, the *Detroit Free Press* took a poll and found that two-thirds of Detroiters thought the People Mover was a bad idea. One joker quipped, "It will make a nice-looking Stonehenge."[30] At its opening in 1987, the *New York Times* reported that the People Mover, "once considered a possible savior of the city's dying center, is now being greeted with both the excitement of a child trying out a new toy and the dread of a parent wondering what will go wrong first."[31] The People Mover was a white elephant from day 1.

When construction began, it was projected to carry 55,000 passengers per day.[32] People Mover ridership, in actual service, averaged about 11,000 in its first year and slowly dropped thereafter. 4,600 passengers rode it on an average weekday in 2019. The People Mover has been the object of ridicule the entire time. *Gizmodo* called it "the world's most pointless transit system" in 2015.[33]

Détente

Half a century has passed since Metro Detroit turned down President Ford's $600 million. It has been 36 years since the People Mover opened. Coleman Young and L. Brooks Patterson have died. Some of the earlier enmity has seemingly been buried with them. For the first time in decades, the city and the suburbs are finally beginning to work together on their shared issues.

For example, the People Mover was joined in 2017 by a 3.3-mile streetcar called the QLine (fig. 7.5). The streetcar runs from downtown Detroit to the old General Motors headquarters on Grand Boulevard, and it's meant to promote center-city development. Initially, the federal government refused to fund the QLine because the city and suburbs hadn't agreed to establish a regional transit authority.[34] Thus, in 2013, the same year the city of Detroit declared bankruptcy, the Regional Transit Authority opened. Metro Detroit's first regional transit plan came out two years after.[35] Other improvements have happened as well. The city and suburban bus authorities now share a ticketing sys-

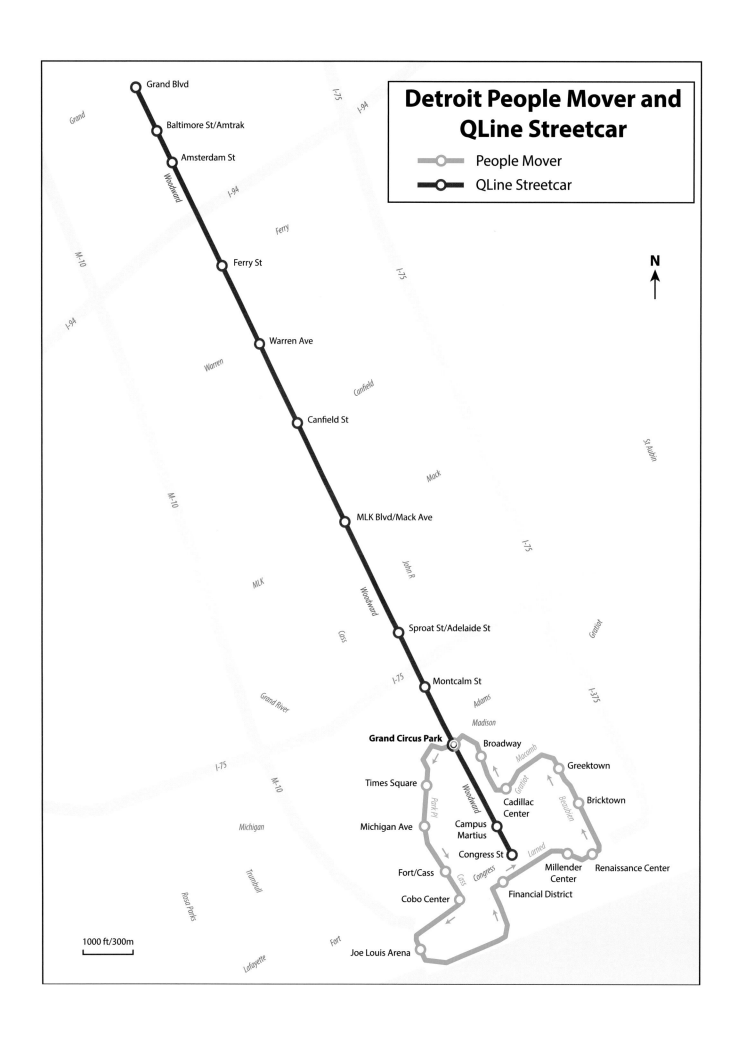

Detroit People Mover and QLine Streetcar

People Mover
QLine Streetcar

N

Grand Blvd
Baltimore St/Amtrak
Amsterdam St
Ferry St
Warren Ave
Canfield St
MLK Blvd/Mack Ave
Sproat St/Adelaide St
Montcalm St
Grand Circus Park
Broadway
Times Square
Greektown
Cadillac Center
Bricktown
Michigan Ave
Campus Martius
Congress St
Fort/Cass
Millender Center
Renaissance Center
Cobo Center
Financial District
Joe Louis Arena

1000 ft/300m

7.5 Detroit People Mover and QLine streetcar, 2022.

tem. The suburban bus authority started running a frequent regional express bus service that serves both the city and the suburbs. This collaboration is an encouraging sign that the decades-long regional feud is finally ending.

Streetcars,
1895

METRORail, 2022

GRAND PARKWAY

BELTWAY 8

HOUSTON

N

Houston

10 miles/16 km

8

HOUSTON

The City of Organic Growth

8.1 Map of greater Houston.

Transport and land use policy are closely related, because suburban sprawl isn't just about the physical transport infrastructure that's in place. It's also about the land use laws that restrict what can be built. Before World War II, neighborhoods generally developed organically around transport infrastructure. Land was used more efficiently as land values increased. This isn't the norm anymore. In the early 21st century, it's distressingly common for expensive, high-capacity mass transit investments to go to waste for lack of appropriate land use policies. The problem is especially egregious in expensive coastal cities with low housing supply, like Los Angeles. This pattern is also present in other postwar metropolises, like Dallas. But there's one city where more organic development still happens: Houston (fig. 8.1). This is because Houston has kept its old-fashioned land use laws. In broad strokes, Houston loosely regulates the shape of the built environment, but not building uses or densities.[1] This makes Houston a useful counterexample to typical growth patterns.

Due to its hot, humid weather, Houston was relatively small for most of its history, and it didn't enter the first rank of American cities until residential air conditioning became universal after World War II. When Houston came of age, it embraced the freeway, the automobile and the parking lot. The city is notably sprawling as a result.

But Houston's land use policy also illustrates how organic densification can result when large quantities of parking aren't required, transit investments are made, and appropriately liberal land use laws are in place.

79

When Houston opened a light rail system in its city core in the early 21st century, the city allowed homes to be built on smaller lots than in the past and partially eliminated its mandatory minimum parking law. As a result, the areas around light rail stations began to undergo the kind of evolution that was common in the past, with apartment buildings and rowhouses replacing old, worn-out tract homes.

Two Big Breaks

Founded in 1836, Houston is today a major city by accident of history. For its first 65 years, Houston was a relatively sleepy town on the banks of the muddy, shallow Buffalo Bayou. Its small streetcar network was typical for a city of its size (fig. 8.2). The city was surrounded by swampy terrain and beset by occasional epidemics of tropical disease. Houston's rival cities had much better natural advantages. Galveston, 50 miles to the southeast on an island in the Gulf of Mexico, had a better natural harbor, a healthier climate, and a bustling port. New Orleans, 300 miles to the east, had a 125-year head start and far better inland connections. New Orleans had been the traditional entrepôt of the Gulf Coast since colonial times, as the Mississippi River and its tributaries all flow through New Orleans.

Houston's first big break came in 1900, when a category 4 hurricane destroyed Galveston, leaving 8,000 dead. (To this day, the Galveston hurricane is still the deadliest natural disaster in the recorded history of North America.) Galveston was seen as too risky a place to invest in after this catastrophe.

Houston took full advantage, dredging the mud of Buffalo Bayou to accommodate oceangoing ships. The city's investment paid off handsomely. By 1930, Houston was Texas's largest city, its busiest port, and the hub of the Texas Gulf Coast's petroleum industry. On the transport front, the city streetcar system was slowly shut down and replaced with buses bit by bit until the streetcar system's final closure in 1940.

Houston's second big break came after World War II, when indoor air conditioning became universal. Houston was already developing rapidly because of its prosperity and its busy port. Air conditioning supercharged that growth. Unusually friendly to Northern transplants and capital, Houston ultimately displaced complacent, old-fashioned New Orleans as the commercial capital of the Gulf Coast. Tellingly, of Houston's postwar mayors, half have been transplants. In contrast, it made national headlines when a Los Angeles native was elected mayor of New Orleans in 2017.[2] President George H. W. Bush, a Connecticut Yankee, began his political career as a Houston congressman. He made Houston his home when not in public service.

The Ultimate Horizontal City

Between 1940 and 1970, Houston's population more than tripled. Greater Houston's population nearly quadrupled. Nearly all that growth was built around freeways and sprawl. In 1975, when asked by the *New York Times* about the potential long-term consequences of sprawl, Mayor Fred Hofheinz remarked simply, "Houston is a very

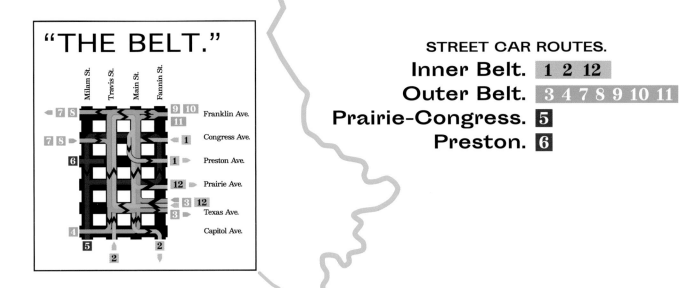

"THE BELT."

STREET CAR ROUTES.
Inner Belt. 1 2 12
Outer Belt. 3 4 7 8 9 10 11
Prairie-Congress. 5
Preston. 6

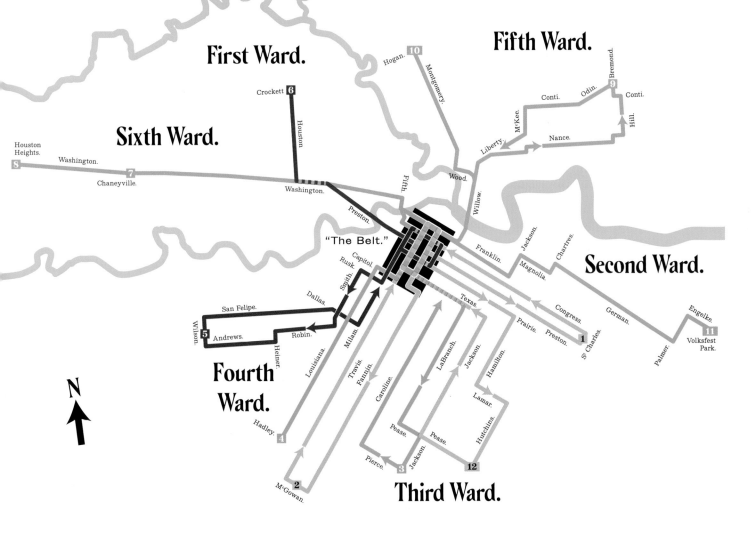

HOUSTON CITY STREET RAILWAY.

STREET CAR ROUTINGS AS OF APRIL 15, 1895.

growth-oriented city, and there is very little opposition to that proposition."[3] By 2020, 2.3 million people lived inside the 640-square-mile city limits, making it one-third as dense as Chicago and half as dense as Los Angeles.

Through all this time, Houstonians steadfastly refused to adopt a traditional zoning law. Voters rejected zoning codes outright in 1929, 1948, 1962, and 1993. Because of this, there are comparatively few restrictions on what can be built, provided that the land isn't subject to restrictive covenants—in a few wealthy neighborhoods, restrictive covenants are included in property deeds to prevent anything except suburban-style houses from being built.[4] Otherwise, Houston's local laws are a throwback to the past. Its laws define the physical shape and scale of buildings using setbacks, height limits, and other regulations on urban form. Uses and densities are not regulated. But Houston turned into sprawl anyway. There are three major reasons for this.

One, Houston's geography is sprawl-friendly. The entire area is flat as a pancake, and the only real limitations on further horizontal expansion are commuter patience and the Gulf of Mexico.

Two, Houston invested early and often in freeways, and continues to do so at an incredibly rapid pace. Greater Houston has already outgrown Interstate 610, its original 38-mile freeway loop, which encircles an area twice the size of San Francisco. It has also outgrown Beltway 8, its 88-mile second freeway loop, which encircles an area larger than New York City. Work is underway to build a third freeway loop, 170 miles long, called the Grand Parkway, which will encircle an area the size of Delaware. For comparison, the entire metropolis has only 23 miles of light rail, nearly all contained within the I-610 inner loop. Transit has simply not been a priority.

Three, until very recently, Houston's minimum parking law and its lot size law generally required development to follow a suburban paradigm.

Suburbia without Zoning

First, the minimum parking law. Houston's minimum parking law requires a lot of parking by default. Office buildings are legally required to be 50 percent parking lot by square footage. Apartments and condominiums are required to be 30 percent–55 percent parking, depending on the number of bedrooms and square footage. Supermarkets are required to be 66 percent parking. Nightclubs are required to be 85 percent parking.[5] To illustrate, Houston requires 14 parking lot spaces for every 1,000 square feet of "bar, club or lounge," and those spaces take up 5,600 square feet, on average.[6] Thus, the ratio is 85 percent parking lot, 15 percent nightclub.

This requirement was set in stone citywide until 2013, when the city abolished the minimum parking law downtown.[7] The city did the same for two other neighborhoods close to downtown and the light rail lines in 2019, but the rest of Houston is still filled with vast acres of mandatory parking lots.[8]

The other major issue was the minimum lot size law. Before 1998, Houston required all parcels used for single-family homes to be at least 5,000 square feet. All parcels used for townhouses had to be at least 2,500 square feet. These requirements made it impossible to build the kinds of rowhouse neighborhoods that characterize most older cities, even though the demand for that type of neighborhood existed. For comparison, cities like Philadelphia and San Francisco have rowhouses on lots that range from 600 to 2,500 square feet. The 1998 lot size reform reduced the minimum size for single-family homes to 1,400 square feet inside the I-610 loop. The reform was extended to the rest of the city in 2013. As a result, Houston contractors often raze old tract homes to build multiple rowhouses instead of doing house flips.[9]

The ability to build rowhouses is important, because rowhouses are much simpler to build than apartment buildings. In the United States, rowhouses use the looser International Residential Code standard, which is meant for one- and two-family buildings. Multifamily apartment buildings use the more complex International Building Code. Most general contractors have the skills to construct a building to International Residential Code standards, but the International Building Code requires more specialized knowledge. (North of the border, Canadian building codes make similar distinctions.)

Old-Fashioned Development, Houston Style

Houston's land use reforms roughly coincided with the opening of the city's light rail system in the '00s. These changes enabled Houston to take full advantage of the new transit infrastructure. Houston built its new light rail system in the city core, almost entirely inside the I-610 loop (fig. 8.3). Significant development has sprung up around the rail stations, especially in formerly rundown Midtown Houston on the southern part of the Red Line. The Midtown stations are now surrounded by new mid-rise and high-rise condominium buildings, making effective use of this transit access and proximity to jobs. The Midtown population grew by 50 percent between 2012 and 2022.[10] The same process is happening on the Purple and Green Lines southeast of the city center, near the campuses of Texas Southern University and the University of Houston.

Houston's results contrast favorably with those of its 21st-century rivals, Dallas and Los Angeles. Dallas built long suburban light rail lines that miss major destinations and employment centers, and largely failed to adjust its land use laws to accommodate the new infrastructure. The result is expensive mass transit that serves few passengers and compares poorly to Houston's city-center light rail. Per mile of track, Houston's light rail has twice as many riders as Dallas's.

Compared to Los Angeles, the supply of housing is plentiful and relatively inexpensive, and the homelessness rate is unusually low.[11] In 2019, Greater Houston had 3,938 homeless in a metro area of 7.1 million. Los Angeles County had 56,247 homeless—14 times the homeless population—out of a total population of 9.8 million.[12] The ample supply of housing also means that Houston's gentrification is slower than other wealthy cities. With the abundance of housing options, affluent Houstonians tend to move to existing prosperous neighborhoods. In cities like Los Angeles, those same types of families tend to move to more marginal neighborhoods and price out the working

METRORail System Map

N

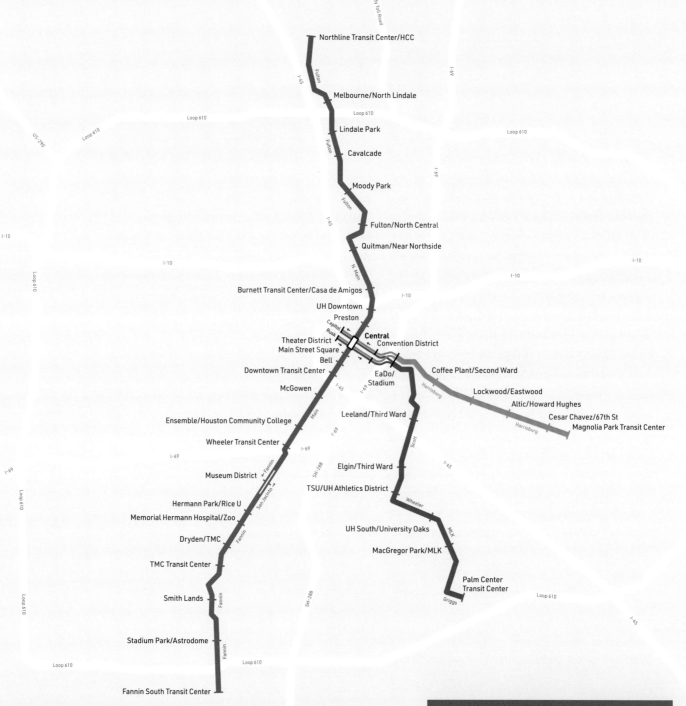

Northline Transit Center/HCC

Melbourne/North Lindale

Lindale Park

Cavalcade

Moody Park

Fulton/North Central

Quitman/Near Northside

Burnett Transit Center/Casa de Amigos

UH Downtown

Preston

Capitol

Rusk

Theater District

Main Street Square

Bell

Downtown Transit Center

McGowen

Ensemble/Houston Community College

Wheeler Transit Center

Museum District

Hermann Park/Rice U

Memorial Hermann Hospital/Zoo

Dryden/TMC

TMC Transit Center

Smith Lands

Stadium Park/Astrodome

Fannin South Transit Center

Central

Convention District

EaDo/Stadium

Coffee Plant/Second Ward

Lockwood/Eastwood

Altic/Howard Hughes

Cesar Chavez/67th St

Magnolia Park Transit Center

Leeland/Third Ward

Elgin/Third Ward

TSU/UH Athletics District

UH South/University Oaks

MacGregor Park/MLK

Palm Center Transit Center

Red Line
Northline TC to Fannin South

Purple Line
Theater District to Palm Center TC

Green Line
Theater District to Magnolia Park TC

1 mile/1600m

8.3 Harris County METRORail, 2022.

classes.[13] The average Los Angeles house costs 3.6 times what it does in Houston.[14]

Houston has a well-justified reputation as the ultimate horizontal city. But it did the right things inside the I-610 loop. Houston built light rail lines within the city core, where the people and the jobs are, rather than extending lines into faraway suburbs. At the same time, its loose land use regulations allow extensive private development near high-capacity mass transit, supplying plentiful housing at minimal cost to the public. This approach is exactly right if the goal is to have people near train stations, and train stations near people.

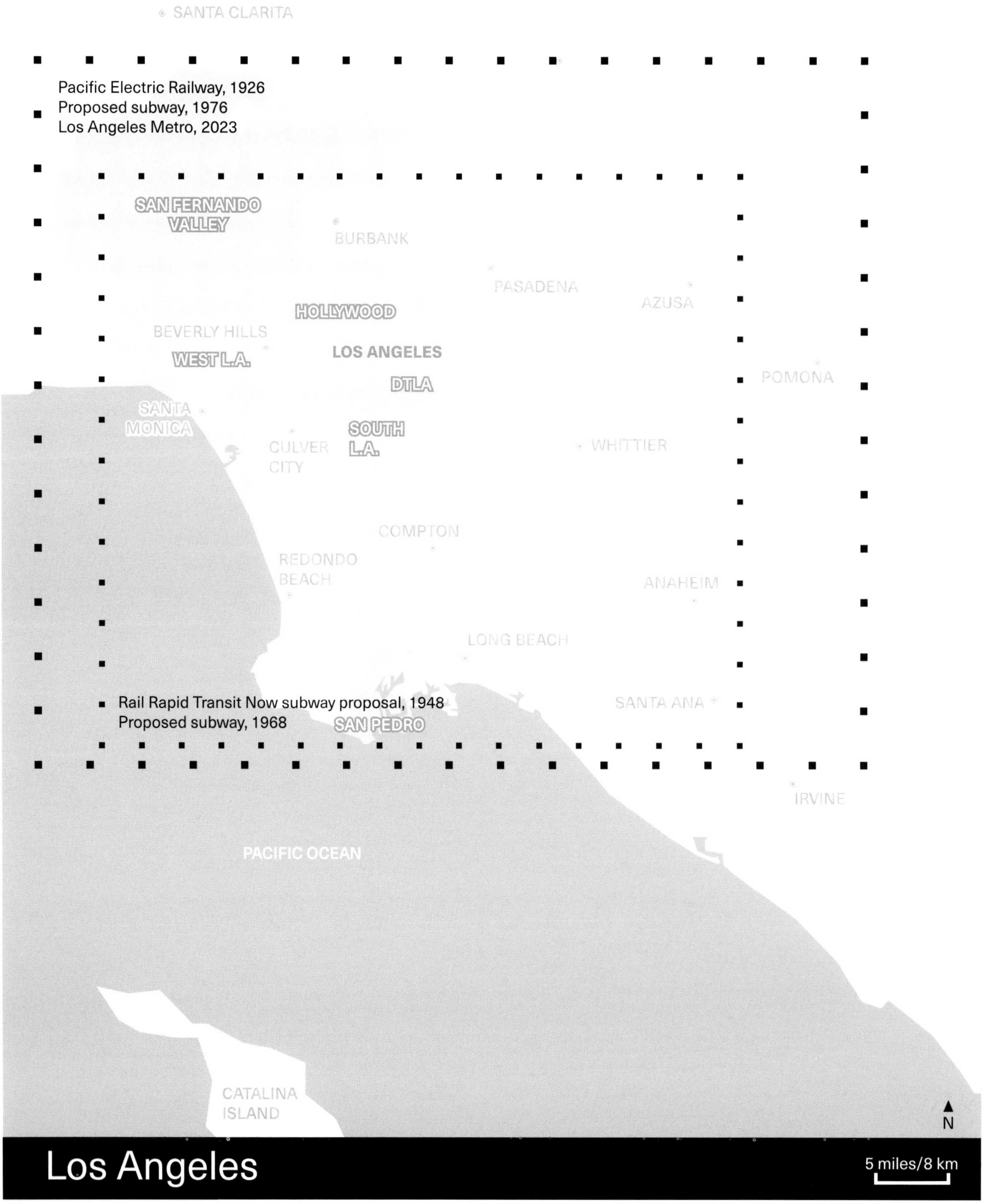

SANTA CLARITA

Pacific Electric Railway, 1926
Proposed subway, 1976
Los Angeles Metro, 2023

SAN FERNANDO
VALLEY

BURBANK

PASADENA

AZUSA

HOLLYWOOD

BEVERLY HILLS

LOS ANGELES

WEST L.A.

DTLA

POMONA

SANTA
MONICA

CULVER
CITY

SOUTH
L.A.

WHITTIER

COMPTON

REDONDO
BEACH

ANAHEIM

LONG BEACH

Rail Rapid Transit Now subway proposal, 1948
Proposed subway, 1968

SANTA ANA

SAN PEDRO

IRVINE

PACIFIC OCEAN

CATALINA
ISLAND

N

Los Angeles

5 miles/8 km

9

LOS ANGELES

72 Suburbs in Search of a City

9.1 Map of greater Los Angeles.

Los Angeles is synonymous with urban sprawl, interminable traffic jams, and smog.[1] This is popularly blamed on the General Motors streetcar conspiracy, which allegedly brought LA's interurbans, the Pacific Electric Railway's Red Cars, to ruin. The truth is more complicated, as it reveals a much more complex relationship between Angelenos and their cars than is commonly understood. Los Angeles (fig. 9.1) has always meant sprawl. It was designed that way.

Even in the early 20th century, Los Angeles was quite decentralized, in no small part because of the Pacific Electric itself. The Pacific Electric's original owner, Henry Huntington, used the Red Cars to promote his real estate interests, not the other way around. His modus operandi was simple: buy cheap farmland far from downtown, build subdivisions, and connect them to downtown with fast, frequent interurbans. The Pacific Electric didn't just benefit from suburban sprawl—the company encouraged it. For the first quarter of the 20th century, the Pacific Electric reigned supreme.

This monopoly was not to last.

As early as the 1920s, suburbs like Santa Monica and Culver City were setting up municipal bus services to compete with the Red Cars. The Pacific Electric was rapidly running out of real estate to develop. When the first freeways opened in 1940, Angelenos were hooked. Over the course of the next four decades, Los Angeles's mass transit system steadily fell apart as Angelenos refused to use tax dollars to support public transportation. LA built freeways instead. Even an oil crisis wasn't enough to convince Angelenos of the 1970s that rapid transit was necessary. Exacerbating the problem, late 20th-century Angelenos decided there was little need to build enough housing to meet demand. Los Angeles's most desirable inner-ring suburbs, initially settled during the Red Car era, were particularly eager to preserve their exclusivity.

As time passed, traffic got worse, and the memory of the Pacific Electric's lousy service faded. A revisionist myth began to take hold: that a nefarious conspiracy had brought the beloved Red Car system to grief. The theory, famously expounded in the film *Who Framed Roger Rabbit*, became a part of Los Angeles's collective consciousness. By the 1980s and 1990s, a sea change was underway as Los Angeles began to reexamine its commitment to the automobile.

The process was slow and painful. Nevertheless, in the 21st century, Los Angeles has undergone a revolution in its thinking about public transportation. An unprecedented regional consensus has formed in favor of rapid transit. But Los Angeles has not adapted its land use laws to take advantage of its newly built transit system. Nor has it built sufficient housing to match its job growth. Even today, significant sections of the modern Los Angeles Metro system run through neighborhoods of suburban single-family homes—many of which were built during the days of the Red Cars.

Water, Oil, and Rail

After its founding in 1781, Los Angeles languished in obscurity for a century. Then everything changed. Los Angeles was linked to San Francisco by rail in 1876, and the transcontinental Santa Fe Railroad reached Los Angeles in 1885. During the early years, the biggest driver of development was petroleum. A prospector named Edward Doheny struck oil in 1892, and forests of oil derricks sprung up all over the Los Angeles Basin. By 1920, California was the world's largest oil producer. Wells were scattered around the region in the unlikeliest of places. (The Long Beach municipal cemetery, the upscale Beverly Center mall, and Beverly Hills High School all played host to oil derricks at one time or another.) Between 1890 and 1920, Los Angeles County's population grew from 101,000 to over 900,000.

The oil boom and resulting population growth put enormous stress on Los Angeles's water supply. The sunny, dry climate simply doesn't provide enough water to support a large city. Los Angeles's solution was to engage in a series of delightfully sketchy dealings to secure a reliable water supply in the Owens Valley, 200 miles away. Through a mix of bribery, fearmongering, and self-dealing, city engineer William Mulholland and former mayor Fred Eaton acquired the rights to the Owens Valley water and used it to fill the city-controlled San Fernando Valley aquifer.

Huntington, owner of the Pacific Electric, got wind of Mulholland and Eaton's scheme. Before long, Huntington was part of a realty syndicate called the San Fernando Mission Land Company, established in 1904. Huntington bought into the syndicate for $15,000 ($458,000 in 2022 dollars). Using Huntington's inside information, the cabal quickly and quietly bought up wide swaths of arid San Fernando Valley semidesert.[2] Los Angeles voters approved bonds to build the aqueduct in 1905. Huntington and his coconspirators made a killing. With the water supply secured, Huntington's San Fernando Mission Land Company sold its newly valuable lands for orchards and townships, and Huntington's Pacific Electric provided the transportation. Huntington sold his shares in the syndicate for $130,000 (nearly $4 million in 2022 dollars) in 1912, shortly before the megalithic Los Angeles Aqueduct opened. Huntington got an 866 percent return on his original investment.[3]

The Largest Electric Railway System in the World

The Los Angeles Aqueduct put the final piece in place for the City of Angels to enter the first rank of American cities. The Pacific Electric used its control over transportation to build the first fully realized horizontal city. Even in its early days, Los Angeles was far less dense than its counterparts farther east. This policy was deliberately encouraged by the city's land use laws. At the time, Chicago was building 400-foot skyscrapers, and New York was building 700-foot towers. Los Angeles banned all buildings taller than 150 feet, about 13 stories. The city's goal was "preventing undue concentration of traffic," according to a 1925 engineering report commissioned by the City Council.[4] Angelenos knew that this policy of decentralization was an obstacle to functional mass transit. That same 1925 report noted: "The desire of the average citizen to own his own home has caused the

single-family dwelling to predominate and the absence of large apartment buildings is noticeable. Such a condition is very desirable, but it is one of the prime factors that makes the construction and operation of rapid transit lines on a self-sustaining basis, a difficult financial problem."[5]

The Pacific Electric recognized this, too. Even during the heyday of the Red Cars (fig. 9.2), passenger traffic was a consistent money loser, and the company made up the difference by carrying freight and developing real estate.[6] But by the mid-1920s, this system was beginning to falter. There was less cheap land left to develop. Politicians and the public expected the Pacific Electric to keep running passenger trains. The company had a huge amount of legacy rail infrastructure to maintain. The system was overextended.

The Red Cars faced two problems, one technological and one political. The technological problem was straightforward: autos, buses, and trucks drastically cut into the Pacific Electric's long-term financial viability. The unprofitable passenger business continued to shrink. Angelenos wealthy enough to buy cars bought them and took to the roads. By 1925, Los Angeles had the highest level of car ownership in the United States, with one car for every four residents.[7] Train service got steadily worse, as the Red Cars usually had to share lanes with motorists.

Buses also posed a major challenge, as they allowed even small operators to take on the mammoth Pacific Electric at a low cost. The first city bus companies in California opened in 1928, when the then-small communities of Santa Monica and Culver City bought buses in response to a Red

89

PACIFIC ELECTRIC RAILWAY SYSTEM

IN

Southern California

9.2 "Red Car" system of the Pacific Electric
     Railway, 1926.

Car fare hike. These municipal buses bypassed the Red Car lines entirely and instead connected with the Yellow Cars, which provided local streetcar service in central Los Angeles. Bus networks generally have lower capacities than rail networks, but buses do have certain advantages. In particular, bus networks are cheaper to build and maintain than rail because they use existing roads. Buses can also detour around obstructions that would block a train. To make trains really shine, trains have to have dedicated rights-of-way—that is, their own lanes or trackways—which the Red Cars generally didn't have in the congested areas of central Los Angeles. While the Pacific Electric also bought buses, the competition wasn't saddled by the cost of operating the largest electric railway system in the world.

The other problem was political. The company couldn't raise fares enough to cover their costs or make infrastructure improvements, because regulators denied the requests. (A similar dynamic was present in other cities, most notably in New York's decades-long battle with its transit operators over the five-cent subway fare.) When the Pacific Electric requested public dollars for infrastructure upgrades, Los Angeles voters were unwilling to pay. Angelenos rejected a proposal to convert the busiest Red Car lines, including the one-mile LA Subway, for rapid transit use in the mid-1920s.[8]

So Los Angeles did nothing. The region continued to sprawl, its transit infrastructure declined, and commuters turned increasingly to the car.

Victory of the Highway Men

Debates between supporters of rapid transit and supporters of the automobile continued apace through the Great Depression and World War II. Los Angeles's first freeways opened in 1940. The Hollywood Freeway included Pacific Electric tracks in the median when it opened. The other, the Arroyo Seco Parkway, did not. These freeways were immediately popular. Soon there was general accord that Los Angeles should build more of them. The pressing questions were who should get freeways, and whether those freeways should include rapid transit. The war put the issue on the back burner, but it returned with a vengeance after the war's end in 1945.

The war years were a bright spot for Los Angeles public transportation. Gasoline and rubber were rationed, so Angelenos took transit in record numbers. The Pacific Electric's passenger services turned a profit for the first time in 20 years.[9] But the Red Cars' situation was dire by the end of the war. The system was under heavy strain from wartime passenger loads and decades of deferred maintenance. The Pacific Electric was threatening to abandon all passenger service in favor of its still-profitable buses, as the company lost $3.4 million ($45 million, adjusted for inflation) running passenger trains in 1947 alone.[10] It was well known that the government would have to make large investments to keep the Red Cars viable.

Simply put, the trains were slow, unreliable, and old. To illustrate, Culver City is seven and a

RAIL RAPID TRANSIT *now!*

As proposed by the Los Angeles Chamber of Commerce, February 1948

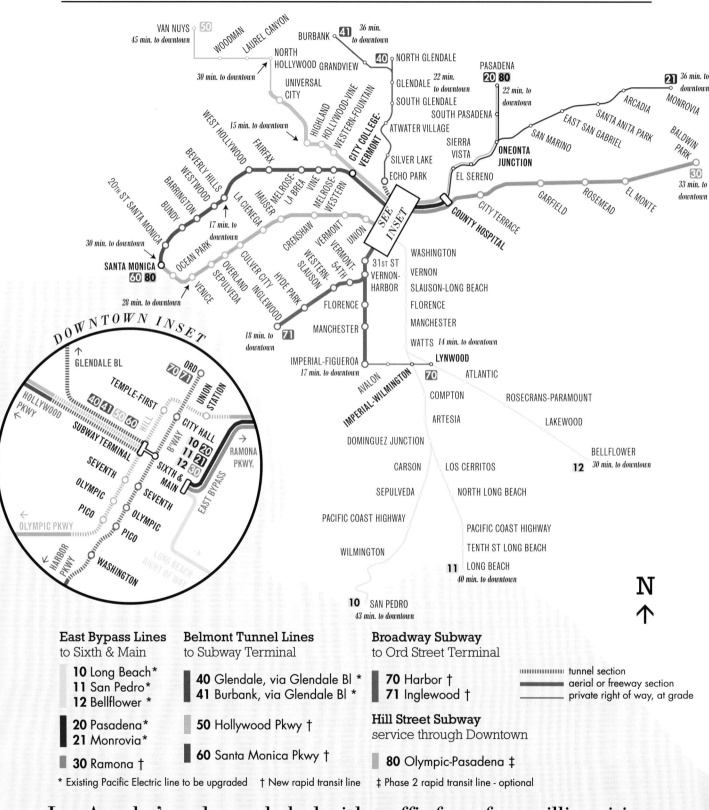

N ↑

East Bypass Lines
to Sixth & Main

10 Long Beach*
11 San Pedro*
12 Bellflower*

20 Pasadena*
21 Monrovia*

30 Ramona †

Belmont Tunnel Lines
to Subway Terminal

40 Glendale, via Glendale Bl *
41 Burbank, via Glendale Bl *

50 Hollywood Pkwy †

60 Santa Monica Pkwy †

Broadway Subway
to Ord Street Terminal

70 Harbor †
71 Inglewood †

Hill Street Subway
service through Downtown

80 Olympic-Pasadena ‡

⦙⦙⦙⦙⦙⦙ tunnel section
▬▬▬ aerial or freeway section
——— private right of way, at grade

* Existing Pacific Electric line to be upgraded   † New rapid transit line   ‡ Phase 2 rapid transit line - optional

Los Angeles' roads are choked with traffic from four million citizens.
What happens when we have six million?

9.3 Rail Rapid Transit Now proposal, 1948. I've re-
constructed stop locations based on contemporary
practices. They are of necessity an estimate.

half miles from Downtown Los Angeles as the crow flies. The modern Metro E Line light rail covers that distance in 30 minutes. In the late 1940s, that same trip took 39 minutes by bus and 43 minutes by Red Car.[11]

These circumstances led to the creation of the 1948 plan *Rail Rapid Transit Now*, assembled by a committee known as the Rapid Transit Action Group (RTAG) (fig. 9.3). RTAG was composed of an array of Downtown Los Angeles power brokers, including Mayor Fletcher Bowron, the city attorney Ray Chesebro, and the Chamber of Commerce president James Beebe. *Rail Rapid Transit Now* proposed a total overhaul of the Red Car system. Street-level Red Car lines would be relocated to freeway medians. Red Car lines with dedicated rights-of-way would be upgraded with crossing gates, like today's light rail. New tunnels and viaducts would completely remove trains from downtown streets. The government would take over the system from the failing Pacific Electric. By doing so, Los Angeles would kill three birds with one stone. In theory, the new freeways and transit would alleviate traffic congestion by taking cars off the streets, reduce urban blight by demolishing inner-city neighborhoods of poor people and minorities, and provide an alternative to the automobile.

Bowron assembled a coalition of powerful downtown interests in favor of the plan, but the Los Angeles City Council narrowly killed it in a dramatic vote.[12] It was Los Angeles's ascendant powers that defeated *Rail Rapid Transit Now*. Fearful of losing shoppers to Downtown LA, an unlikely alliance of suburban cities, shopping mall developers, and outlying commercial districts

blasted the plan.[13] In the view of RTAG's opponents, it would be an un-American interference in the free market to take over and modernize the failing rail operations of the Pacific Electric.

With *Rail Rapid Transit Now* defeated, the freeways went forward without rapid transit tracks. Both the Pacific Electric and the local Yellow Cars entered their final death spiral. The Red Car tracks in the median of the Hollywood Freeway were removed in the 1950s, the LA Subway closed in 1955, the last Red Cars ran in 1961, and the last Yellow Car lines closed in 1963.

A Genuine, Bona Fide, Electrified Six-Car Monorail

Los Angeles's freeway system was intended to put an end to the metropolis's traffic congestion. Ironically, the new freeways only made traffic worse. The phenomenon, known as induced demand, is laconically summed up as follows: "Traffic expands to fill the road space available."[14] The experience of the pioneering Hollywood Freeway is instructive: one year after its opening, the road was carrying twice its design capacity.[15]

The freeways opened up Los Angeles's surrounding farmland to new development, and the suburbs boomed. Whites, in particular, left central Los Angeles en masse, enabled by federal housing subsidies and freeway policies that encouraged suburban homeownership and excluded minorities. (These policies are discussed in greater depth in the Detroit chapter.) Flush with tax dollars, Los Angeles demolished much of its city core to make it easier for the new suburbanites to drive down-

93

town. Downtown offices, homes, and stores were replaced with surface parking lots. Whole neighborhoods were razed for freeways.

Making matters worse, Angelenos were unwilling to pay for better transit. The generation that had grown up with the slow, rickety Red Cars was in no mood to fund the kinds of fixes necessary to make trains run on time. Mayor Bowron's experience with *Rail Rapid Transit Now* had shown as much. With traditional rail out of fashion, Los Angeles's leadership tried a space-age solution to the problem of urban congestion: the monorail. The Los Angeles County Board of Supervisors released a monorail plan in 1954, seven years before the end of the Red Cars.[16]

The monorail push treated Los Angeles's longstanding antipathy to rapid transit as a marketing problem. Monorails had a much better public image in the 1950s than traditional elevated and subway lines, even though they don't have that many technological advantages. Monorails are somewhat cheaper to build than traditional elevated lines and marginally better at climbing slopes, but their disadvantages are fairly serious. First, monorail technology isn't standardized. If a city buys, say, a Lanley Monorail (with apologies to *The Simpsons*), trains and maintenance equipment can't be bought off the rack. Rather, the equipment must be ordered from the original manufacturer or custom built. Second, monorail junctions are much more expensive and mechanically complex than traditional rail junctions, limiting transit operators' ability to build branch lines.[17] This is because monorail junctions require the entire track to bend. In contrast, the technology for traditional

rail junctions is cheap, mechanically simple, and has been in use since the 1830s. Third, monorails require substantially more vertical clearance than traditional subway systems. Construction savings from cheaper aerial structures are usually offset by the bigger tunnels required.[18]

But in the end, it wasn't the technical or marketing aspects that doomed the 1954 monorail plan. It was the money. Contemporary planners, politicians, and the public assumed that rapid transit systems could finance themselves through fares, once built. Transit was still considered a business, not a public service.[19] But Los Angeles's investment in auto infrastructure was causing Southern California's sprawl to rapidly expand, and public transportation ridership was collapsing. The financials just didn't add up. The plan went in the dustbin.

The Best Laid Plans

The remnants of the Red Car and Yellow Car systems were taken over by the Metropolitan Transit Authority, which dismantled the surviving lines between 1958 and 1963. Around the same time, rival San Francisco voted to build the high-tech BART subway. Los Angeles felt obligated to jump on the bandwagon, but the Metropolitan Transit Authority was hamstrung by its inability to levy taxes.[20] This led the state legislature to reorganize the Metropolitan Transit Authority as the Southern California Rapid Transit District (RTD), a replacement agency with authority to tax. RTD tried to convince the voters three times to build a subway system. The first attempt went on the ballot in 1968 (fig. 9.4).

94

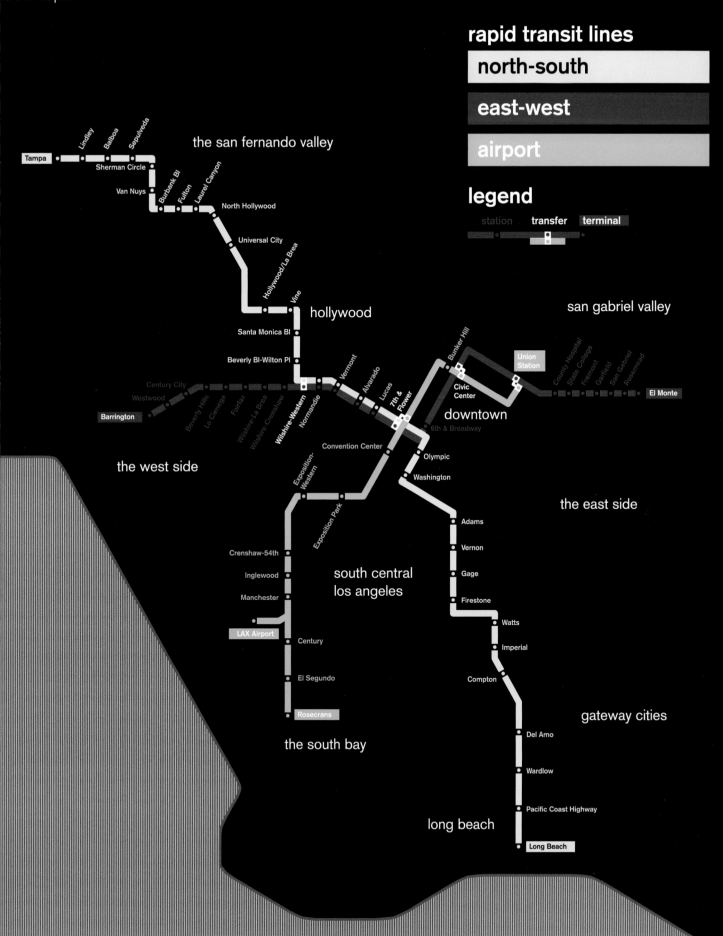

Baxter Ward's subway proposal, 1976. The specific services of the Baxter Ward plan are an educated guess on my part. The original plans include copious information on track layouts, station locations, and alignments, but I haven't seen a proposed service guide.

At the time, Los Angeles was still amenable to large-scale urban changes, although the region's center of gravity was slowly shifting away from traditional downtown interests to the wealthy West Los Angeles suburbs along Wilshire Boulevard. The single largest landowner to take advantage of the shift was none other than movie studio 20th Century Fox. Fox was nearly bankrupt after spending $31 million ($290 million in 2022 dollars) to make *Cleopatra*, starring Elizabeth Taylor and Richard Burton. To save the company, executives decided to sell off the 200-acre studio backlot and fill it with skyscrapers.[21] Other developers soon joined the push. By the end of this era, modernist high-rises stretched all the way from Downtown Los Angeles to Westwood, 11 miles away.

Given the circumstances, it made perfect sense to build a subway to serve the new towers. Voters weren't convinced. In particular, the expense of the new system and its perceived focus on serving trips to and from Downtown Los Angeles was unpopular.[22] RTD's 1968 subway plan needed a 60 percent supermajority to pass, but only 45 percent of voters supported it.[23] RTD floated two more rapid transit proposals were floated over the next 10 years. Neither convinced a skeptical electorate that rapid transit was necessary.

The second plan, from 1974, tried to square the circle. This plan focused more on suburban routes and less on downtown. For example, dedicated busways were planned for the 710 and Century Freeways through the suburbs northeast and south of downtown. Los Angeles had a good demonstration of the possibilities of busways after the El Monte busway in the median of the 10 Freeway—

the eastern half of today's J Line—had proved a success. Voters still weren't convinced. Even in the middle of the 1970s oil crisis, the measure fell eight percentage points short of a majority.

The third and final plan of this era came two years later, in 1976 (fig. 9.5). RTD's 1976 proposal was the brainchild of county supervisor and sometime TV anchor Baxter Ward. Ward believed that voters would be sold on a rapid transit plan if it blanketed the entire county, not just the densest corridors of central Los Angeles. Ward also wanted to include special excursion cars, with luxuries like transparent roofs, so surfers and cyclists could carry their boards and bikes to the beach.[24] He envisioned 230 miles of subway and elevated lines, mostly in freeway medians or following existing railways. This enormous system was the length of the London Underground. Ward dubbed it "the new Red Cars."[25] Voters were offered the possibility to pay extra to put their local line underground, to defuse neighborhood opposition to the visual impact of elevated lines. So, for example, the Central Line, planned for the 110 Freeway median, could be shifted to parallel tunnels under busy Vermont Avenue with voter approval. None of these gestures had any effect. 60 percent of voters opposed the plan on election day.[26]

In the meantime, a marked transformation was happening in local politics. In earlier decades, growth was treated as a necessity. This changed drastically in the 1960s and 1970s, as the newly built suburbs altered their land use policies to restrict further development.[27] This was driven by a mix of self-interest and fear. After the courts voided racial restrictions on home sales, the largely

white suburbs turned to zoning policy to achieve the same result. This process only accelerated after the Watts riots of 1965 turned the poor, mostly black neighborhoods of South Los Angeles into a war zone for six days.

To illustrate, at the 1960 census, Los Angeles proper had 2.5 million inhabitants, and its zoning laws were designed to accommodate 10 million. By 1990, the city's population had grown to 3.5 million—but the city had reduced its housing capacity to 3.9 million residents by reducing the land available for apartments.[28] The suburbs were even more aggressive. Many placed de facto moratoriums on new housing construction. Particularly wealthy suburbs like Beverly Hills and Santa Monica stopped gaining residents around 1970 and began to shrink thereafter. Beverly Hills' population at the 1970 census was 33,416, and by 1990, it had shrunk to 31,971. Santa Monica shrank from 88,289 to 86,905.

These new housing supply restrictions caused home prices to skyrocket. Increased property values meant higher property taxes. In previous eras of development, homeowners who couldn't afford their taxes would sell to apartment developers and pocket the cash. But 1970s homeowners wanted both high home values and low property taxes. A man named Howard Jarvis tapped into that desire.

Howard Jarvis had come onto Los Angeles's political scene in the 1930s. He was a failed mayoral candidate and consummate gadfly, the kind of man who showed up to city council meetings to rant. Jarvis had little traction until he started assembling a political campaign to reduce local property taxes. He hit paydirt. Jarvis's signature initiative, Proposition 13, rewrote the state property tax law so that property taxes would be based on initial purchase price, not current market value. Property tax assessments are effectively frozen in time.[29] While Proposition 13's current iteration allows annual valuation increases up to 2 percent, inflation has averaged 3.5 percent per year since 1978. In other words, property owners pay less tax every year relative to inflation.[30]

In spite of opposition from nearly every major political group in California, Proposition 13 passed in a landslide in June 1978.[31] It has been the poster child for unintended consequences ever since. Local property tax revenues dropped by 60 percent. Local governments were forced to rely on sales tax revenue and development fees to stay afloat. Worse, Proposition 13 also applied to commercial and industrial property, not just residential. This completely changed local governments' revenue calculations.

Before Proposition 13, when new housing was built, increased property taxes would offset increased demand for city services. After Proposition 13, new housing would be a net negative for city coffers over time. New residents would need more city services, but over time those residents' tax payments would slowly decrease relative to inflation. In contrast, shopping centers and hotel complexes would produce sales tax and be a net positive. This created a strong financial incentive for local governments to prevent new housing from being built.[32] New housing was thus pushed to the edge of the metropolis, instead of densifying the existing urban core.* As of 2020, greater Los Angeles covered an area twice the size of Switzerland.[33]

Old Habits Die Hard

At the beginning of the 21st century, Los Angeles's car-oriented suburbia and its restrictive land use laws have created a crisis. There's little land left for new freeways or suburban subdivisions, unless you want to live in the Mojave Desert or in a wildfire zone. This combination neatly defines modern Los Angeles's two biggest challenges: bad traffic and expensive real estate, a situation that the coronavirus pandemic has only made worse.

On the transportation side, Angelenos slowly came to recognize the problem. Voters approved the LA Metro in 1980, but the system was controversial in the beginning. (The first line, the A Line light rail, followed the route of the old Red Car line to Long Beach.) Wealthy neighborhoods looked to preserve their exclusivity by canceling or stripping down voter-authorized rapid transit. The D Line subway initially fell victim to these shenanigans. In March 1985, a freak methane gas explosion destroyed a Ross Dress for Less store near the subway's route. West Los Angeles residents, never big fans of the subway to begin with, successfully lobbied Congressman Henry Waxman to ban tunneling along the corridor. This ban would last until 2005.[34] Similarly, the G Line busway in the San Fernando Valley was originally meant to be an ele-

vated extension of the B Line subway, but the Valley's representatives in the state legislature banned elevateds and light rail there in 1991. Subway tunnels cost too much to build, leaving a busway as the only possibility. One reporter later remarked that it was "the days of LA's anti-rail zealotry."[35]

Nevertheless, the LA Metro system has gradually expanded in fits and spurts to include 110 miles of light rail and subway and 60 miles of busway (fig. 9.6). This expansion has only gained steam over the decades. Twenty-first-century Angelenos have assembled an unprecedented consensus in support of better mass transit, voting twice to expand the LA Metro with new taxes. This includes the long-awaited D Line subway extension to the towers of West Los Angeles, a rail connection to Los Angeles International Airport, an expansion of the busway network, and a tunnel allowing the A Line and E Line to run through downtown. It's the largest, most ambitious set of transit projects since the Pacific Electric.

The real estate question is more challenging, because it involves changing Los Angeles's conception of itself. Since the days of the Red Cars, the metropolis has defined itself as a horizontal city, not a vertical one. Modern Los Angeles's laws still reflect this ambivalence about building upward. Even in 2022, the City of Los Angeles still bans apartments in three-quarters of its residential land. In the areas where apartments are legal, development is hobbled by sclerotic bureaucracy, city council meddling, and byzantine laws like the California Environmental Quality Act. (The California Environmental Quality Act is discussed in

---

* Proposition 13 also applied to the San Francisco Bay Area and had similar effects there.

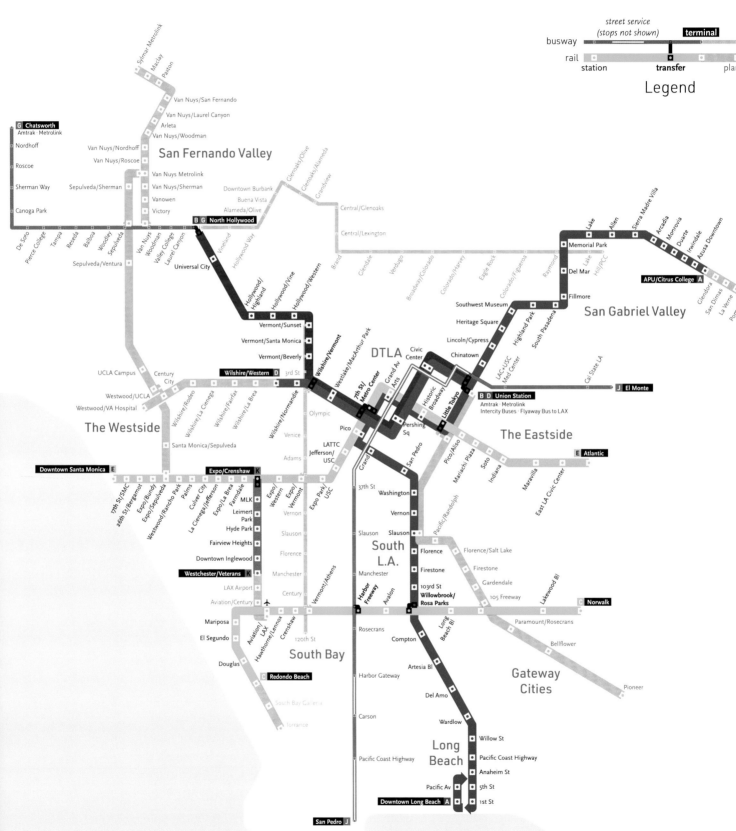

LOS ANGELES METRO

RAIL AND BUSWAY SYSTEM MA

street service
(stops not shown)
busway   terminal
rail
station   transfer   plann

Legend

San Fernando Valley

G Chatsworth
Amtrak · Metrolink
Nordhoff
Roscoe
Sherman Way
Canoga Park

Sylmar/Metrolink
Maclay
Paxton
Van Nuys/San Fernando
Van Nuys/Laurel Canyon
Arleta
Van Nuys/Woodman
Van Nuys/Nordhoff
Van Nuys/Roscoe
Van Nuys Metrolink
Van Nuys/Sherman
Vanowen
Victory

Sepulveda/Sherman

Downtown Burbank
Buena Vista
Alameda/Olive

B G North Hollywood

Central/Glenoaks
Central/Lexington

De Soto
Pierce College
Tampa
Reseda
Balboa
Woodley
Sepulveda
Van Nuys
Woodman
Valley College
Laurel Canyon

Sepulveda/Ventura

Universal City

Hollywood/Highland
Hollywood/Vine
Hollywood/Western

Glenoaks/Olive
Glenoaks/Alameda
Grandview

Central/Glenoaks

Brand
Glendale
Verdugo
Broadway/Colorado
Colorado/Harvey
Eagle Rock
Colorado/Figueroa
Raymond

Lake
Allen
Sierra Madre Villa
Arcadia
Monrovia
Duarte
Irwindale
Azusa Downtown

Memorial Park
Del Mar
Lake Hill/PCC

APU/Citrus College A

Glendora
San Dimas
La Verne
Pomona

San Gabriel Valley

Southwest Museum
Heritage Square
Lincoln/Cypress
Chinatown

Highland Park
South Pasadena

Fillmore

LAC+USC Med Center

Cal State LA

J El Monte

Vermont/Sunset
Vermont/Santa Monica
Vermont/Beverly

UCLA Campus
Century City

Wilshire/Western D   3rd St

Wilshire/Vermont
Westlake/MacArthur Park
7th St/Metro Center
Grand Av Arts
Civic Center

DTLA

Historic Broadway
Little Tokyo

B D Union Station
Amtrak · Metrolink
Intercity Buses · Flyaway Bus to LAX

The Eastside

Westwood/UCLA
Westwood/VA Hospital

The Westside

Wilshire/Rodeo
Wilshire/La Cienega
Wilshire/Fairfax
Wilshire/La Brea
Wilshire/Normandie

Olympic
Venice
Pico
Adams

Pershing Sq

Pico/Aliso
Mariachi Plaza
Soto
Indiana
Maravilla
East LA Civic Center

E Atlantic

Santa Monica/Sepulveda

Downtown Santa Monica E

Expo/Crenshaw K

17th St/SMC
26th St/Bergamot
Expo/Bundy
Expo/Sepulveda
Westwood/Rancho Park
Palms
Culver City
La Cienega/Jefferson
Expo/La Brea
Farndale
MLK
Leimert Park
Hyde Park
Fairview Heights
Downtown Inglewood

Westchester/Veterans K

LAX Airport

Aviation/Century

Expo/Western
Expo/Vermont
Expo Park/USC

Vermont
Vernon
Slauson
Florence
Manchester
Century

37th St
Washington
Vernon
Slauson

LATTC Jefferson/USC
Grand
San Pedro

Jefferson/USC

South L.A.

Manchester

Harbor Freeway
Avalon

Florence
Firestone
103rd St
Willowbrook/Rosa Parks

Florence/Salt Lake
Firestone
Gardendale
105 Freeway

Pacific/Randolph

Paramount/Rosecrans
Bellflower
Pioneer

C Norwalk

Gateway Cities

Lakewood Bl

Mariposa
El Segundo
Douglas

Aviation/LAX
Hawthorne/Lennox
Crenshaw
120th St

Vermont/Athens

Rosecrans
Compton
Artesia Bl

Long Beach Bl

South Bay

Redondo Beach

South Bay Galleria

Torrance

Harbor Gateway
Carson

Pacific Coast Highway

Del Amo
Wardlow
Willow St
Pacific Coast Highway
Anaheim St
5th St
1st St

Long Beach

Pacific Av

Downtown Long Beach A

San Pedro J

N

9.6 Los Angeles Metro, 2023.

greater detail in the chapter on San Francisco.) The metropolis's zoning is still designed for the car. In the parts of Los Angeles with frequent transit, huge amounts of parking was legally required until the State of California overrode local minimum parking laws in 2022.[36] For example, the upscale Beverly Center mall in West Los Angeles is an eight-story building. The first story is half parking, half shops, the second through fifth stories are all parking, and the sixth through eighth stories are all shops.[37]

Some reforms have been made, like the City of Los Angeles's Transit Oriented Communities program. Transit Oriented Communities allows developers to exceed zoning restrictions if below-market-rate housing is included in new transit-adjacent buildings. But these changes have been inadequate to address the scale of the crisis. Between 2011 and 2017, Los Angeles County added 5.5 new jobs for every new unit of housing.[38] The most desirable places to live in the metropolis—places like Beverly Hills, Santa Monica, and Manhattan Beach—have had little or no population growth since 1970.[39] The result has been pollution, a horrific housing shortage, tent cities of the homeless, and an exodus of working-class Angelenos for Nevada and Texas. Over half of homes in Los Angeles sold for above the asking price in early 2021.[40]

It's not an exaggeration to say that the status quo threatens the region's long-term prosperity. As a technical matter, the problem is fixable: streamline the land use bureaucracy, require less parking, and put more jobs and apartments near transit. It's not brain surgery. Even in California, other cities have figured out how to do this.[41] But the politics are hard, even though the average Los Angeles home is unaffordable to anyone making less than $200,000 per year.[42] In early 2020, Los Angeles's representatives in the California state senate provided the decisive votes against a bill that would have allowed six-story apartments near all LA Metro stations.[43] Old habits die hard—especially the ones that have been part of Los Angeles's DNA since the Pacific Electric.

101

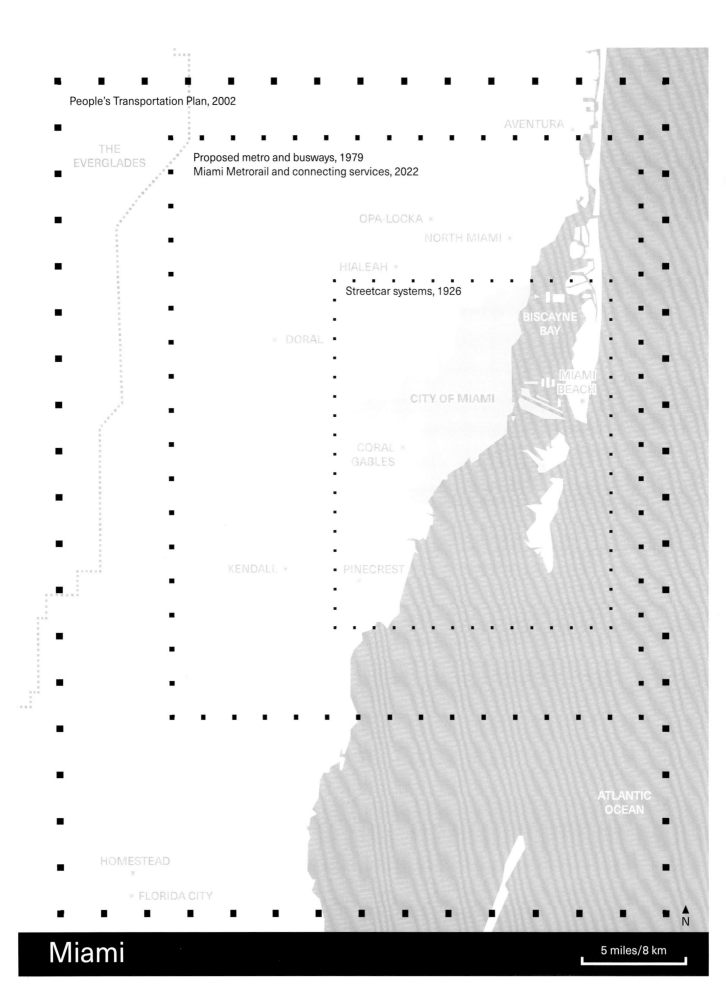

People's Transportation Plan, 2002

THE
EVERGLADES

AVENTURA

Proposed metro and busways, 1979
Miami Metrorail and connecting services, 2022

OPA-LOCKA

NORTH MIAMI

HIALEAH

Streetcar systems, 1926

BISCAYNE
BAY

DORAL

MIAMI
BEACH

CITY OF MIAMI

CORAL
GABLES

KENDALL   PINECREST

ATLANTIC
OCEAN

HOMESTEAD

FLORIDA CITY

N

Miami

5 miles/8 km

MIAMI

Overpromise, Underdeliver

10.1 Map of greater Miami.

Miami is renowned for its fine weather, its boom-bust economy, and its sketchy real estate deals. Greater Miami (fig. 10.1) has a reputation as a region for salesmen and hucksters, an environment where the natural state of things is to overpromise and underdeliver. This tradition goes all the way back to the city's founding. The mother of Miami was a Cleveland real estate speculator named Julia Tuttle, who used her inheritance to buy a square mile of orange groves along the Miami River in 1891. The city got its big break when Tuttle convinced robber baron Henry Flagler to extend his railway to Miami.

Flagler's railway arrived in 1896. He invested early and often in Miami real estate and infrastructure to maximize the value of his railway assets. But even with Flagler's investments, Miami remained a small winter colony for the rich until the 1920s, when brilliant marketing and cheap land turned South Florida into a real estate speculator's paradise. During this first great Florida land bubble, the developers who built Miami Beach and Coral Gables opened their own streetcar companies to attract home buyers and vacationers. The bubble burst in 1926, taking South Florida's economy with it. Between 1926 and 1940, the streetcar system was slowly dismantled.

After World War II, South Florida built expressways like its contemporaries, and real estate promoters moved out into the suburbs. Developers replaced orange groves and avocado plantations with suburban subdivisions and, later, condominiums. In the 1970s, South Florida's leaders decided that the next urban amenity to add was a rapid transit system, like the

103

then-fashionable ones built in San Francisco and Washington, DC. Miami's leadership successfully pitched Uncle Sam. The result was the elevated Metrorail system.

Metrorail fell victim to South Florida's usual combination of overpromising and underdelivering. Only half of the system was built due to mismanagement. Metrorail didn't go where the people or the jobs were, and South Florida's authorities didn't concentrate new development near the stations. Metrorail's lack of utility has continued into the 21st century, despite South Florida's persistently bad traffic, high population density, and the addition of a downtown people mover. In the early 2000s, Miami-Dade's voters approved a massive system expansion. The benefits were oversold and the costs were lowballed, once again. Only 3 percent of the promised new Metrorail mileage was built.

Miami has delightful beaches, the fourth-tallest skyline in North America, and a population density greater than Chicago or Philadelphia. Unlike those other dense cities, Miami is still utterly dependent on the car. In the 2020s, three in four Miamians drive alone to work, and only one in 20 takes public transportation. This figure is broadly in line with United States averages, but is anomalous considering Miami's sheer density.[1]

It's June in Miami

Miami was not a natural location for a city in the 19th century. Lethal tropical diseases were endemic. The uninhabitable swamp of the Everglades came within six miles of Biscayne Bay. The bay was too shallow to provide a good port.[2] Thus, when oil magnate Flagler extended his Florida East Coast Railway down the Atlantic coast in the 1880s and 1890s, he initially had no intention of extending it to Biscayne Bay. Despite Tuttle's repeated offers of free land on Biscayne Bay in exchange for a rail connection, Flagler saw no reason to extend the line beyond his chosen terminus in West Palm Beach, 70 miles to the north.

That all changed over the winter of 1894–1895, when the worst winter in living memory wiped out nearly the entire Florida orange crop—except along the shores of Biscayne Bay. Tuttle renewed her offer to Flagler, sending him orange blossoms as evidence that her oranges had not frozen. Flagler took the hint. He promptly expanded the railway line south, marking the beginning of the era when Miami was a winter colony for the wealthy.[3] Over the course of 20 years, Flagler did for Miami what nature did not: he paid to dredge a deepwater port and drain the fever swamps of the Everglades. Flagler successfully sold Miami as a winter destination to the rich Northerners of the Gilded Age. By 1910, Miami was a small resort town of 5,471, with its promoters calling it "the Magic City."[4]

After World War I ended in 1918, South Florida boomed, and Miami transitioned from resort town to boomtown, aided by cheap credit, loose Prohibition enforcement, and aggressive marketing. During the prosperous 1920s, speculative new housing developments were all the rage, and South Florida was at the center of the nationwide bubble. Carl Fisher, father of Miami Beach, bought a billboard in Manhattan's Times Square in midwinter that read "It's June in Miami."[5] George Merrick, founder of Coral Gables, hired William Jennings Bryan, former secretary of state and three-time Democratic presidential nominee, to be his pitchman.[6]

A Sunny Place for Shady People

During the 1920s land rush, much of the land being bought and sold didn't even exist.[7] When it didn't exist, developers created it, building artificial islands in the middle of Biscayne Bay and draining mangrove swamps to build up as subdivisions. But the vast amount of money needed to make these infrastructure improvements virtually required developers to sell lots before the subdivisions were built. Frederick Lewis Allen's classic *Only Yesterday* captured the spirit of the age quite well: "The men whose fantastic projects had seemed in 1923 to be evidences of megalomania were now coining millions: by the pragmatic test they were not madmen but—as the advertisements put it—inspired dreamers. Coral Gables, Hollywood-by-the-Sea, Miami Beach, Davis Islands—there they stood: mere patterns on a blue-print no longer, but actual cities of brick and concrete and stucco; unfinished, to be sure, but growing with amazing speed, while prospects stood in line to buy and every square foot within their limits leaped in price."[8] This modus operandi worked so well in Miami Beach and Coral Gables that Fisher and Merrick even built streetcar lines on speculation, so commuters and tourists could connect with downtown Miami and

the all-important Florida East Coast Railway depot. Fisher's Miami Beach Railway also operated the City of Miami's municipal streetcars (fig. 10.2).

By mid-1925, the bubble was ready to burst, as the supply of new developments was rapidly outpacing the demand. The trigger for the collapse was Florida's overloaded transport infrastructure. All through the summer, railway lines into Florida were far over capacity, causing food shortages in Miami. By October, railway congestion had gotten so bad that the rail companies jointly embargoed all shipments except absolute necessities.[9] Over 10,000 railway cars full of goods were left idle. As November and December came and went, the expected wave of Northern buyers never arrived. Cooler heads began to prevail, and the market began to drop.

Then, on January 10, 1926, the *Prinz Valdemar*, an Imperial German blockade runner turned floating hotel, capsized at the entrance to Miami's port. While the crew escaped unharmed, the ship completely blocked the channel.[10] For 25 days, nothing bigger than a rowboat could get in or out. The accident, combined with the railway embargo, brought Miami's commerce to a virtual halt. By March 1926, a full-blown crash was underway: the average value of new construction in Miami had dropped by 50 percent in four months. News reports began to appear that speculation had gone too far. Promoters had sold enough lots to house 2 million people in a county with 111,332 inhabitants.[11]

The worst was yet to come, in the form of two eternal Florida hazards: indictments and hurri-

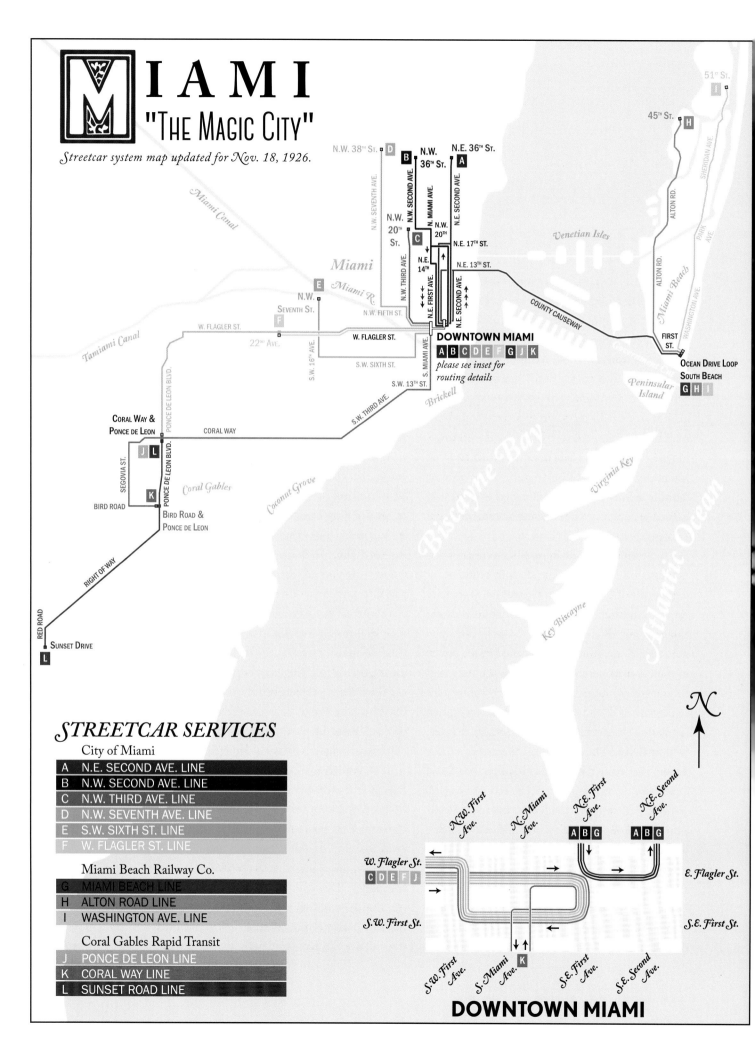

MIAMI
"THE MAGIC CITY"
Streetcar system map updated for Nov. 18, 1926.

51ST ST. — I
45TH ST. — H

N.W. 38TH ST. — D
B — N.W. 36TH ST.
A — N.E. 36TH ST.

N.W. 20TH ST.
C
N.W. 20TH
N.E. 17TH ST.
N.E. 14TH
N.E. 13TH ST.

Miami
Miami R.

N.W. SEVENTH AVE.
N.W. SECOND AVE.
N.W. MIAMI AVE.
N.E. SECOND AVE.
N.W. THIRD AVE.
N.E. FIRST AVE.
N.E. SECOND AVE.

Venetian Isles

COUNTY CAUSEWAY

E — N.W. SEVENTH ST.
F — W. FLAGLER ST.
22ND AVE.
W. FLAGLER ST.
N.W. FIFTH ST.
S.W. SIXTH ST.
S.W. 16TH AVE.
S. MIAMI AVE.

ALTON RD.
SHERIDAN AVE.
PARK AVE.
Miami Beach
Washington Ave.

DOWNTOWN MIAMI
A B C D E F G J K
please see inset for routing details

FIRST ST.
OCEAN DRIVE LOOP
SOUTH BEACH
G H I

Tamiami Canal
Miami Canal

S.W. 13TH ST.
S.W. THIRD AVE.
Brickell
Peninsular Island

CORAL WAY & PONCE DE LEON
J L
SEGOVIA ST.
PONCE DE LEON BLVD.
CORAL WAY
Coral Gables
Coconut Grove

K
BIRD ROAD
BIRD ROAD & PONCE DE LEON

Biscayne Bay
Virginia Key
Key Biscayne
Atlantic Ocean

RIGHT OF WAY
RED ROAD
SUNSET DRIVE
L

N ↑

STREETCAR SERVICES
City of Miami
A	N.E. SECOND AVE. LINE
B	N.W. SECOND AVE. LINE
C	N.W. THIRD AVE. LINE
D	N.W. SEVENTH AVE. LINE
E	S.W. SIXTH ST. LINE
F	W. FLAGLER ST. LINE

Miami Beach Railway Co.
G	MIAMI BEACH LINE
H	ALTON ROAD LINE
I	WASHINGTON AVE. LINE

Coral Gables Rapid Transit
J	PONCE DE LEON LINE
K	CORAL WAY LINE
L	SUNSET ROAD LINE

N.W. First Ave.
N. Miami Ave.
N.E. First Ave.
N.E. Second Ave.

W. Flagler St.
C D E F J
E. Flagler St.

A B G
A B G

S.W. First St.
S.E. First St.

S.W. First Ave.
S. Miami Ave.
S.E. First Ave.
S.E. Second Ave.
K

DOWNTOWN MIAMI

10.2 Early Miami streetcar systems, 1926.

canes. When investors and prosecutors began to look hard at some developments, it was clear that many of these proposed developments were scams. For example, Merle Tebbetts, developer of Fulford-by-the-Sea (today's North Miami Beach), was arrested in April 1926 on fraud charges. Tebbetts had created a fictional university and lied about his investments in the town to attract homebuyers.[12] Prices continued to fall. By the late summer of 1926, the average value of new construction in Miami was 77 percent below its peak of only 10 months before. Merrick used the opportunity to offload his unprofitable streetcar system on Coral Gables.[13]

The final blow came in mid-September, when a category 4 hurricane scored a direct hit, leaving a third of the county's population homeless. An 11.5-foot storm surge tore through downtown Miami. Most of the remaining land developers were ruined, including Fisher and Merrick. Some of the planned developments from the 1920s are still empty in the 2020s. Aladdin City, 20 miles southwest of downtown Miami, has reverted to farmland. The ruins of the unfinished artificial island Isola di Lolando are still visible from the Julia Tuttle Causeway between Miami Beach and the mainland.[14]

With the collapse, the local economy entered a prolonged depression. Henry S. Villard, writing for the *Nation* in 1928, described the scene: "Dead subdivisions line the highway, their pompous names half-obliterated on crumbling stucco gates. Lonely white-way lights stand guard over miles of cement side-walks, where grass and palmetto take the place of homes that were to be."[15] The last

streetcars ran in 1940, one year before the United States entered World War II.

Cheaper to Buy a Limousine

World War II was the catalyst for South Florida's renewal. The military turned Miami Beach's hotels and apartments into training stations. In total, half a million personnel trained in South Florida. Postwar population growth was nothing short of explosive. Returning servicemembers settled down in large numbers. In addition, hundreds of thousands of Cuban refugees settled in Miami after Havana fell to Fidel Castro's communist rebels in 1959. In 30 years, Miami-Dade County added a million residents, including over 400,000 Cuban refugees and their families. New expressways were built to handle the new residents. Unsurprisingly, many of these expressways were built through existing black neighborhoods in the center city. I-95 was built through the black neighborhood of Overtown, the "Harlem of the South," to open up new land for the expansion of Downtown Miami.[16] But by the late 1960s, the fervor for new expressways was beginning to wane, and traffic kept getting worse.

The fashionable solution at the time was to take after San Francisco and Washington, DC. Both cities had opted for new metro systems to compete with the automobile. Miami's elevated Metrorail system was designed to do the same. (Metrorail was designed as all elevated because the high water table makes an underground metro impractical.) According to a county planning study, the fully built 50-mile Metrorail system would attract 280,000 riders per day by 1985 (fig. 10.3).[17]

The federal government agreed to fund 80 percent of the system's construction on that basis. While planning was underway, the need for better transportation and more housing was becoming ever more urgent. Fidel Castro, who had declared himself president of Cuba, allowed Cubans to leave the island by boat, triggering an exodus known as the Mariel Boatlift. Between April and October 1980, 125,000 refugees arrived in Miami. At the same time, Miami was flooded with dirty money from the cocaine trade. Much of that money was invested in South Florida real estate.[18]

An appropriate solution would have been to tackle both issues at once and put the new condos near the new train stations. This did not happen. Miami-Dade County assumed that commuters would drive to Metrorail and park. Thus, multi-level garages were built instead—a common error in American transit planning.[19] Compounding the problem, Metrorail construction was so poorly managed that money ran out halfway through. The project was drastically scaled down. When Metrorail opened in 1984, the half-built system had fewer than 10,000 riders a day. President Ronald Reagan quipped, "It would have been a lot cheaper to buy everyone a limousine."[20]

The Best Bait and Switch

In the decades after Metrorail's opening, ridership remained far below original projections, despite ever-denser development and South Florida's persistent traffic woes. Downtown Miami also added the Metromover during this period, the only downtown people mover in North America

to gain significant ridership. (The other two, the Jacksonville Skyway and the Detroit People Mover, are rightfully considered expensive mistakes.)[21] In 2002, a tax measure called the People's Transportation Plan was put to a vote to extend Metrorail and make it useful (fig. 10.4). The People's Transportation Plan included an 89-mile expansion of the Metrorail system to cover the whole county. The plan's backers bought the punchy website TrafficRelief.com. Voters enthusiastically agreed, approving the 0.5 percent sales tax increase. But there wasn't nearly enough money. And the county knew it at the time.

The grand promises quickly began to vanish. By 2004, it had become clear that the people of Miami-Dade County had been taken for a ride. That year, the *Miami New Times*, in its end-of-year Best of Miami awards, gave the People's Transportation Plan its "Best Bait and Switch" award: "The shiny lure that voters went for was the promise that a half-penny sales tax would buy them an expanded and improved mass transit system. Now they're told the tax was in reality designed to win matching state and federal funds, not to actually build a new system—never mind the ballot language. According to a county study, absent these outside funds (and none of them are guaranteed), the PTP could literally bankrupt the transportation system."[22] Of the 89 miles promised, only a three-mile airport extension was built.

The More Things Change, the More They Stay the Same

Thirty-five years after it opened, Miami's urban rail

10

METRORAIL *Miami*

Proposed System - March 1979

RAPID TRANSIT

1 Cutler Ridge to Opa-Locka

2 Dadeland South to West 8th Ave Hialeah

3 Airport to Douglas Rd

4 Airport to Miami Beach Convention Center

BUSWAY

5 Golden Glades to Government Center

Station   Transfer   Terminal

Jake Berman '19

SCALE IN MILES

0   1   2   3   4   5

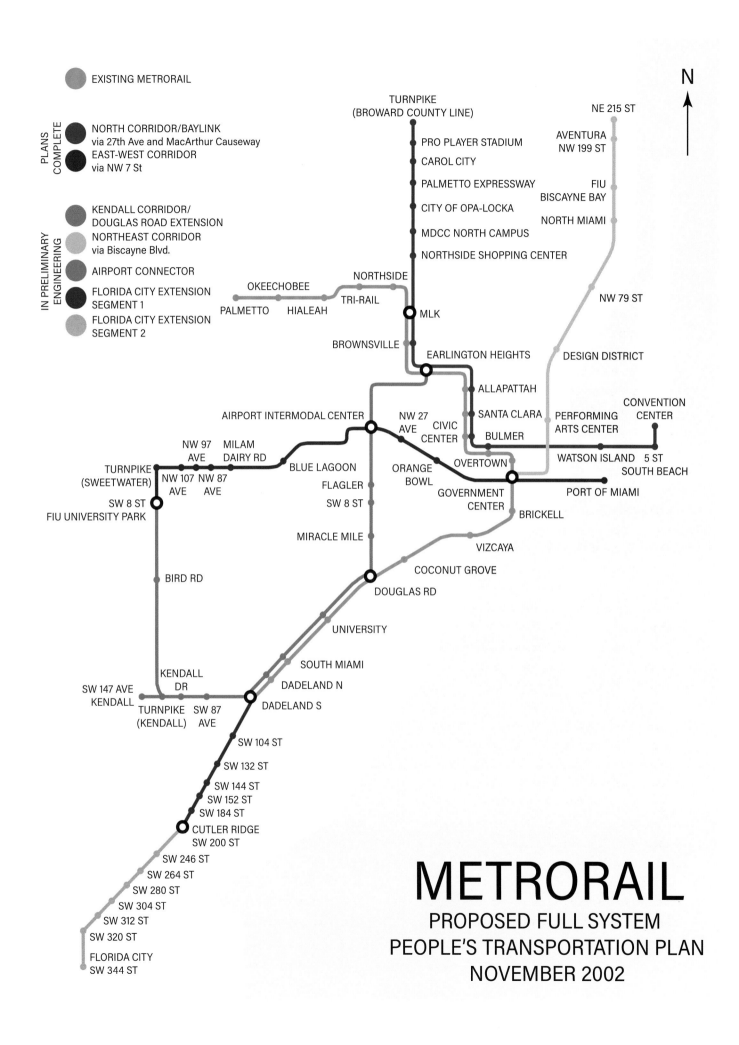

10.4 The People's Transporation Plan, 2002.

network hasn't done much to affect Miami's development pattern. This is atypical, because greater Miami is geographically constrained in a way that most Sun Belt cities aren't. The entire urban area is squeezed into a flat strip of land 20 miles wide and 120 miles long, stretching from Jupiter in Palm Beach County to Florida City at the end of Florida's Turnpike. The metropolis is stuck between the Everglades and the Atlantic, so there's nowhere else to build. With sea levels rising due to climate change, the amount of available land is slowly decreasing. Despite this, Miami has continued to be totally dependent on the automobile. Metrorail is a marginal part of the region's transport picture. Before the coronavirus pandemic hit, the entire Metrorail system carried 66,200 passengers per weekday. For comparison, Downtown Los Angeles's Seventh Street station carried 54,000 passengers per weekday all by itself.[23]

To its credit, Miami has tried hard to encourage more vertical development, to the point of rewriting its land use law from scratch. The principal goal of Miami's old law was to separate commercial, residential, and industrial uses, like most traditional zoning codes. The old code made allowable uses relatively clear, but it did a poor job of defining which types of buildings were permissible. The new land use law junked all those rules and instead put restrictions on the sizes and forms of buildings, with an emphasis on promoting street-level activity. The reforms have made development significantly more predictable, because developers know in advance which types of buildings are allowed. Under the old system, a fight with the

neighbors was virtually assured.[24] These reforms won the city the American Planning Association's 2011 National Planning Excellence Award for Best Practice.

But these new land use laws haven't translated into higher transit ridership. Metrorail and Metromover ridership collapsed between 2013 and 2019. Things only got worse when the coronavirus pandemic hit in spring 2020. Ridership is weak for a very prosaic reason: the trains don't go where the passengers are. Miami's busiest bus routes cross Biscayne Bay to and from the towers of Miami Beach, in addition to a few other routes which run to downtown Miami.[25] Metrorail doesn't serve these corridors.

Perhaps it is unsurprising that Miami-Dade County planned to close the Biscayne Bay gap in the early 2020s by relying on real estate development schemes. Rather than pay for the rail connection itself, Miami-Dade County initially cut a deal with a Malaysian gambling conglomerate. The conglomerate agreed to build a monorail from the mainland to Miami Beach, in exchange for county permission to build a $3.2 billion casino complex (fig. 10.5).[26] The mainland monorail terminal would connect with a new Metromover station.

A rail transit link over Biscayne Bay is a sound idea, but the execution was lacking. The proposed mainland terminal was Herald Plaza, 15 blocks northeast of Miami's historic central business district (CBD). Most Metromover passengers have destinations in the CBD or in the Brickell neighborhood immediately south of the CBD, so most riders would have to transfer to reach their final

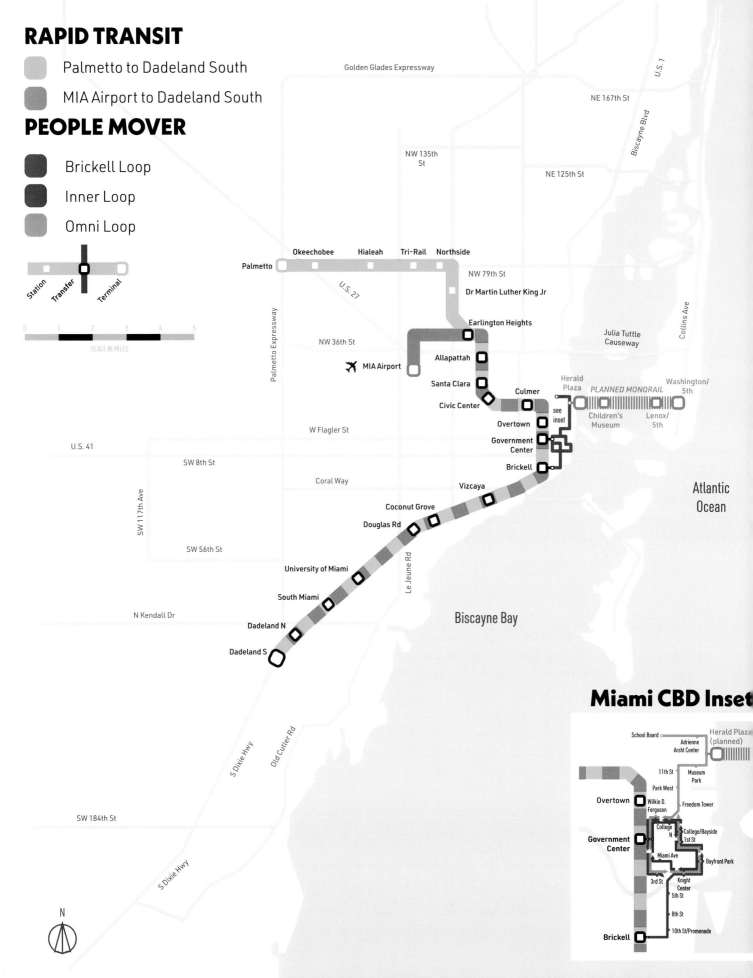

RAPID TRANSIT

Palmetto to Dadeland South

MIA Airport to Dadeland South

PEOPLE MOVER

Brickell Loop

Inner Loop

Omni Loop

Station Transfer Terminal

0 1 2 3 4 5
SCALE IN MILES

Golden Glades Expressway

NE 167th St

Biscayne Blvd

U.S. 1

NW 135th St

NE 125th St

Okeechobee Hialeah Tri-Rail Northside

Palmetto

NW 79th St

U.S. 27

Dr Martin Luther King Jr

Palmetto Expressway

NW 36th St

Earlington Heights

Julia Tuttle Causeway

Collins Ave

MIA Airport

Allapattah

Santa Clara

Culmer

Herald Plaza

PLANNED MONORAIL

Washington/ 5th

Civic Center

Children's Museum

Lenox/ 5th

W Flagler St

Overtown

see inset

U.S. 41

Government Center

SW 8th St

Brickell

SW 117th Ave

Coral Way

Vizcaya

Atlantic Ocean

SW 56th St

Coconut Grove

Douglas Rd

Le Jeune Rd

University of Miami

South Miami

Biscayne Bay

N Kendall Dr

Dadeland N

Dadeland S

SW 184th St

S Dixie Hwy

Old Cutler Rd

Miami CBD Inset

School Board

Herald Plaza (planned)

Adrienne Arsht Center

11th St

Museum Park

Overtown

Park West

Wilkie D. Ferguson

Freedom Tower

College N

College/Bayside

Government Center

1st St

Miami Ave

Bayfront Park

3rd St

Knight Center

5th St

8th St

10th St/Promenade

Brickell

S Dixie Hwy

N

MIAMI METRORAIL AND METROMOVER

10.5 Miami Metrorail and connecting services, 2022.

destinations.[27] No direct connection was planned between the monorail and Metrorail, so Metrorail passengers would have to transfer twice to get to Miami Beach.

The monorail project fell through in the end. After an initial development contract was signed in 2020, the monorail's price tag rapidly ballooned from $770 million to $1.3 billion. The sticker shock led the county to pull the plug. The Miami Beach monorail is, in a way, a microcosm of Miami's history: flashy real estate deals and half-cocked mass transit plans.

MINNEAPOLIS

ST. PAUL

Twin City Rapid Transit, 1948
Metro Transit, 2022

Minneapolis-St. Paul

N

2 miles/3.2 km

MINNEAPOLIS– ST. PAUL

The Mob Takeover of Twin City Rapid Transit

11.1 Map of the Twin Cities.

There is a common folk theory that the North American streetcar disappeared because of a conspiracy. Allegedly, industrialists connected with the auto and motor bus industries bought up North American streetcar lines, then shut them down to sell more cars and buses. Made famous by the film *Who Framed Roger Rabbit*, the theory has little truth to it. By the 1930s, the for-profit streetcar business was rapidly becoming unprofitable. The industrialists weren't hawks swooping in on healthy prey—they were vultures picking the bones clean. In fact, there was only one place in North America where a relatively healthy streetcar system fell victim to a conspiracy. That place was the Twin Cities: Minneapolis and St. Paul (fig. 11.1). The Twin City Rapid Transit Company was a national model for how to run a transit system until a Wall Street financier, a crooked lawyer, and the Minneapolis mob came along and stripped it for parts. High-capacity rail transit wouldn't return to the Twin Cities until the 21st century.

A Family Business

Twin City Rapid Transit had its origins in the 1886 merger of the Minneapolis Street Railway and St. Paul City Railway. This merger brought nearly all urban transport in the Twin Cities under the control of one company. The first head of the company was Thomas Lowry, the largest real estate developer in the Twin Cities. Lowry used his control over the streetcar system to promote his real estate interests. Lowry was an enthusiastic civic booster, a founder of Minneapolis's Library Board, and part of the founding group of the Soo Line Railroad, which still exists today. The Lowry Hill neighborhood, Lowry Park, and Lowry Avenue in greater Minneapolis are all named in his honor. Although Twin City Rapid Transit was publicly traded, the company on the whole remained a family business. Thomas Lowry died in 1909 and was succeeded by his son-in-law, Calvin Goodrich, who ran the company until his death in 1915. Goodrich in turn was succeeded by Thomas Lowry's son Horace, who ran the company until 1931.

During the Lowry family's management of Twin City Rapid Transit, the company was an innovator, establishing a reputation for long-term vision and quiet, unflashy management. This included building its own advanced, lightweight, "noiseless" streetcars in-house, which it exported to places like Chicago, Seattle, and Nashville. This culture of competence continued even after the Lowry family no longer managed the company. By the 1940s, as the streetcar era was beginning to end, the management of Twin City Rapid Transit continued to take the same long-term view to maintaining a sustainable business. While lightly used streetcar lines were replaced with buses, Twin City Rapid Transit, under company president D. J. Strouse, undertook major investments to modernize the busiest streetcar lines and bought brand-new, state-of-the-art President's Conference Committee (PCC) streetcars to serve those lines (fig. 11.2).[1]

Sold for Scrap

Strouse's decision to modernize the core of the streetcar system set events in motion that would ultimately lead to the system's collapse. In 1948, a New York financier named Charles Green bought 6,000 shares of company stock, expecting a large stock dividend and to make a quick buck. No dividend was forthcoming. Twin City Rapid Transit, to his surprise, was reinvesting revenues back into the streetcar system as it always had. Furious, Green started a hostile takeover attempt. To succeed in this endeavor, Green had to gain the support of other shareholders. One of those shareholders was Isadore Blumenfield, alias Kid Cann, the godfather of the Minneapolis mob.

Kid Cann had gotten his start as a bootlegger during Prohibition, importing industrial alcohol from Canada and selling it in speakeasies. In the 1940s, he still controlled the by then legal liquor trade and had his fingers in all sorts of vice, including labor racketeering, illegal gambling, and prostitution. He was the untouchable king of the Minneapolis underworld. When he was indicted on federal money-laundering charges in Oklahoma, the Minneapolis police chief personally traveled to Oklahoma and testified in Kid Cann's defense.[2] Kid Cann was acquitted. Arrested twice for murder, he beat the charges both times.

Green and Kid Cann met at the Club Carnival nightclub in 1948 to plot the takeover of Twin City Rapid Transit, and they decided to bring lawyer Fred Ossanna into the scheme. Ossanna was a criminal lawyer, not a specialist in transit

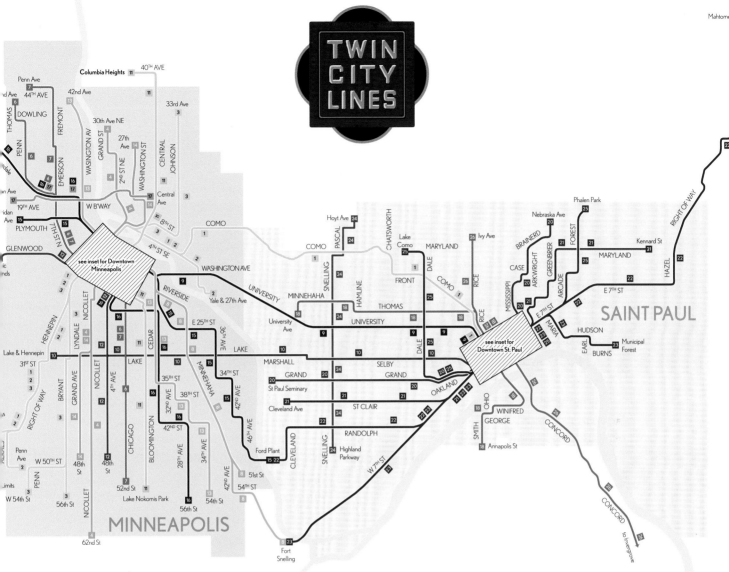

TWIN CITY LINES

MINNEAPOLIS

SAINT PAUL

see inset for Downtown Minneapolis

see inset for Downtown St. Paul

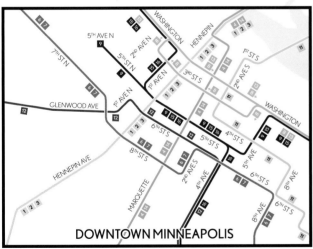

DOWNTOWN MINNEAPOLIS

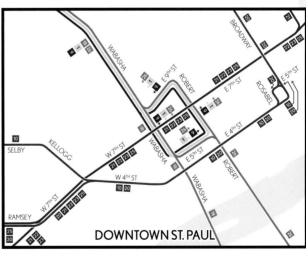

DOWNTOWN ST. PAUL

STREETCAR ROUTES

1 Como–Harriet–Hopkins Interurban	10 Selby–Lake Interurban
2 Oak–Harriet	20 Grand–Mississippi
3 Bryant–Johnson	11 Bloomington–Columbia Heights
4 Nicollet–2nd St NE	12 Glenwood–4th Ave S
14 Grand–Monroe	17 Broadway Crosstown
6 Chicago–Penn	18 Hamline–Cherokee Heights
7 Chicago–Fremont	26 Rice–South St Paul
8 Minnehaha–Ft Snelling	21 St Clair–Payne
13 34th Ave S–N Washington Ave	22 Randolph–Hazel Park–N St Paul–
9 Interurban	Willernie–Mahtomedi
15 Plymouth–E 25th St	23 Fort Snelling–Maria
16 28th Ave S–Robbinsdale	25 Dale–Forest
	24 Snelling Crosstown

EFFECTIVE AUGUST 1, 1948.

law. Ossanna had become wealthy representing the interests of the liquor business, in the process becoming a political fixer for Minnesota's Democratic-Farmer-Labor Party. Kid Cann was one of Ossanna's clients.

The triumvirate of Green, Kid Cann, and Ossanna went to war with Strouse and his allies over the course of 1948 and 1949, charging that Strouse's fiscal prudence was costing shareholders money. Green masterminded a letter-writing campaign to shareholders, contacting each one individually to make his case. This campaign had legs. When Green went to Minneapolis for the annual Twin City Rapid Transit shareholder meeting in the spring of 1949, he presented an ultimatum to company management: fire Strouse and allow Green to pick half of the board of directors, or else. Strouse rejected the ultimatum, chose to fight, and lost. By late 1949, a majority of the shareholders had been won over to the plotters' cause. That November, Strouse was out, and Green was in.

As company president, Green brought his brash, pushy New York City style to a state known for its "Minnesota nice." Ossanna was appointed as the company's chief lawyer. Green immediately took an ax to Twin City Rapid Transit, promising that all streetcars would be replaced with buses by 1958. He fired 800 workers without explanation.[3] He slashed feeder service to outlying areas, leaving commuters stranded. Maintenance was reduced to the bare minimum. At the very same time, Green had the audacity to ask the state railroad commission to approve a fare hike. He told the press: "The public be damned! I intend to force a profit out of this company! If necessary, I'll auction off all the streetcars and buses and sell the rails for scrap iron!"[4] Green acted so aggressively that the state railroad commission had to get a court order stopping him from closing any more lines without the commission's explicit permission.

Ridership dropped like a brick. Transit ridership was already falling nationwide because of the resumption of civilian auto sales and the end of gas rationing, but Green's management was running the company into the ground. In 1946, with Strouse in charge, Twin City Rapid Transit carried 201 million passengers. By 1950, with Green in charge, ridership had dropped to 144 million—a 28 percent decrease.[5] For comparison, New York City subway ridership decreased by 13 percent over the same period.[6] Green's heavy-handedness didn't endear him to the people of the Twin Cities, either. Frustrated commuters and their elected representatives made a major political issue out of Green's ham-fisted cuts. Twin City Rapid Transit became a political target.

The troika of Green, Ossanna, and Kid Cann didn't last long. Green's combative style alienated the other two. By May 1950, Ossanna and Kid Cann had turned against Green and stripped him of his power to speak for the company. All official Twin City Rapid Transit pronouncements were to come from Ossanna. By September, Green and his allies on the board of directors had agreed to resign. Green was forced to sell his shares. In the process, he profited $100,000 (over $1 million in 2022 dollars). Through intermediaries, Kid Cann, Ossanna, and other mob associates snapped up

11

Green's stock, giving them enough shares to control the company outright.[7] The following year, Ossanna became company president.

Crime and Punishment

Ossanna, like his predecessor, had little interest in running a competent transit operation. He only accelerated the streetcar shutdown. Ossanna soon signed a sweetheart deal to buy 525 buses from General Motors on credit, and hired a bus expert from Los Angeles to oversee the bus transition. By 1954 the entire streetcar system had closed, four years faster than Green's original timeline. Twin City Rapid Transit's modern PCC streetcars were sold to other cities. (Eleven of these PCCs are still in daily use in San Francisco, seven decades later.) The streetcars that Ossanna couldn't sell were set on fire and sold for scrap metal. Ossanna turned the destruction of the streetcars into a photo op. The newspapers ran pictures of a grinning Ossanna being presented with a check in front of a burning streetcar.

Ossanna and Kid Cann used the system's closure as an opportunity to corruptly line their pockets. When the duo sold off Twin City Rapid Transit's streetcar equipment, they sold off company assets at below-market rates and took kickbacks. These dealings defrauded Twin City Rapid Transit out of about $1 million (nearly $10 million in 2022 dollars). A few years later, in 1959, federal agents found out. The feds arrested Ossanna, Kid Cann, and several others on fraud and conspiracy charges. In August 1960, after a four-month trial, the jury found Ossanna guilty. His sentence was four years in prison.[8] The Minnesota Supreme Court promptly stripped Ossanna of his law license.[9]

Kid Cann was acquitted yet again but remained in federal crosshairs. Later that year, prosecutors finally caught up with Kid Cann. He didn't go to prison for decades of murder, bootlegging, extortion, bribery, fraud, embezzlement, or racketeering, though. Federal prosecutors finally jailed him for transporting a prostitute across state lines.[10] This, and a later jury tampering charge, finally put Kid Cann behind bars.

After the Minneapolis–St. Paul streetcar system closed, Twin Cities transit entered the same death spiral that can be seen all over this book. Mass transit soon faded away as the freeway and the automobile became favored. The regional Metropolitan Council eventually took over Twin City Rapid Transit's bus system. Attempts to build a subway in the 1960s and a pod-car system in the 1970s never came to anything.

It wasn't until the 21st century that rapid transit of any kind returned to the Twin Cities (fig. 11.3). The two modern METRO light rail lines operated by the Metropolitan Council run parallel to old Twin City Rapid Transit lines. The Blue Line, which opened in 2004, runs parallel to Twin City Rapid Transit's old Hiawatha Line. The Green Line, which opened in 2014, follows the same route as the old Twin City Rapid Transit Interurban Line, which ran from 1890 to 1953. These two light rail lines have proved unexpectedly popular, beating initial ridership projections and creating the political conditions for further expansion. Per

mile of track, METRO is notably busier than its contemporaries like Phoenix's Valley Metro Rail and the Hudson-Bergen Light Rail in New Jersey. This success is due in part to good planning. The Green and Blue Lines both follow busy streets, have major destinations at both ends of the line, and have dedicated lanes so the trains don't get stuck in traffic.

But none of these modern attempts at rebuilding high-capacity mass transit have the comprehensiveness of the old streetcar system. In the 1940s, Twin City Rapid Transit had the right ideas: replace streetcars with buses on marginal routes but keep the higher-capacity, more comfortable streetcars running on busy routes. This approach prefigured the consensus that developed in favor of light rail in the 1980s. Of course, Minnesotans never got to see how things would've played out, because of the meddling of a corporate raider, a crooked lawyer, and the king of the Minneapolis underworld. One is left to wonder what might have been.

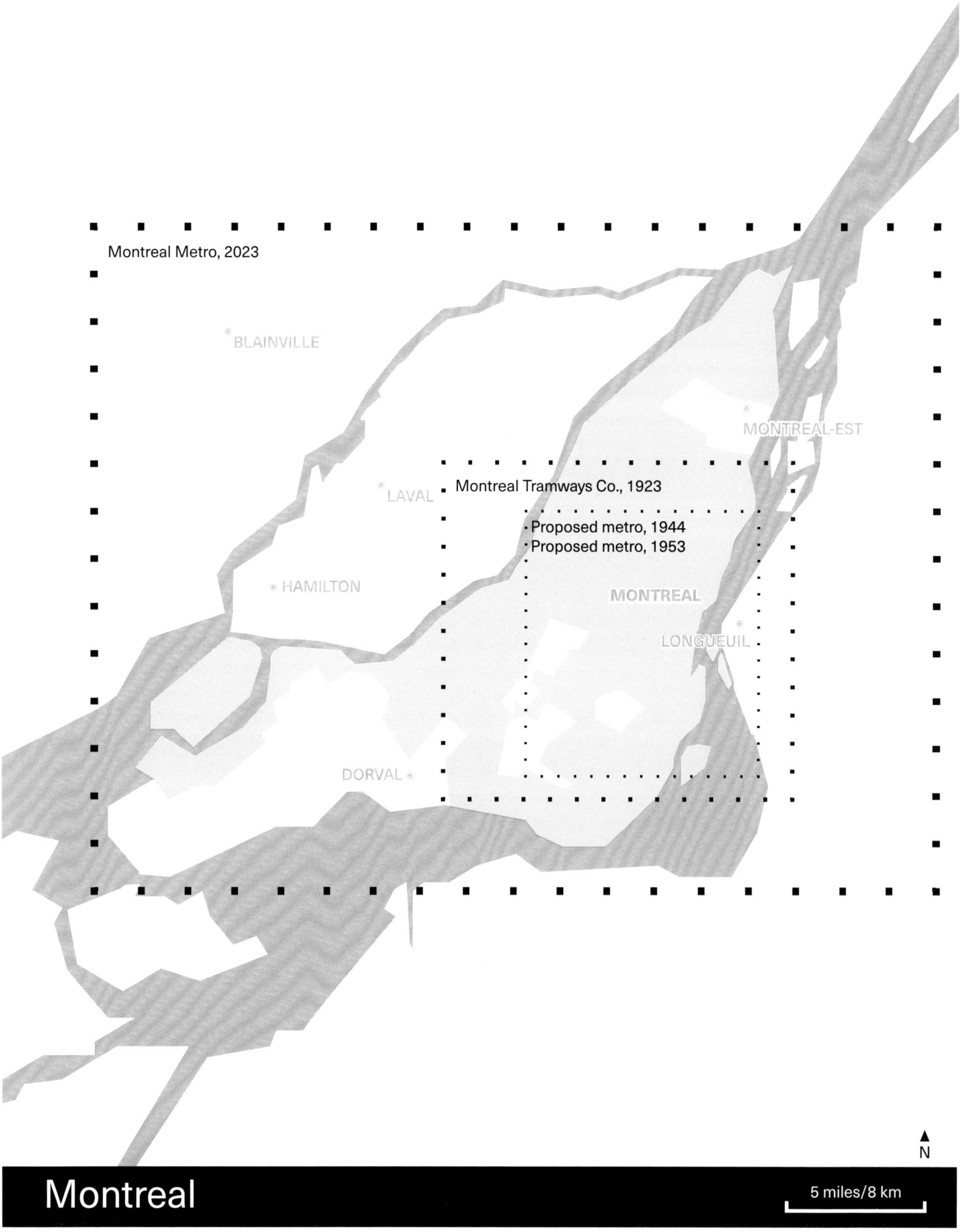

Montreal Metro, 2023

BLAINVILLE

MONTREAL-EST

LAVAL

Montreal Tramways Co., 1923

· Proposed metro, 1944
· Proposed metro, 1953

HAMILTON

MONTREAL

LONGUEUIL

DORVAL

Montreal

5 miles/8 km

N

MONTREAL

The Metro as Showcase Megaproject

12.1 Map of greater Montreal.

Not every transit system is built to solve a transport problem. Los Angeles's Red Cars were built to sell real estate. Cleveland's Waterfront Line light rail was built to keep up with the Joneses. The Detroit QLine was built to promote the development of downtown. The Montreal Metro wasn't built to solve a transport problem, either. This is ironic, since the Metro is one of the great success stories in North American mass transit. Per mile of track, Montreal (fig. 12.1) has the second-busiest metro system in North America, following only New York. In absolute terms, it's the fourth-busiest metro, behind New York, Mexico City, and Toronto. The Metro really has resolved many of Montreal's transport issues. But at the time it was built, that was considered an incidental benefit to the city's powers that be. Jean Drapeau (rhymes with *chateau*), mayor of Montreal, wanted a megaproject to show off Montreal to the world, and the Metro was it.

Proposals to build a metro had been advanced multiple times since 1910. Nothing ever came of them. The eternal enemy of public infrastructure raised its head over and over: no money. The city government simply wasn't willing to pay, and private enterprises couldn't find funding either. As mayor in the mid-1950s, Drapeau, of the Quebec Liberal Party, opposed building a metro. At the time he believed metro systems were outdated. This all changed dramatically in 1960. But when Drapeau returned to the mayor's office that year after a three-year interregnum, he did an about-face. He went all-in on the metro.

In the intervening half decade between Drapeau's first mayoralty and his second, the transportation picture had changed little. The political picture, however, had changed in three critical ways. First, Quebec itself was different. From the mid-1930s to 1959, Quebec was dominated socially by the Catholic Church, controlled economically by an English-speaking business class, and governed corruptly by a conservative political machine. That machine collapsed in 1959. The Liberals filled the power vacuum, marking the start of Quebec's secular, prosperous, and unabashedly Francophone modern era. Second, the choice of technology mattered. Drapeau refused to back a system that used conventional steel-wheeled trains and steel rails. (Instead, during his first mayoralty, he preferred American-style expressways.) But when the opportunity presented itself in 1960, Drapeau was happy to import an ultramodern Parisian metro system with rubber-tired trains and concrete tracks. Third, rival Toronto had opened Canada's first metro in 1954. If Toronto had one, Montreal needed one too, preferably in time for the city's Expo '67, Canada's first world's fair. Expo '67 was Montreal's coming-out party on the global stage, and Drapeau was determined to show off the city to the world.

The Great Darkness

The earliest proposal to build a metro in Montreal dates to 1910. At the time, Montreal was the largest city and commercial capital of Canada. Montreal suffered from the same streetcar congestion issues as Boston and Philadelphia (fig. 12.2 shows the

system near its height in the early 20th century). Thus, early metro proposals were originally envisioned as streetcar tunnels. The 1910 plan was put forth by the Montreal Tramways Company, which operated the streetcar system, and had gained a monopoly over the city's transport.[1] Like its contemporaries elsewhere in North America, the company was aggressively protective of this monopoly. It freely used corruption to get its way. When the company's franchise was up for renewal in 1915, the Tramways paid a city councilman named Napoléon Hébert $200,000 (US$3.6 million in 2022) for his support.[2]

Other metro plans in the late 1920s, as well as the company's metro plan of 1944 (fig. 12.3), never came to anything, due to the Great Depression and the World War II. During this period, the Tramways declined like the other privately owned streetcar operators in this book, due to bus and car competition. The government-owned Montreal Transportation Commission took over the remnants of the system in 1951. (The Société de Transport de Montréal, which runs the Metro and the city buses today, is the commission's descendant.) Something of the flavor of the era can be gathered from the maps themselves. In early 20th-century Montreal, most major planning documents were in English. Maps and other documents meant for public consumption were bilingual. The figures in this book follow that same convention.

The last half of this era was defined by a singular politician, Premier Maurice Duplessis. In power from 1936 to 1939 and 1944 to 1959, Duplessis stood for a conservative vision of Quebec nationalism. His party, the Union Nationale (Na-

COMPAGNIE DES TRAMWAYS DE
MONTRÉAL
TRAMWAYS CO.

Trajets pendant la journée sur semaine, août 1923
Regular weekday service, August 1923

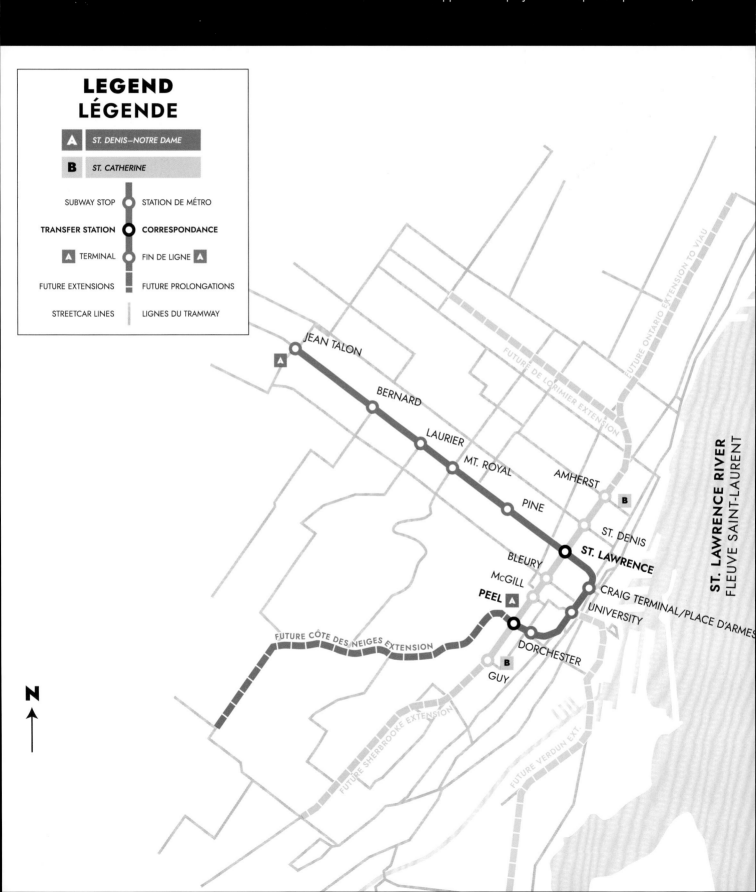

LEGEND
LÉGENDE

A *ST. DENIS—NOTRE DAME*

B *ST. CATHERINE*

SUBWAY STOP	○	STATION DE MÉTRO	
TRANSFER STATION	◉	**CORRESPONDANCE**	
A TERMINAL		FIN DE LIGNE **A**	
FUTURE EXTENSIONS		FUTURE PROLONGATIONS	
STREETCAR LINES		LIGNES DU TRAMWAY	

JEAN TALON

BERNARD

LAURIER

MT. ROYAL

PINE

AMHERST

B

ST. DENIS

BLEURY

ST. LAWRENCE

McGILL

CRAIG TERMINAL/PLACE D'ARMES

PEEL A

UNIVERSITY

FUTURE CÔTE DES NEIGES EXTENSION

DORCHESTER

B

GUY

FUTURE ONTARIO EXTENSION TO VIAU

FUTURE DE LORIMIER EXTENSION

ST. LAWRENCE RIVER
FLEUVE SAINT-LAURENT

FUTURE SHERBROOKE EXTENSION

FUTURE VERDUN EXT.

N

12.3 Proposed metro system, 1944.

tional Union Party), was fervently Catholic, anti-communist, and at peace with Quebec's social and economic status quo. The Union Nationale urged the French-speaking majority to know and accept their place in society. The province was largely under the economic control of a Montreal-based, English-speaking commercial elite. For example, English-speaking Quebecers made up 13 percent of the population in 1961 but owned 74 percent of Quebec's large financial services companies.[3] French speakers' wages were 51 percent of their English-speaking counterparts.[4] The Catholic Church controlled provincial schools, hospitals, and other social services. Labor unrest was put down with brute force. In 1949, Duplessis broke a four-month miners' strike with mass arrests, tear gas, and threats to shoot strikers. He governed by running a shamelessly corrupt patronage machine, rewarding his allies and stiffing his enemies. In this era, the Union Nationale's rallying cry was a line attributed to a 19th-century Catholic bishop: "Le ciel est bleu, l'enfer est rouge" (Heaven is blue, hell is red). Blue was the Union Nationale's color. The rival Liberals used red.

Duplessis's opponents tarred this period as La Grande Noirceur, or "The Great Darkness."[5] When Duplessis died in 1959, he left no worthy political heir. The Union Nationale was left in chaos. The Liberals—including Jean Drapeau—took full advantage leading into the 1960 election.

He Just Yells Louder

The Liberal vision was the opposite of the Union Nationale. The Liberal rallying cry at the time was "Maîtres chez nous" (Masters in our own house), a call for major changes to Quebec's socioeconomic fabric. At the provincial level, this meant nationalizing Quebec's electric companies, ending Church control over hospitals and schools, opening the workplace to women, and creating a provincial pension system.[6] At the local level, Jean Drapeau dreamed big after his 1960 return to the mayor's office. One would not expect such ambition from the man, at first glance. A contemporary profile described Drapeau as a man who "has few visible manifestations of leadership. He is small, pear-shaped, with a high-domed bald head. Behind the thick lenses of his glasses his eyes are red-rimmed and pouchy from lack of sleep. There is no sparkling wit, little animation. no sense of presence. His voice is shrill, without expression. To make a point, he just yells louder."[7]

But behind the scenes, Drapeau was notable for his intense work ethic and his desire to showcase Montreal to the world. When it came to transportation, Drapeau was all over the place before he became the prime mover of the Metro. In the 1950s, he pushed for a full-scale American-style expressway system. He oversaw the final shutdown of the city's streetcars. He opposed the metro system proposed in 1953 (fig. 12.4). Drapeau even pushed a monorail, before political considerations led him to run for mayor in 1960 on a promise to build a metro system.[8] Drapeau's running mate, Lucien Saulnier, had made a metro a precondition of Saulnier's support.[9] (All the same, the Montreal expressway network significantly expanded during

127

Drapeau's mayoralty. When he left office in 1986, Montreal had the most expressway mileage per capita of any major Canadian city.)[10]

Drapeau fully committed to the Metro because of his deal with Saulnier. Once Drapeau threw his support behind a metro, he was not content to build any ordinary system—it had to have flash. The Metro, built out competently, would be the perfect Expo '67 megaproject. Thus, the Montreal Metro would run on concrete tracks using rubber tires instead of normal steel wheels and steel rails. This system, imported from Paris, allowed for "whisper-quiet" operation. Each station would have distinctive architecture and public art. This practice was common in the communist bloc, but nearly unheard of in the West. On top of all these innovations, Drapeau encouraged real estate developers to connect their new downtown towers directly to Metro stations. Called RESÓ in French and the Underground City in English, this network of city-center commercial tunnels is still the largest in the world. But the advanced technology and architectural grandeur weren't there for their own sake. It was also to one-up Toronto.

Humble and Very Envious Salutations

Six years before, Toronto had opened Canada's first metro line, and that fact stung Montrealers. While the Toronto Subway was under construction in 1950, *MacLean's* reported "that Toronto has copped the honor of becoming America's fifth, and the world's 14th subwayed city is particularly galling to Montrealers. The *Montreal Herald* sent 'humble and very envious salutations,' while the *Montreal*

Star headed an editorial on the subway 'Happy Toronto,' and sighed, 'The *Montreal Gazette* has been celebrating its 172nd birthday. We hope it will live to celebrate its 200th, coincidental with a start on the Montreal subway.'"[11]

Comparisons between Montreal and Toronto were inevitable when the Montreal Metro opened on October 14, 1966. Those comparisons were usually favorable to Montreal. Saulnier had proved a magnificent project manager. The Toronto Subway was a stripped-down, utilitarian beast, decorated with uniform white ceramic tiles. Wags called the stations "bathrooms without plumbing."[12] These "bathrooms without plumbing" paled in comparison to Montreal's bold architecture, state-of-the-art technology, and colorful public art. Drapeau and Saulnier had achieved the impossible. They had delivered an ambitious megaproject on time, under budget, and with panache. It was a stunning political and technological success.[13] Nine days after the Metro opened, Montrealers went to the polls. Drapeau won reelection with 94 percent of the vote.

A Legacy

The Metro's success brought Drapeau accolades and a virtual monopoly on power in Montreal. He freely pursued more megaprojects. But Drapeau's magic touch seemed to wane as the years passed. Expo '67 is still a Canadian cultural touchpoint half a century later, but it cost two and a half times its original estimate. The skyscraper complexes and the Underground City received acclaim for their ability to stabilize downtown Montreal at a time

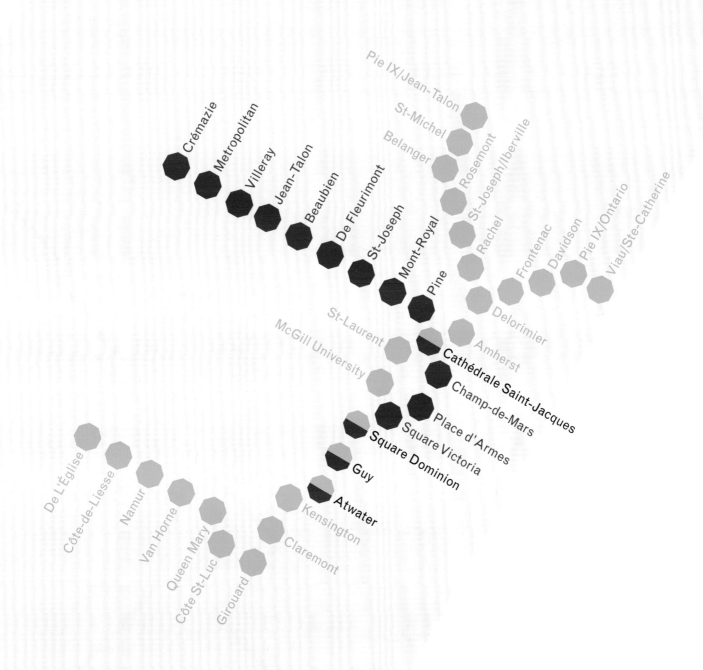

Crémazie
Metropolitan
Villeray
Jean-Talon
Beaubien
De Fleurimont
St-Joseph
Mont-Royal
Pine

Pie IX/Jean-Talon
St-Michel
Belanger
Rosemont
St-Joseph/Iberville
Rachel
Frontenac
Davidson
Pie IX/Ontario
Viau/Ste-Catherine
Delorimier
Amherst

St-Laurent
McGill University
Cathédrale Saint-Jacques
Champ-de-Mars
Place d'Armes
Square Victoria
Square Dominion
Guy
Atwater

De L'Église
Côte-de-Liesse
Namur
Van Horne
Queen Mary
Côte St-Luc
Girouard
Claremont
Kensington

Commission de Transport de Montréal
Plan du projet de Métro pour Montréal
Juin 1953

METRO DE MONTRÉAL
ET RÉSEAU EXPRESS MÉTROPOLITAIN

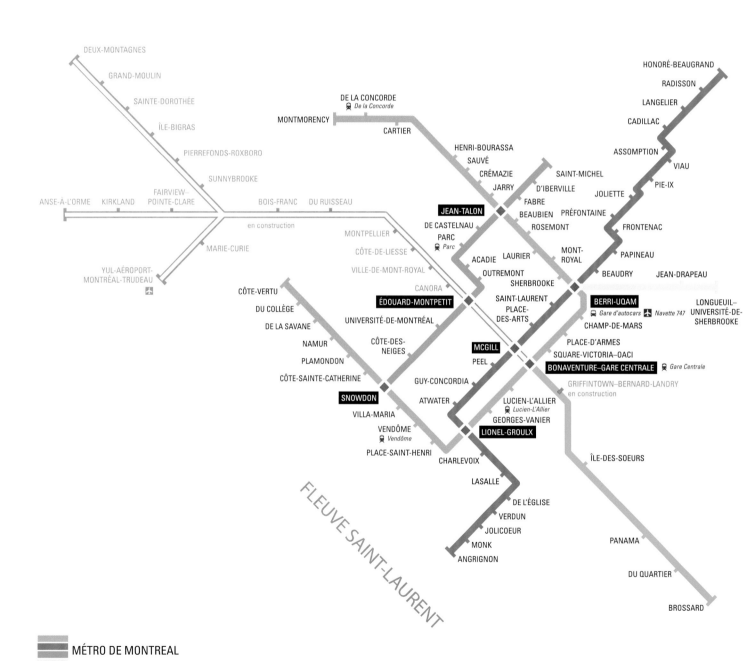

DEUX-MONTAGNES

GRAND-MOULIN

SAINTE-DOROTHÉE

ÎLE-BIGRAS

PIERREFONDS-ROXBORO

SUNNYBROOKE

ANSE-À-L'ORME   KIRKLAND   FAIRVIEW–POINTE-CLARE   BOIS-FRANC   DU RUISSEAU

en construction

MARIE-CURIE

YUL-AÉROPORT-MONTRÉAL-TRUDEAU

DE LA CONCORDE
De la Concorde

MONTMORENCY

CARTIER

HENRI-BOURASSA
SAUVÉ
CRÉMAZIE
JARRY

MONTPELLIER

CÔTE-DE-LIESSE

VILLE-DE-MONT-ROYAL

CANORA

SAINT-MICHEL

D'IBERVILLE
FABRE
BEAUBIEN
ROSEMONT

JEAN-TALON
DE CASTELNAU
PARC
Parc

ACADIE   LAURIER
OUTREMONT
SHERBROOKE

PRÉFONTAINE

MONT-ROYAL

HONORÉ-BEAUGRAND
RADISSON
LANGELIER
CADILLAC
ASSOMPTION
VIAU
JOLIETTE   PIE-IX
FRONTENAC
PAPINEAU
BEAUDRY   JEAN-DRAPEAU

CÔTE-VERTU

DU COLLÈGE

DE LA SAVANE

NAMUR

PLAMONDON

CÔTE-SAINTE-CATHERINE

ÉDOUARD-MONTPETIT

UNIVERSITÉ-DE-MONTRÉAL

CÔTE-DES-NEIGES

SAINT-LAURENT
PLACE-DES-ARTS

MCGILL
PEEL

BERRI-UQAM
Gare d'autocars   Navette 747

CHAMP-DE-MARS

PLACE-D'ARMES
SQUARE-VICTORIA–OACI

BONAVENTURE–GARE CENTRALE   Gare Centrale

LONGUEUIL–UNIVERSITÉ-DE-SHERBROOKE

SNOWDON

VILLA-MARIA

VENDÔME
Vendôme

PLACE-SAINT-HENRI

GUY-CONCORDIA

ATWATER

LUCIEN-L'ALLIER
Lucien-L'Allier
GEORGES-VANIER

LIONEL-GROULX

GRIFFINTOWN–BERNARD-LANDRY
en construction

ÎLE-DES-SOEURS

CHARLEVOIX

LASALLE

DE L'ÉGLISE
VERDUN
JOLICOEUR
MONK
ANGRIGNON

PANAMA

DU QUARTIER

BROSSARD

FLEUVE SAINT-LAURENT

MÉTRO DE MONTREAL

RÉSEAU EXPRESS MÉTROPOLITAIN

STATION   **CORRESPONDANCE**
Pôles d'échanges multimodaux

12.5 Montreal Metro, 2023.

when most North American downtowns were collapsing. All the same, Canada's economic center of gravity shifted from Montreal to Toronto.

Drapeau's beloved 1976 Summer Olympics were an unmitigated financial disaster, with a 720 percent cost overrun. Drapeau had promised that "the Montreal Olympics can no more lose money than a man can have a baby," a statement that subjected him to public ridicule for decades afterward.[14] The Olympics left Montreal US$1.5 billion in debt, a sum that took 30 years to repay.[15] Major League Baseball's Expos started play in 1969 after Drapeau courted an expansion team. In 2004, the Expos left for Washington DC after a decade of anemic attendance and on-field mediocrity. Since

then, the old Olympic Stadium where the Expos played has lacked a permanent team. The Montreal Alouettes of the Canadian Football League and Major League Soccer's CF Montreal use the stadium when a very-high-capacity venue is needed, but these occasions are infrequent.

Sixty years after it opened, the Metro (fig. 12.5) still stands as Drapeau's signature achievement, one that still plays a large role in the daily lives of Montrealers. The Metro truly revolutionized Montreal's transport. The great irony is, that revolution wasn't the reason for its existence. Jean Drapeau wanted Montreal to be a global sensation. The Metro was the means to that end.

LAKE PONTCHARTRAIN

JEFFERSON PARISH

MISSISSIPPI R.

NEW ORLEANS

Streetcar system, 2022

Streetcar system, 1945

Planned monorail, 1958

ST. BERNARD PARISH

N

New Orleans

2 miles/3.2km

13

NEW ORLEANS

How a Big City Grew into a Small Town

13.1 Map of greater New Orleans.

New Orleans once seemed destined for perpetual greatness. Founded by the French in 1718 as a trading post on the swampy portage between Lake Pontchartrain and the Mississippi River, the city sits on the last piece of dry land suitable for a major port before the Mississippi enters the Gulf of Mexico. For the first two centuries of its existence, New Orleans (fig. 13.1) was "Queen City of the South." New Orleans even caught a break during the Civil War, because the city fell to federal troops early in the war and survived intact. As late as 1940, New Orleans was still the largest Southern city. This mix of geographical accident and traditional dominance allowed New Orleans to retain its unique culture and historical continuity. It's not for nothing that the famed St. Charles streetcar line has been running continuously since Andrew Jackson was president. Even when the rest of the continent was busy tearing out streetcar tracks, the St. Charles streetcar survived without much fuss.

New Orleans has survived plagues, a British invasion, the Civil War, two world wars, and a direct hit from Hurricane Katrina, but it lost its former preeminence in the years after World War II. When air conditioning became universal, most Sun Belt cities boomed. Not New Orleans. The old-fashioned gentry of New Orleans were quite satisfied with the status quo. This proved a double-edged sword. While this stubborn conservatism kept the St. Charles Avenue streetcar running and prevented the wholesale destruction of the city to build expressways, it also meant that New Orleans thought small and got small. The city's postwar growth was sluggish.

133

Transplants took their money to Houston and Atlanta instead. To this day, New Orleans's insular, complacent attitude persists, and its mass transit priorities reflect that.

Queen City of the South

New Orleans's streetcar history begins on September 26, 1835, when the New Orleans and Carrollton Rail Road began running steam locomotives out St. Charles Avenue to what was then the suburb of Carrollton. The line was a success. Steampowered street railroads and horsecars soon fanned out as the city developed. All lines were converted to use electric power in the 1890s after abortive experiments with mechanical cables, superheated water, and ammonia-powered locomotives.[1] From the 1920s onward, all streetcars were run by New Orleans Public Service Inc. (NOPSI), the local electric company.

Until World War II, New Orleans had no serious municipal competitor. No other port on the Gulf of Mexico could even come close to New Orleans's reach. Because of its position near the mouth of the Mississippi, New Orleans is the natural ocean port for most of the Midwest and the South. As might be expected, when New Orleans went to war in 1941, the city boomed. But after the war, New Orleans went to war with itself. (Near the war's end, the NOPSI streetcar system appeared as it did in fig. 13.2.)

Rust Belt in the Sun Belt

Postwar prosperity and the advent of cheap home air conditioning led to a major southward shift in where Americans lived. Air conditioning made subtropical cities like Houston, Atlanta, and New Orleans truly inhabitable during the long, sweltering, and humid Southern summers. Houston and Atlanta, open to outside influences, capitalized. New Orleans did not.

In this era, New Orleans's parochially minded, conformist white upper class was extremely insular. It was a place where the accidents of birth mattered far more than mere money. (New Orleans, after all, disdains arrivistes.) The routes into this closed world all ran through New Orleans's Carnival krewes—social clubs whose official purpose was to organize revelry during Mardi Gras season.[2] To join a krewe, one had to be a native. Well into the 1970s, the New Orleans gentry worried more "about who will be Queen of the Comus [Krewe] Ball than about systematic economic expansion—shaking themselves out of their dozy clubbiness only every 30 or 40 years to go along with some grandiose scheme that is touted as the miraculous rejuvenator of the economy."[3]

This is not to say that there was universal accord in New Orleans about this point. Quite the opposite. At the time, New Orleans was engaged in internecine warfare over what the city should be. On one side were reformers like Mayor deLesseps "Chep" Morrison, New Orleans's press, and much of its city council, which wanted to drag New Orleans kicking and screaming into the 20th century.[4] On the other side was much of the New Orleans gentry, who were perfectly fine with things as they were, thank you very much. These types of political fights were not the norm in postwar America.

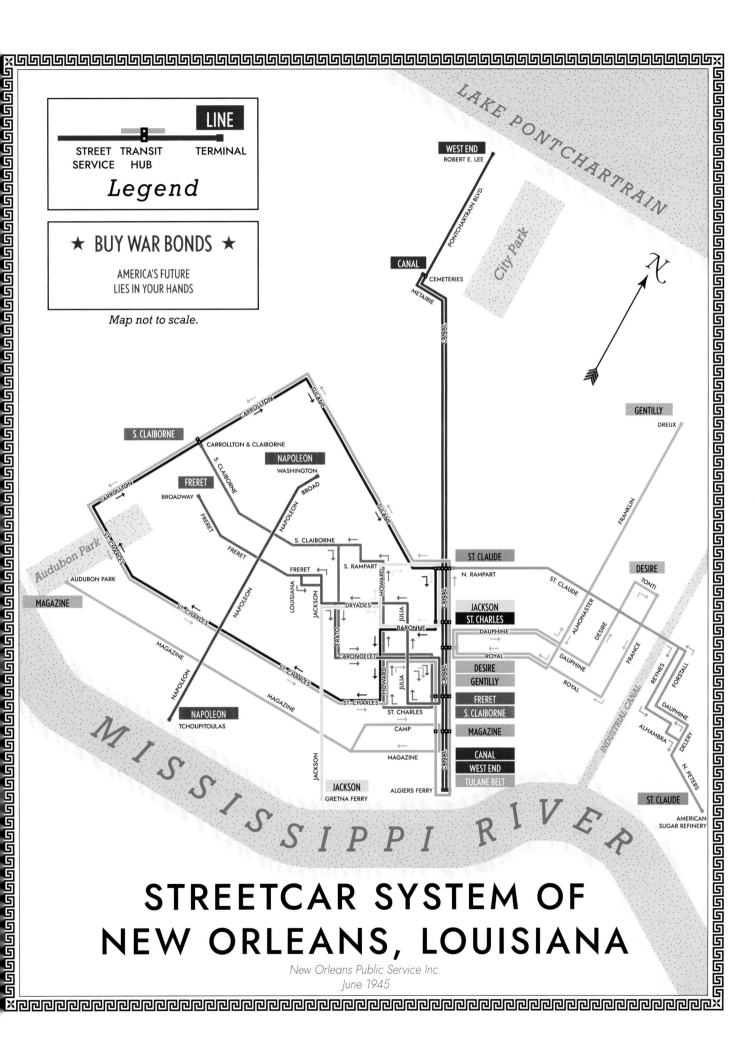

LAKE PONTCHARTRAIN

City Park

LINE

STREET | TRANSIT | TERMINAL
SERVICE | HUB

Legend

★ BUY WAR BONDS ★

AMERICA'S FUTURE
LIES IN YOUR HANDS

Map not to scale.

WEST END
ROBERT E. LEE

CANAL
CEMETERIES
METAIRIE

PONTCHARTRAIN BLVD.

CANAL

GENTILLY
DREUX

S. CLAIBORNE

CARROLLTON

TULANE

CARROLLTON & CLAIBORNE

NAPOLEON
WASHINGTON

FRERET
BROADWAY

S. CLAIBORNE

FRERET

NAPOLEON

BROAD

TULANE

FRANKLIN

CARROLLTON

AUDUBON PARK

ST. CHARLES

Audubon Park

MAGAZINE

FRERET

NAPOLEON

S. CLAIBORNE

FRERET

LOUISIANA

JACKSON

DRYADES

S. RAMPART

HOWARD

JULIA

BARONNE

CARONDELET

ERATO

ST. CHARLES

HOWARD

JULIA

CAMP

ST. CHARLES

ST. CHARLES

CANAL

N. RAMPART

ST. CLAUDE

ST. CLAUDE

DESIRE
TONTI

ALMONASTER

DESIRE

DAUPHINE

ROYAL

DAUPHINE

ROYAL

FRANCE

REYNES

FORSTALL

DAUPHINE

ALHAMBRA

DELERY

N. PETERS

INDUSTRIAL CANAL

JACKSON
ST. CHARLES

DESIRE
GENTILLY

FRERET
S. CLAIBORNE

MAGAZINE

CANAL
WEST END
TULANE BELT

MAGAZINE

NAPOLEON

MAGAZINE

NAPOLEON
TCHOUPITOULAS

JACKSON

JACKSON
GRETNA FERRY

MAGAZINE

ALGIERS FERRY

ST. CLAUDE

AMERICAN
SUGAR REFINERY

N

MISSISSIPPI RIVER

STREETCAR SYSTEM OF
NEW ORLEANS, LOUISIANA

New Orleans Public Service Inc.
June 1945

Elsewhere, opposition to the utopian schemes of the period was disorganized and ineffective. But New Orleans was resistant to such plans, especially when they negatively affected New Orleans's traditional upper classes.

The most obvious example of this process in action is the survival of the St. Charles streetcar. To provide context, NOPSI shut down all but two of New Orleans's streetcar lines in quick succession between 1945 and 1953. The Canal and St. Charles lines were the only survivors. NOPSI closed the Canal line in 1964, over the desultory objections of historical preservationists.[5] This was broadly in line with the historical trend. Streetcar systems were being dismantled across the continent. For the average passenger, air-conditioned buses were a major improvement over the streetcar. New Orleans averages 91 degrees Fahrenheit and 79 percent humidity in August, and NOPSI's old streetcars only had fans.[6] Even so, the St. Charles streetcar remained untouched, and NOPSI never made any serious attempt to replace it with buses. When the Canal line was in the process of closure, NOPSI took out large newspaper ads reassuring the public: "We do not intend now or in the foreseeable future to recommend any changes in the St. Charles streetcar operation."[7]

The St. Charles streetcar is a glaring outlier. There was nothing technically special about the line that would justify its preservation. Most surviving North American streetcar lines ran through tunnels or used dedicated, specialized rights-of-way.[8] The St. Charles line had neither. But politically, there was plenty of reason to keep it. The line runs in the grassy, tree-lined median, or

"neutral ground," of St. Charles Avenue, through some of the wealthiest neighborhoods of uptown New Orleans. Uptown's upper classes feared that the neutral ground would be eliminated for more traffic lanes if the streetcar line was replaced with buses.[9] This political dynamic was also responsible for the cancellation of the Vieux Carré Expressway through the French Quarter in 1969 after much debate, even as Interstate 10 through historically black Tremé was built as planned.

New Orleans's gentry remained wedded to these types of parochial concerns. This approach also kept the gates of polite society closed to outsiders. The result was brain drain and economic stagnation, something that became national news in the 1970s. A contemporary *Atlantic Monthly* article remarked that "New Orleans is perhaps the only large city left in America where birth counts for so much."[10]

Because of these factors, postwar New Orleans's demographics and development pattern were more typical of the Rust Belt than the Sun Belt. Between 1960 and 1990, the city lost a fifth of its total population, the same proportion as Baltimore. The metropolitan population stagnated. Over that time, the city went from being 68 percent white, 32 percent black, to 35 percent white, 62 percent black, as whites left the city for the suburbs. Even the rise of black political power in the city never really changed much. No matter whether the politicians in power were black or white, Catholic or Protestant, aristocrat or arriviste, New Orleans remained poor, unsafe, and resistant to change.

This starkly contrasts with what happened in Houston and Atlanta. Houston's power brokers welcomed successful Yankees like future president George H. W. Bush. Over the same period, the city of Houston's population grew by 73 percent. Atlanta sold itself as "the city too busy to hate" and made a conscious effort to court northern transplants and their money. Atlanta's metropolitan population more than doubled.

Monorail Inc.

As Houston grew larger and richer than New Orleans, New Orleans's leaders felt the need to engage Houston in a game of civic tit-for-tat. When Houston built the Astrodome, the world's first domed stadium, Louisiana governor John J. McKeithen ordered up a "Superdome" twice the size. When Houston built One Shell Plaza, a 50-story high-rise, New Orleans got a 51-story high-rise, One Shell Square. And after a Houston company called Monorail Inc. installed a demonstration track at the 1956 Texas State Fair, Mayor Morrison proposed to one-up Houston by building a monorail between New Orleans's new international airport and its new central railway station (fig. 13.3). For its part, Monorail Inc. argued that it could finance the project without public subsidy.

Monorail Inc.'s pitch to the city council was tantalizing. Fiberglass monorail trains would arrive every four minutes at rush hours. It would take 26 minutes from the city's central business district to the airport, about as fast as driving Interstate 10. Councilmembers were unsurprisingly excited. However, out of an abundance of caution, they needed an independent analysis before final approval. An engineer named S. H. Bingham was engaged to review the proposal in May 1958. Bingham was the former general manager of the New York Subway. He reported back the next January that the project was technically feasible.[11] But the problem wasn't the technology. It was the money. Monorail Inc.'s financial projections were optimistic, to say the least. Morrison was forced to admit that public financing might be needed. The City Planning Commission and other bastions of old New Orleans soon soured on the project. Even some monorail supporters balked at the potential financial exposure. In the end, the City Council bowed to the pressure. The council canceled the monorail in August 1959.

The monorail, the Superdome, and the new high-rises illustrated the city's curse: no matter how the skyline changed, the culture stayed the same. As writer Michael Lewis wrote of his childhood in 1960s New Orleans: "No one of importance ever seemed to move in, just as no one of importance ever moved away. The absence of any sort of movement into or out of the upper and upper-middle classes was obviously bad for business, but it was great for what are now called family values."[12]

Back to the Future

Modern New Orleans's complacency persists. Even before the catastrophic direct hit from Hurricane Katrina in 2005, the city was shrinking and metropolitan population growth was minimal. Tourism had become the principal driver of the economy, overtaking petroleum and shipping. In the early 2020s, the economy of greater New Orleans is one-sixth the size of greater Houston, and one-fifth the

137

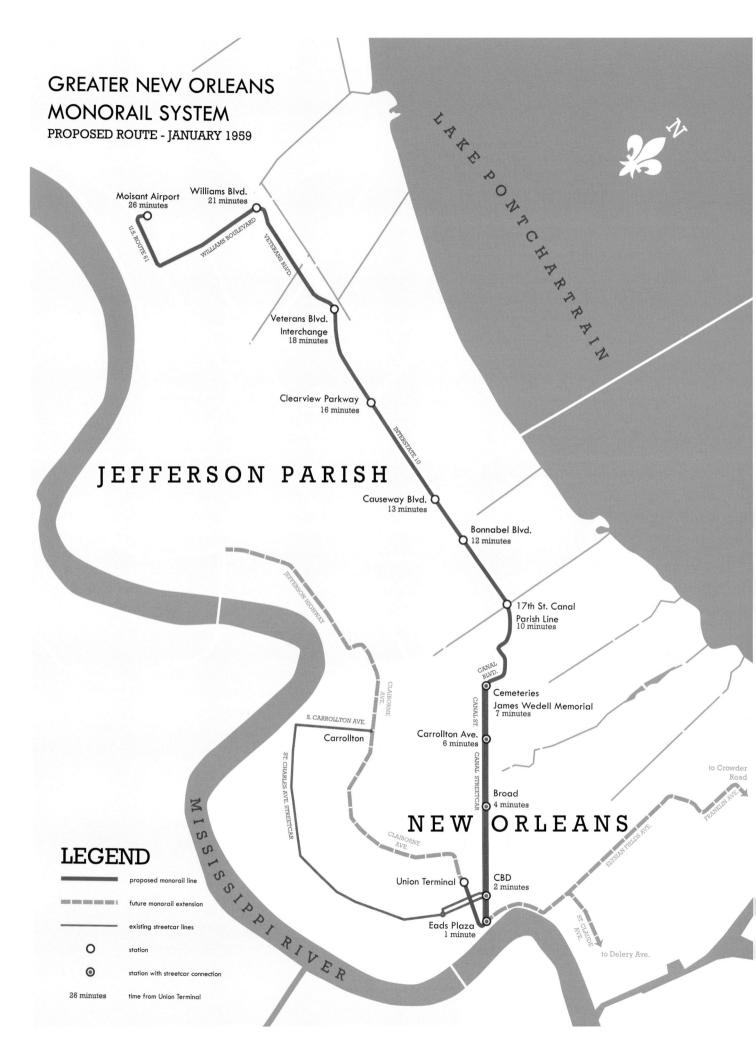

13.3 Monorail Inc.'s monorail proposal, 1959.

size of greater Atlanta. Houston is home to two dozen Fortune 500 companies; Atlanta is home to a dozen; New Orleans has only Entergy, NOPSI's successor company. The third-tallest building in Houston, the Williams Tower, hosts the Williams Companies, a Fortune 500 natural gas business; the third-tallest building in Atlanta, One Atlantic Center, is home to international law firm Alston & Bird; the third-tallest building in New Orleans, the 45-story Plaza Tower, is abandoned, filled with asbestos and toxic mold, and has been rotting, empty, for 20 years.[13]

New Orleans's approach to rebuilding its streetcar system over the past 50 years reflects this lack of ambition. Between 1964 and 1988, the only operational streetcar line was the venerable St. Charles line. Since then, three new lines have opened. The Riverfront line opened in 1988, a new Canal line opened in 2004, and the Rampart–St. Claude line opened in 2013 (fig. 13.4). Although it's superficially encouraging to see New Orleans revive its rail system, the improvements don't stand up to close inspection.

The Riverfront and Rampart–St. Claude lines don't serve any pressing urban transport need. These lines run infrequent, slow service through New Orleans's major tourist attractions, with stops every two blocks. They are meant for visitors. Unsurprisingly, these lines use hand-built, wheelchair-accessible, air-conditioned replicas of the century-old streetcars that still grace the St. Charles Avenue line. The Riverfront and Rampart–St. Claude lines also employ the same operational practices as the St. Charles line.

More illuminating of New Orleans's problem is the new Canal line, which opened in 2004, 40 years after its 1861-vintage namesake was shut down. This line is emphatically *not* for tourists. Canal Street is New Orleans's main thoroughfare, and has been the backbone of the city's transit network since the Civil War. Nine in 10 riders are locals.[14] In short, Canal Street is the perfect place to put a modern high-capacity transit line. New Orleans did nothing of the sort. When the Canal line was rebuilt in the late 1990s and early 2000s, the New Orleans Regional Transit Authority (RTA) made few gestures to modernity. (RTA, a public agency, took over transit operations from NOPSI in 1983.) Aside from air-conditioned trains and wheelchair accessibility, RTA ignored best practices when rebuilding the Canal line.

An explanation of why requires a slight detour into the technical details of mass transit planning. Generally speaking, light rail and streetcar lines use similar equipment.[15] The differences between the two are in operational procedures. To provide fast, high-capacity service, the best practice is to provide the trains dedicated lanes, run multiple-car trains, sell tickets to passengers before they board, give trains priority at intersections, and build stops every half mile or so when operating in dense areas.[16] This exploits rail's largest advantage: the ability to move large numbers of people quickly. Houston's Red Line, which opened in 2004, does all these things. It averages 17 miles per hour.

New Orleans's Canal line has dedicated lanes for the trains, but none of the remaining operational features. RTA runs single-car trains, the trains have to wait for cross traffic, passengers pay the fare onboard one by one, and the line has stops

139

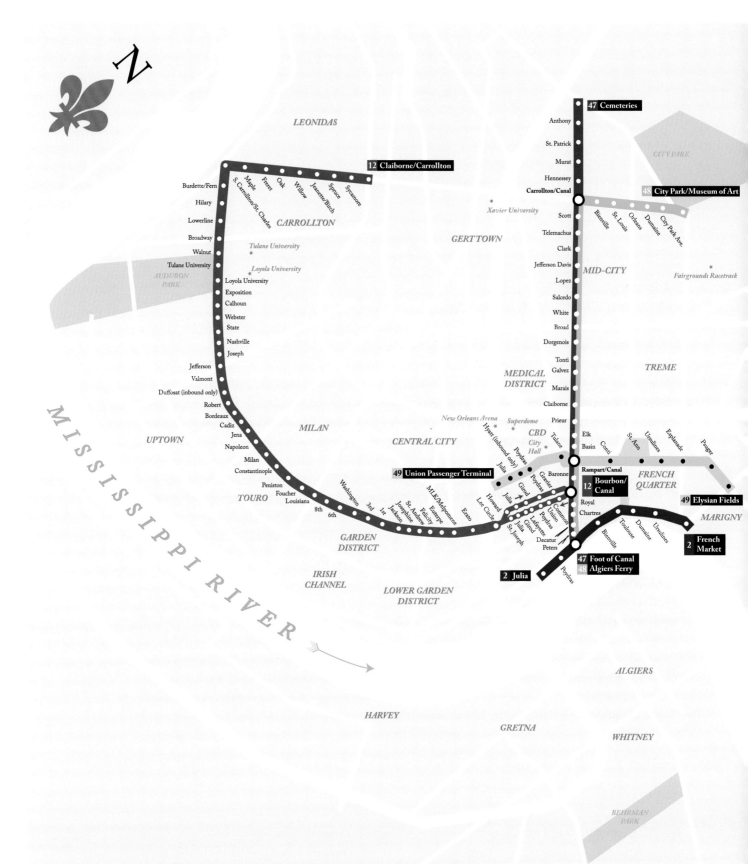

New Orleans Streetcars

| 2 | *Riverfront* | 47 | *Canal-Cemeteries* | 49 | *Rampart-St. Claude* |
| 12 | *Saint Charles* | 48 | *Canal-Museum* | | |

13.4 RTA streetcar system, 2022.

every 600 feet.[17] (To its credit, RTA rolled out smartphone tickets in the late 2010s, but uptake has been slow.[18]) It takes over half an hour to cover the 2.75 miles between Carrollton Avenue and the Mississippi River, an average speed of 5.5 miles per hour.[19] That is to say, it isn't much faster than walking.

New Orleans didn't just make mistakes in the initial construction of the new Canal line. The city has also stubbornly opposed any attempts to fix it. In February 2019, RTA released a pilot plan with most of the best practices already described. RTA wanted to start selling tickets at stations, give trains priority at intersections, close certain motor vehicle U-turn locations across the tracks, and reduce the number of round-trip stops from 49 to 19— that is, a station every three-tenths of a mile instead of every 600 feet. This plan would have cut travel times by a third, reduced accidents by a quarter, and dramatically improved on-time performance. It would have brought New Orleans more in line with the standards used in the rest of the country.

And yet, a firestorm of criticism from riders and politicians caused RTA to put the plan on ice. The complaints were, drivers would have fewer places to make U-turns, passengers would have to walk further to take the train, and the proposal relied on objective computer modeling rather than the subjective experiences of riders. By March, opposition had mounted so much that RTA's executive director had to apologize to the City Council. The pilot was dropped.[20]

In a sense, RTA's streetcar improvement pilot is representative of New Orleans's metropolitan experience since World War II. As with so many other things in the Big Easy, New Orleans's residents simply preferred the status quo—an unreliable, slow streetcar—to the possibility of change. This conservative attitude has been a major factor in the preservation of New Orleans's historic, quirky neighborhoods and culture. This attitude also saved the French Quarter from an expressway. Likewise, the St. Charles streetcar is still there, running the same route it did when James Madison was still alive, the Union had 24 states, and California was part of Mexico. Urban continuity is important, a lesson that New Orleans knows better than most. But New Orleans's approach to urban continuity also makes adaptation difficult when there are lessons that can be learned from the outside.

141

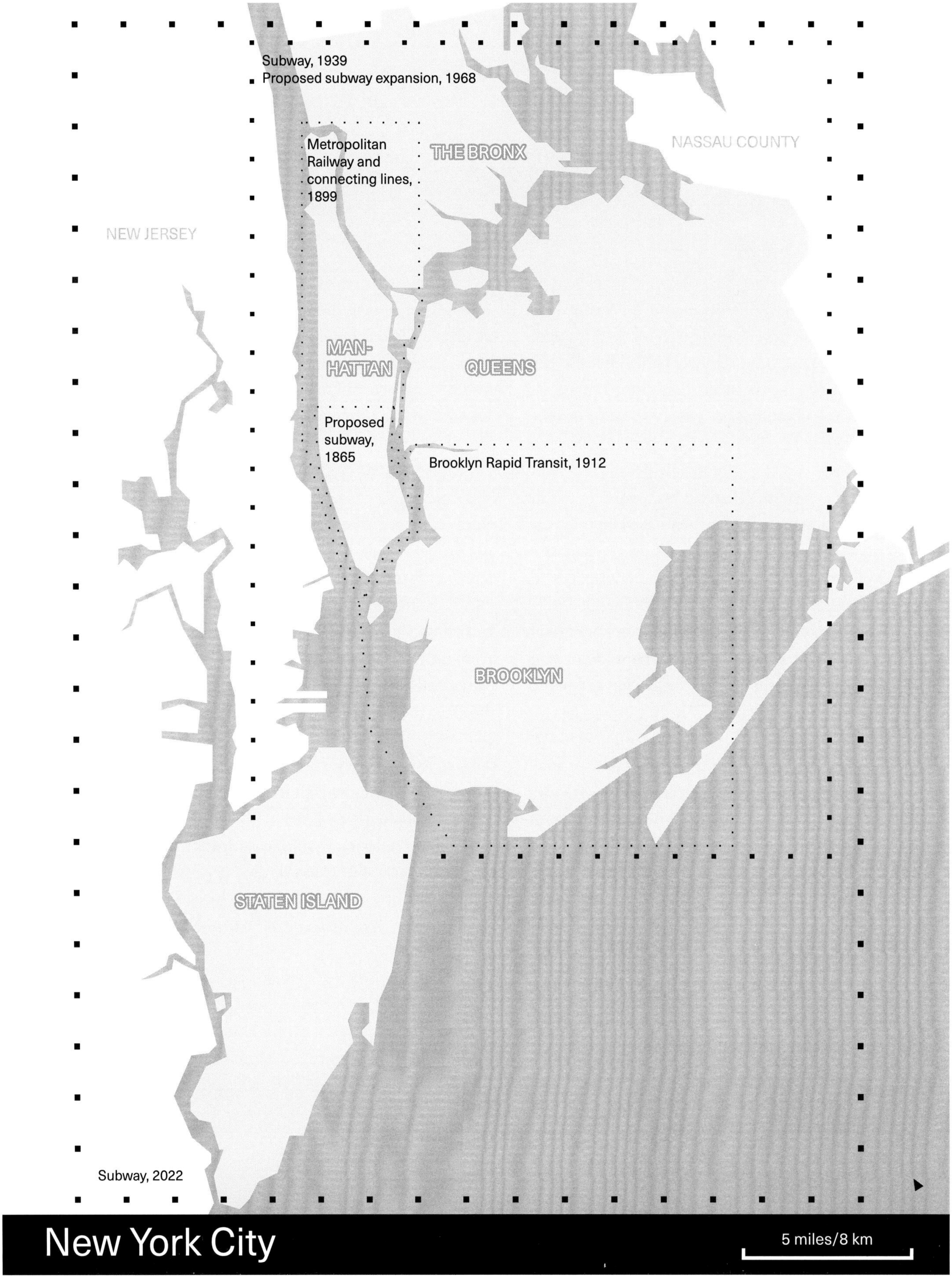

Subway, 1939
Proposed subway expansion, 1968

Metropolitan
Railway and
connecting lines,
1899

THE BRONX

NASSAU COUNTY

NEW JERSEY

MAN-
HATTAN

QUEENS

Proposed
subway,
1865

Brooklyn Rapid Transit, 1912

BROOKLYN

STATEN ISLAND

Subway, 2022

▶

New York City

5 miles/8 km

NEW YORK CITY

The Tortured History of the Second Avenue Subway

14.1 Map of New York
City.

In 1939, my grandmother, a young doctor from Kansas, had just moved
to New York City to train at the great Bellevue Hospital complex on the
East Side of Manhattan. To get to Bellevue, she would've taken the Second
Avenue Elevated, a decrepit, 60-year-old steel eyesore on its last legs. At the
time, the Second Avenue Elevated was still using wooden train cars built in
the late 1880s. The City of New York had made plans in 1929 to demolish
the Second Avenue Elevated and replace it with a modern subway line.
At the time, few doubted that a Second Avenue Subway was in the near
future. After all, New York (fig. 14.1) had been on a decades-long subway-
building orgy. The city government had vowed to tear down the noisy, ugly
elevated lines that bathed the streets below in darkness. But the best laid
plans often go astray. Nearly a century later, the Second Avenue Subway
still hasn't been finished.

The city closed the Second Avenue Elevated in stages between 1940
and 1942. Lack of money and America's entry into World War II put the
replacement subway line on hold. Despite this, planning continued. New
York City voters approved funds for a comprehensive subway expansion
in 1951. The city even bought train cars to serve the future Second Avenue
Subway. But the Korean War drove construction costs so high that plans
were scrapped. A second bond measure passed in 1968 to build the Second
Avenue Subway, as part of a 50-mile citywide subway expansion. But New
York City was falling apart in the 1970s, and work on the unfinished line
was halted in 1975. The 1970s and 1980s marked a turning point for New

143

York. From that point on, the city lost its ability to think big.

Prosperity returned in the 1990s and 2000s. New York's ability to plan and execute megaprojects at reasonable cost did not. The modern plans for the Second Avenue Subway reflect this lack of vision and lack of institutional competence. The state Metropolitan Transportation Authority (MTA), which currently runs the subways, approved a 10-mile Second Avenue Subway in the year 2000. Twenty-three years later, only a 1.8-mile stub is complete. No one has been held accountable for this failure, and the MTA's governance structure makes it exceedingly difficult to hold anyone responsible.

A century later, the Second Avenue Subway still isn't done.

Primitive Subways and Elevateds

The New York City Subway's history begins in 1864, when a Michigan businessman named Hugh Willson proposed to build a steam-powered Metropolitan Railroad inspired by the eponymous line in London (fig. 14.2). After all, if London could build an underground railway, why couldn't New York build a tunnel under Fifth Avenue? Willson sponsored a bill to authorize construction in the state legislature, which approved it in April 1865. To the surprise of underground railway supporters, Governor Reuben Fenton vetoed the project in late May. Fenton's veto message cited the possibility that the stations would be an unsightly presence in city streets and parklands.[1] He stated, "I cannot consent, on my part, to such use of these grounds, without feeling that I had violated the trust reposed in me by the people."[2] Later attempts to revive the Metropolitan Railroad project in 1866 fell short. Momentum passed to the backers of the competing elevated proposals. Later that year, Fenton approved the construction of New York's first rapid transit line, the Ninth Avenue Elevated, which opened in 1868. Nine years later, the Ninth Avenue Elevated was extended to the southern tip of Manhattan. A smoky four-track station served by coal-burning locomotives was built in the middle of Battery Park.[3]

After working out the kinks, the Ninth Avenue Elevated was an immense success and tremendously profitable. This kicked off a frenzy of elevated construction in Manhattan and what was then the independent city of Brooklyn. The Second Avenue Elevated opened in 1878. By 1880, four steam-powered elevated lines covered the entire island of Manhattan, and the competing elevated companies had all come under common ownership (fig. 14.3 depicts the Manhattan elevateds at the turn of the 20th century). A similar process took place across the East River in Brooklyn under the Brooklyn Rapid Transit Company, albeit somewhat more slowly. Around the turn of the 20th century, all steam elevated lines were electrified. Elevated lines, while useful and cheap to build, were noisy and ugly, and caused property damage to adjacent buildings. (See the chapter on Vancouver for a discussion of the benefits and drawbacks of elevated construction in more detail.) A subway, while more expensive, would be hidden safely underground, where the earth would absorb much of the noise and vibration.

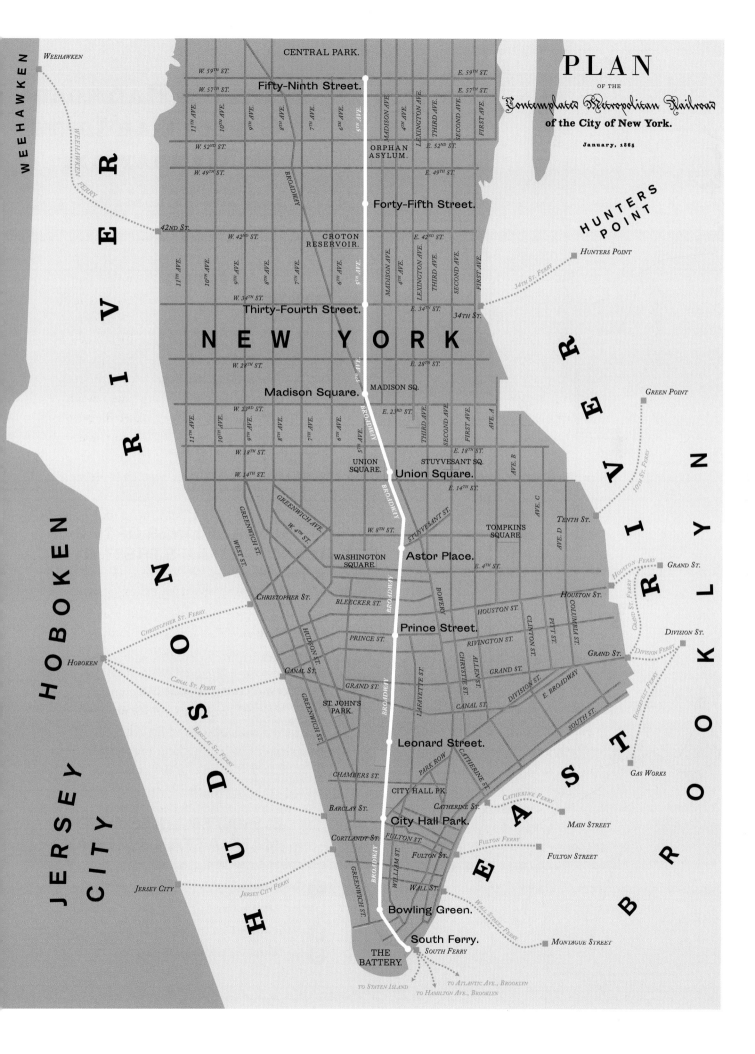

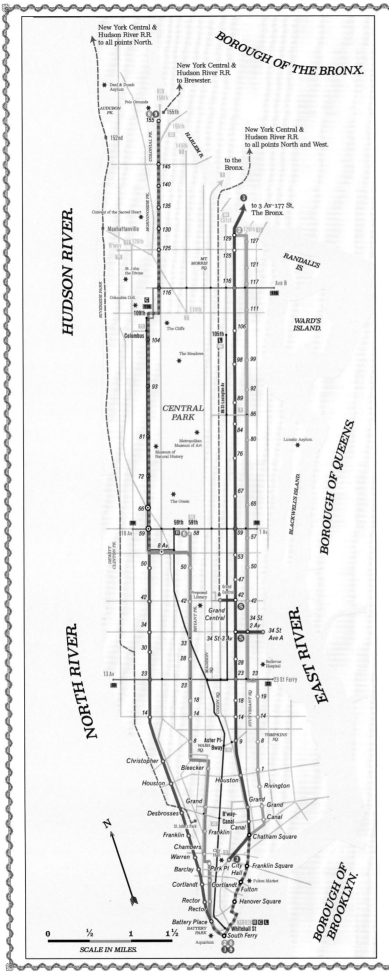

ELEVATED RAILROAD & STREET CAR LINES OF MANHATTAN.

New Map and Guide to the
Elevated Railways of the Manhattan Railway &
the Street Cars of the Metropolitan Street Railway Co.

Updated for 1899.

LEGEND.

Elevated Railways.

Electric Street Cars.

Crosstown Electric Street Cars.

Cable Cars.

Main Line Railroads.

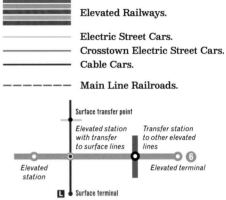

Surface transfer point

Elevated station with transfer to surface lines

Transfer station to other elevated lines

Elevated station

Elevated terminal

L Surface terminal

SURFACE LINES OF THE METROPOLITAN STREET RY. CO.

ELECTRIC STREET CAR.

1 N.Y. & Harlem via Madison Ave.
7 Lenox Ave.
11 Ninth Ave.
15 Second Ave.
20 Eighth Ave.
55 Sixth Ave.
98 Lexington Ave.

CROSSTOWN ELECTRIC STREET CAR.

23 Twenty-Third St.
59 Central Park - 59th St.
116 116th St & Manhattan Ave.

CABLE CAR.

B Broadway Cable.
C Columbus Avenue Cable.
L Lexington Avenue Cable.

ELEVATED LINES OF THE MANHATTAN RAILWAY.

2 Second Avenue Line.
3 Third Avenue Line.
6 Sixth Avenue Line.
9 Ninth Avenue Line.
S 42nd Street and 34th Street Shuttles.

14.3 Elevated lines of Manhattan, with connecting
streetcars and cable cars, 1899. The Second
Avenue Elevated is in light blue.

Suddenly Admitted into Civilization

The first subway opened on October 27, 1904. The
line was built by the city and leased to Interborough Rapid Transit, which had taken control of
Manhattan's elevated lines.* New Yorkers immediately took to the new invention. On the first day of
the subway's operation, the *New York Times* wrote,
with an unfortunate metaphor typical of the time,
"The up-bound Brooklynites and Jerseyites and
Richmondites [from what is today Staten Island]
had boarded the trains with the stolid air of an African chief suddenly admitted into civilization and
unwilling to admit that anything surprised him.
The Manhattanites boarded the trains with the
sneaking air of men who were ashamed to admit
that they were doing something new, and attempting to cover up the disgraceful fact."[4]

The Interborough subway was a smashing
success. Shortly thereafter, in 1913, the city signed
contracts to build new subways with the Interborough, as well as with Brooklyn Rapid Transit.
These lines were called the Dual Contracts (fig.
14.4 depicts Brooklyn Rapid Transit's lines immediately before the Dual Contracts were signed).
The subway quickly became a victim of its own
success. By the early 1920s, the new subways were
overcrowded. The relationship between the private
subway companies and the city soured. World War

I-related inflation had greatly reduced the value of
the dollar, and the city would not allow the Interborough and Brooklyn Rapid Transit to raise
fares.[5] By 1925 the city had started construction on
a third subway system, the Independent Subway
System. The Independent would be city operated.
It would compete directly with the Interborough
and Brooklyn-Manhattan Transit (BMT), successor to Brooklyn Rapid Transit.

Close but No Cigar

In 1929, before a single Independent train had run,
the Independent's planners had already put together a second-phase system expansion with another
hundred miles of subway.[6] This second phase was
meant to put the remaining elevated lines out of
business, enabling the city-owned Independent
to take over the Interborough and BMT. A four-track Second Avenue Subway was included when
this expansion plan became public on September
15, 1929. The plan proved excessively ambitious.[7]
Thirty-nine days later, the stock market crashed
and the Great Depression began. The first phase
of the Independent would open over the course
of the 1930s, vastly over budget. The cost overruns
and the country's economic woes meant that the
second phase never came to anything.

By 1940, the private subway companies were
in deep trouble. Competitive pressure was heavy
due to the Great Depression, the opening of the
Independent, and the city's continuing refusal to
allow fare hikes.[8] Fares had been frozen at five cents
for three and a half decades. The BMT was barely profitable and the Interborough was bankrupt.

---

* The MTA uses three-letter abbreviations for the three
early subway operators: IRT (Interborough Rapid Transit),
BMT (Brooklyn-Manhattan Transit), and IND (Independent
Subway System). To avoid abbreviation confusion, I use Interborough, BMT, and Independent, respectively.

147

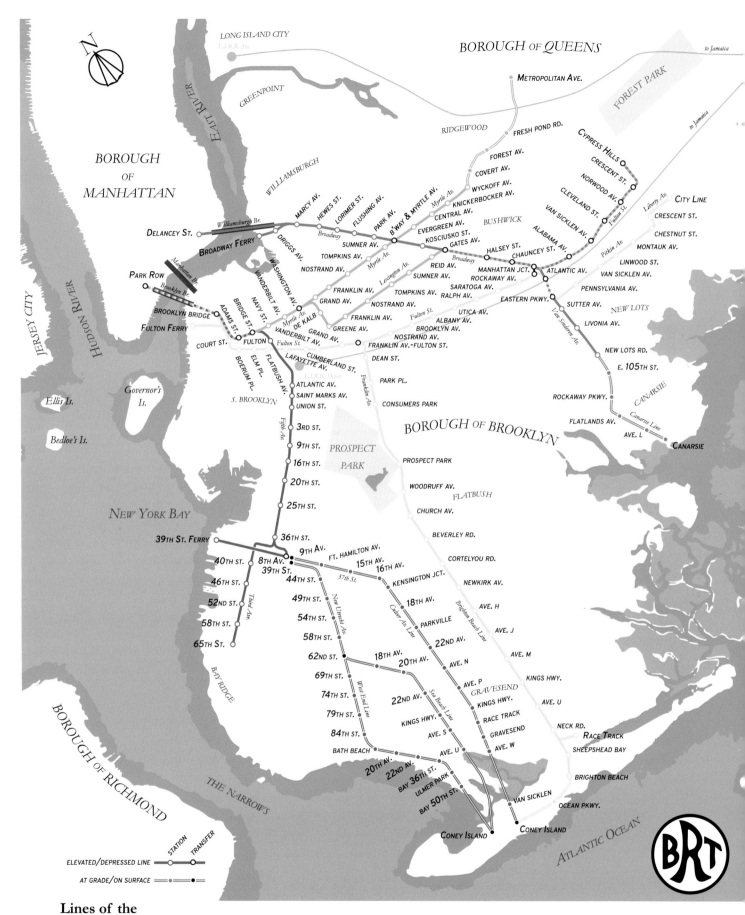

Lines of the
Brooklyn Rapid Transit Co.

rev. Dec. 1912.

Thus, in June 1940, the Independent took over both the BMT and the Interborough, bringing the entire system under city control (fig. 14.5 shows the system immediately before the takeover). The city moved quickly to shut down the Manhattan elevated lines. On June 13, 1942, the last portions of the Second Avenue Elevated were closed. No immediate replacement plans were in the works.

Hope Springs Eternal

There were two more major attempts to build the line in the 20th century. In 1951, New York City voters approved $500 million in bonds ($5.3 billion, adjusted for inflation) to build a Second Avenue Subway, among other things. Construction was supposed to begin in 1957 or 1958, but most of the money went to modernization and rehabilitation of existing subway lines, not to new construction. Compounding this problem, inflation related to the Korean War significantly reduced how much subway $500 million could buy. By 1957, the front page of the New York Times reported, "The Second Avenue subway, envisioned long ago and promised in 1951, probably never will be built."[9]

The final 20th-century attempt to build the Second Avenue Subway came as part of a major 1960s reform package to revive transport in greater New York. The transportation situation at the time was dire. There was no regional coordination. The subway was falling apart. The region's commuter railroads were bankrupt. Meanwhile, the Triborough Bridge and Tunnel Authority, which operated New York City's tolled bridges and tunnels, was flush with cash. Thus, between 1965 and 1968,

a series of shotgun marriages brought the subways, the commuter railroads, and the Bridge and Tunnel Authority under the umbrella of the newly established MTA. The MTA answered to the state governor in Albany. At the same time, state voters approved a $2.5 billion transportation bond ($19.8 billion, adjusted for inflation) to modernize and expand the region's transportation infrastructure.[10] The Second Avenue Subway was the centerpiece of the modernization plan, called the Program for Action. Fifty miles of new routes were proposed, including a 10-mile Second Avenue Subway (fig. 14.6).[11] Things started out well. In October 1972, the Second Avenue Subway made a ceremonial groundbreaking in Spanish Harlem, Manhattan. The MTA was optimistic in 1973 that "the Second Avenue subway, talked about since the 1920's, is becoming a reality in the 1970's."[12]

It did not become a reality.

New York City in the 1970s was rapidly disintegrating. Crime had skyrocketed. Whole neighborhoods were abandoned. Unreliable, graffiti-festooned subway cars were the norm. By 1975, the economy was stagnant, inflation was high, and the city government was in the middle of a fiscal crisis. With the city in danger of defaulting on its debts, Mayor Abe Beame ordered Second Avenue Subway construction to halt in September 1975. The next month, the city had to be saved from insolvency by the city teachers union's pension fund.[13] On October 29, President Gerald Ford rejected a federal bailout, prompting the now-famous newspaper headline "Ford to City: Drop Dead."[14]

Things continued to be bad through the 1980s. Despite slow improvement, the subway was dan-

149

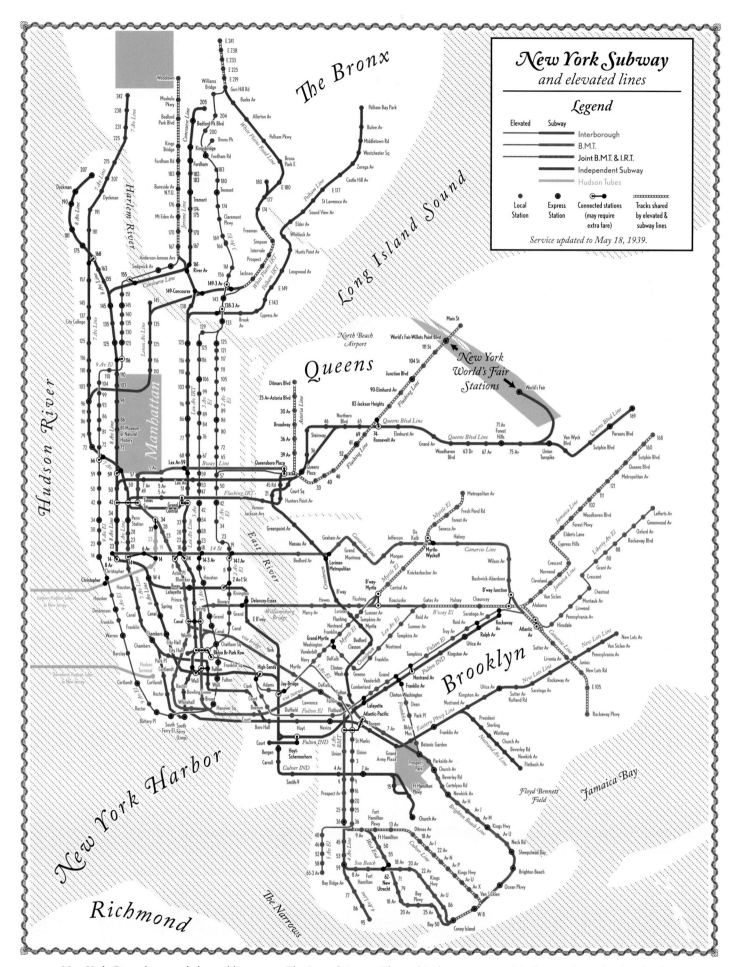

14.5 New York City subway and elevated lines, 1939. The Second Avenue Elevated is the easternmost Manhattan elevated, colored red.

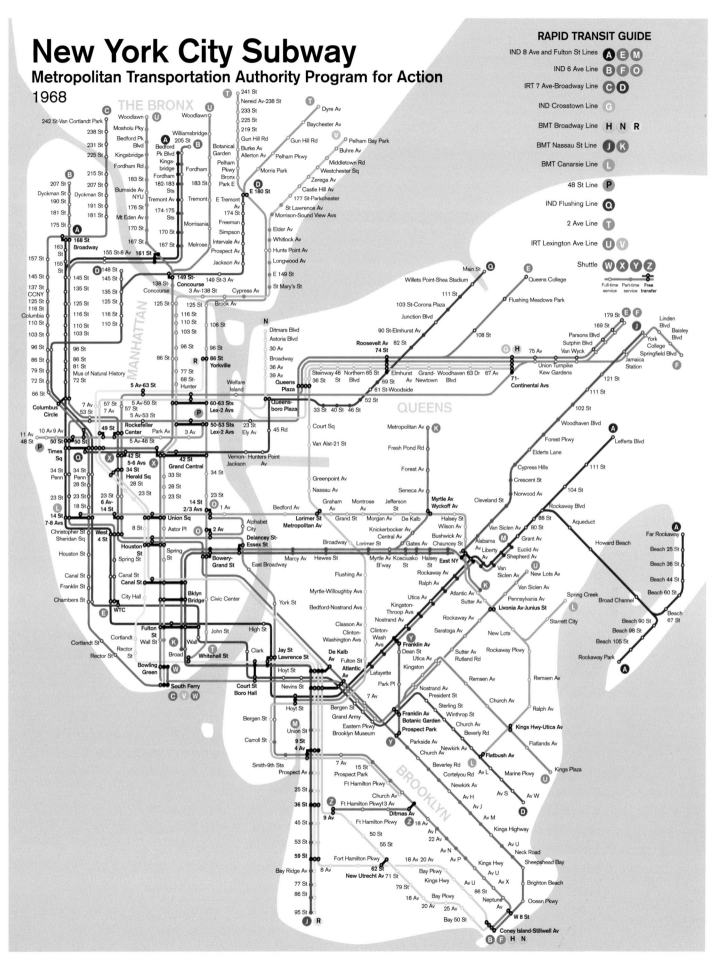

14.6 The MTA's 1968 Program for Action. In the map, I have used the line-naming scheme proposed in 1968, so line designations do not match modern usage. Line T is the Second Avenue Subway.

gerous and dilapidated. In 1984, a white subway passenger named Bernie Goetz became national news for shooting four black teenagers who had tried to rob him. (Amid a media circus, Goetz was acquitted of attempted murder in the criminal trial, but he was held civilly liable.) Money was so short that the MTA considered shutting down 16 percent of its stations in 1986, mostly in poorer sections of the city outside Manhattan.[15] There was no chance of the Second Avenue Subway being built under these conditions.

The Most Expensive Piece of Subway in the World

In the 1990s, crime went down, the economy recovered, and New York City started its comeback. The graffiti was cleaned and the trains began to run on time again. The city and state once again had the financial stability to take another stab at building the Second Avenue Subway. The MTA board committed to building the full 10-mile line in 2000.[16]

Twenty-two years later, only 1.8 miles of two-track subway is complete. The cost has been $2.5 billion per mile, making it the most expensive piece of subway in the world.[17] Per mile, New York paid two and half times what Los Angeles pays, six times what Tokyo pays, and 14 times what Milan pays.[18] The challenges of building subways in New York City are not technological compared to these other cities. Los Angeles has earthquakes. Tokyo has earthquakes and is very densely populated. Milan is 2,500 years old, introducing archaeological and geological complexity. Rather, the challenges of building new subways relate to the MTA's competence as an institution.

The most usual way of addressing governmental institutional failure is to hold officials accountable at the ballot box and through the other mechanisms of electoral politics. But it's extremely difficult for New York City voters to hold officials to account for MTA failures. By law, the suburbs control one-third of votes on the MTA board. New York City's mayor controls another third, and the state governor in faraway Albany controls the remaining third.[19] The governor has effective day-to-day control, as the governor hires senior MTA management. The state legislature in Albany controls the purse strings. This governing structure means that city subway riders are subject to the whims of politicians who aren't accountable to city voters. This encourages political interference and incompetence.

Take the example of Andrew Cuomo. Cuomo took office as Governor of New York in January 2011. In May 2011, 85 percent of subway trains ran on time.[20] In his first six years in office, Cuomo repeatedly diverted the MTA operations budget to fund unrelated projects, in one case using MTA money to bail out bankrupt state-owned ski resorts.[21] Subway on-time performance collapsed, culminating in the 2017 "Summer of Hell," when only 65 percent of trains ran on time, and two trains derailed.[22] Taking political flak, Cuomo hired Andy Byford to run the subways. Byford, foremost transit operations expert in the English-speaking world, was CEO of the Toronto Transit Commission. The soft-spoken Englishman took control of the subways on New Year's Day 2018.

(14)

In Byford's two-year tenure, he began to clear the MTA's decades-long repair backlog, greatly improving subway reliability in the process. But because the press gave Byford credit for this turnaround, rather than Cuomo, Cuomo stripped Byford of his authority in the summer of 2019.[23] By early 2020, Byford was forced out entirely. (The mayor of London seized the opportunity, immediately hiring Byford to run the London Underground.) Cuomo was savaged in the press, but he faced few political consequences for the Byford disaster.[24] Until an unrelated sexual assault scandal forced his resignation in 2021, Cuomo was on track to win a fourth term as governor.

But the MTA's institutional accountability problems aren't just about politicians. The MTA's bureaucracy has little ability to project manage its contractors, and subway construction contracts are vastly overstaffed due to the MTA's work rules. For example, MTA work rules often require up to 25 workers to run a tunnel-boring machine. Other major cities require fewer than 10.[25] The MTA's shorthanded engineering team lacks the in-house expertise and authority to push back against spendthrift consultants, construction companies and politicians. (Similar problems afflict the Massachusetts Bay Transportation Authority in Boston.) As *New York Magazine* put it, "Nobody knows who's in charge, so nobody has to be terrified of taking the blame for obscene costs and endless delays."[26]

The best illustration of this process is to compare two subway stops on opposite sides of Manhattan: 72nd Street station on the Second Avenue Subway, opened in 2017, and 72nd Street–Broadway station, built for the Interborough in 1904 as part of the very first subway line (fig. 14.7).

The 72nd Street station on the Second Avenue Subway is 600 feet long, 50 feet wide, and 100 feet deep, with two subway tracks and high ceilings. The station had to be mined out from within the tunnel at very high cost to keep neighbor complaints to a minimum. A full-length mezzanine with turnstiles and ticket machines is suspended above the platform, inside the tunnel. By comparison, the 1904-vintage station at 72nd Street–Broadway is 520 feet long, 60 feet wide, and 14 feet deep, with space for four subway tracks. To build the station in 1904, the city simply tore up the street and dug a shallow station large enough to fit four tracks. Turnstiles and ticket machines are at street level. In total, the modern station is six times larger, has half the train capacity, and took three times longer to build than the historic station on the other side of Central Park.

The Second Avenue Subway isn't the only public infrastructure in New York suffering from this kind of poor management. The one-stop subway extension of the 7 train from Times Square to 34th Street–Hudson Yards cost $2.1 billion per mile. This is less than the Second Avenue Subway, but several multiples of what Tokyo pays.[27] The Moynihan Train Hall—an expansion of Penn Station, the busiest transportation facility in North America—was originally projected to cost $315 million and take three years to build, with a projected opening date in 2002.[28] Moynihan ultimately opened in 2021 at a cost of $1.6 billion, 19 years after its original estimated completion date and five times its original projected cost.[29]

153

14.7 New York City Subway, 2022. 72nd Street on the
Second Avenue Subway is on the Q train, in yellow;
72nd Street–Broadway from 1904 is on the 1, 2 and
3 trains, in red. The Hudson Yards extension is the
section of the 7 train west of Times Square, in violet.

It's anybody's guess if the Second Avenue Subway will ever get finished, and it's a sign of New York's reduced ambitions. When my grandmother was doing her training at Bellevue Hospital in 1939, New York thought it reasonable to build 100 miles of subway in 10 years, with a four-track, 10-mile main line underneath Second Avenue. A lifetime later, New York considers it a great victory to have the most expensive 1.8 miles of subway ever built.[30]

My grandmother lived a long, fruitful life. After she finished her medical residency at Bellevue Hospital, she married my grandfather, raised five children, practiced medicine for six decades, and passed away quietly in her home in 2008. In her lifetime, she witnessed the entire sweep of the 20th century: the Great Depression, World War II, the civil rights movement, Apollo 11, and the fall of the Berlin Wall. But she did not live to see the completion of the Second Avenue Subway.

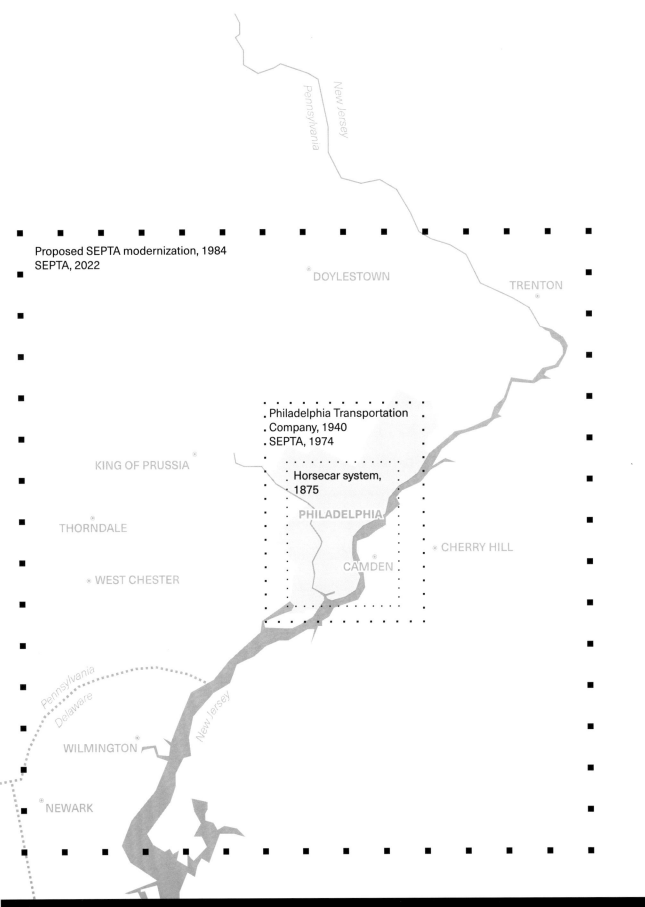

Proposed SEPTA modernization, 1984
SEPTA, 2022

DOYLESTOWN

TRENTON

Philadelphia Transportation
Company, 1940
SEPTA, 1974

Horsecar system,
1875

KING OF PRUSSIA

PHILADELPHIA

THORNDALE

CHERRY HILL

CAMDEN

WEST CHESTER

Pennsylvania

Delaware

New Jersey

WILMINGTON

NEWARK

N

Philadelphia

5 miles/8 km

PHILADELPHIA

How Not to Run a Railroad

15.1 Map of greater Philadelphia.

There's a German saying in transportation planning that a transit operator's order of priorities should be *Organisation vor Elektronik vor Beton*, or organization before electronics before concrete. That is, the most important factor in running effective mass transit is the bread and butter of everyday transit operations, rather than state-of-the-art technology. Building new physical infrastructure should be a last resort. Many public transport agencies and the politicians who oversee them don't think this way. Nowhere is this more apparent than Philadelphia. Philadelphia (fig. 15.1) has all the physical infrastructure in place to have the best regional rapid transit system in the country. But the Southeastern Pennsylvania Transportation Authority (SEPTA) doesn't do that. Instead, SEPTA spends millions of dollars per year running infrequent, expensive regional rail service. It's a useful illustration of how bad organizational policies can get in the way of good transportation.

Like most North American cities, private operators built out the local transit system in Philadelphia, which the newly established SEPTA took over in the 1960s. SEPTA had its hands full from day 1. Money was tight, so SEPTA suspended most surface trolley services one by one and replaced them with buses. But SEPTA did sign on to an innovative plan to fix the regional rail system and support Center City Philadelphia (the city's downtown) against the tectonic forces of suburbanization and deindustrialization.

The plan was to convert the regional rail system into a de facto part of the subway network, modeled after similar projects in the German-speaking world. This would require some concrete, including a new commuter tunnel (officially the Center City Commuter Connection) between Philadelphia's two main railway stations. Some technological upgrades were also included. But most of the substantive changes would happen by reforming SEPTA's organization. Two problems prevented this conversion from happening.

The first problem was labor. SEPTA's relationship with its unions has never been good, and Philadelphia's transit unions are unusually strike-prone.[1] From 1970 to 2022, SEPTA unions have gone on strike 12 times—that is, once every five years or so. For comparison, New York has had only three strikes in that time, and Chicago, two. One of these strikes proved fatal to the plans to reinvent the regional rail system. In 1983, one year before the commuter tunnel was slated to open, a lengthy regional rail strike caused ridership to crater, making it impossible to implement the necessary reforms.

The second problem SEPTA faces is its own bureaucratic inertia. This inertia can be traced to how SEPTA is governed. SEPTA ultimately answers to a board of directors appointed by elected officials from all over the state of Pennsylvania. In theory, this structure allows for a wide range of state and local officials to get a voice in SEPTA's decision-making. In practice, because everyone is responsible, no one is responsible. This governance problem is not unique to SEPTA, but SEPTA's dynamic is unusually dysfunctional because so many politicians have seats at the table. For these reasons, it's difficult to change SEPTA.

The reforms to make SEPTA's regional rail run like a modern rapid transit system aren't rocket science. The system needs the following: automated fare collection (i.e., ticket machines) instead of having train conductors check every ticket; one-person train crews instead of two- to four-person crews; subway fares for trips within the city instead of the more expensive regional rail fares; and train-level platforms, eliminating the need for conductors to manually raise and lower train stairs for passengers to board. Combined, these four reforms would enable SEPTA to run rapid-transit-like service at relatively low cost. But SEPTA hasn't done these things, plagued as it is by its labor issues and its own hidebound ways.

SEPTA and the City of Philadelphia know what the problems are. Ticket machines, less wasteful staffing, equal fares, and level boarding are straightforward solutions to figure out on paper. Their long-range planning departments have put these items on the wish list. If enacted, the changes would allow the system to realize its true potential, decades after the physical infrastructure was built. It's another question entirely whether SEPTA's management will put in the hard work necessary to make these organizational reforms a reality.

Horsecar, Trolley, Subway, Railway, Bus

To explore this history requires going all the way back to the beginning. SEPTA's transit system is the heir to the intricate network of horsecars, long-distance railroads, trolleys, buses, and subways originally operated by private companies in the 19th and early 20th centuries. This transit network was one of the largest in the world. A sense of the chaotic nature of the early system can be gathered from figure 15.2, which depicts Philadelphia's horsecars and urban steam railroads in 1875.

By 1940, transit operations within greater Philadelphia had been largely consolidated under four roofs: the Philadelphia Transportation Company (PTC), which ran trolley, subway, elevated, and trolleybus service within the city; the Red Arrow Lines, which ran most local trolley and bus service in the suburbs; the Pennsylvania Railroad (Pennsy), which ran a large network of electric regional rail lines, among other things; and the Reading Railroad (*Reading* rhymes with *bedding*), which was the Pennsy's direct competitor (fig. 15.3).

All four of these companies' transit operations became financially unsustainable after World War II. At the same time, with the construction of the Interstate Highway System, the city was in the midst of a major demographic shift, although Philadelphia's white flight and attendant population loss was never as severe as in cities like Detroit. The newly formed SEPTA took effective control between 1965 and 1970, during the era of white flight but before the onset of the deindustrialization of the 1970s.

SEPTA's response to this crisis was basically reactive. The old private transit operators had left their physical infrastructure in a dreadful state. At one point, three-quarters of subway cars were judged unusable.[2] Most trolley services were deactivated for lack of funds, even for busy lines that could have been salvaged with enough money. SEPTA lacked a dedicated source of funding in its early years, making it extremely difficult for the agency to plan for long-term maintenance and operations.[3] This, in turn, meant that bus substitution was the order of the day. By 1974, the enormous trolley network had greatly withered (fig. 15.4). In the tumultuous 1970s, SEPTA's unions went on strike three times over wages and pensions, culminating in a 44-day walkout in 1977 that permanently reduced SEPTA ridership by 10 percent.[4] As the largest SEPTA workers' union put it in its official history, "Labor relations in Philadelphia were never conventional, and rarely peaceful," a dynamic examined in more detail in the following section.[5]

The Commuter Tunnel

Center City Philadelphia was in decline during this period. The City of Philadelphia and SEPTA proposed strengthening Center City with infrastructure improvements to attract commuters and shoppers back from the suburbs. The keystone of the improvements was to revamp the regional rail network. At the time, the regional rail system was bifurcated, as the Pennsy and the Reading used separate dead-end terminals half a mile apart on opposite sides of City Hall. This system was inconvenient and difficult to use. A subway transfer was required for Reading passengers to reach the office core in Center City West, or for Pennsy passengers to reach the city's main shopping district in Center City East. Suburb-to-suburb travel was impractically slow.

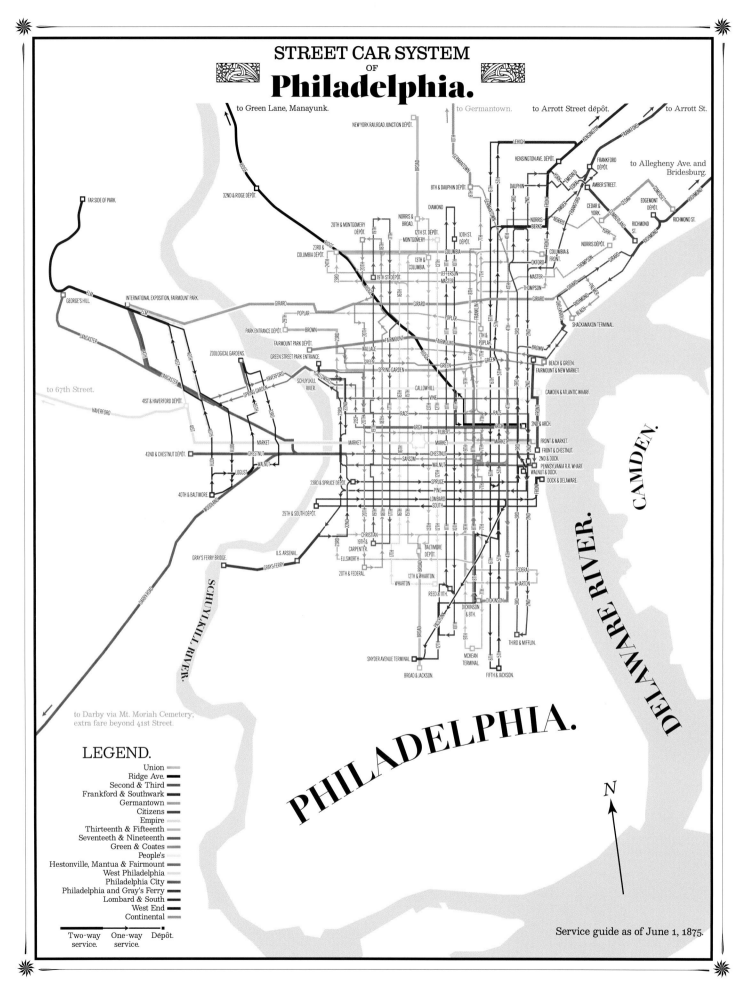

15.2 Horsecars and urban steam railroads of Philadelphia, 1875.

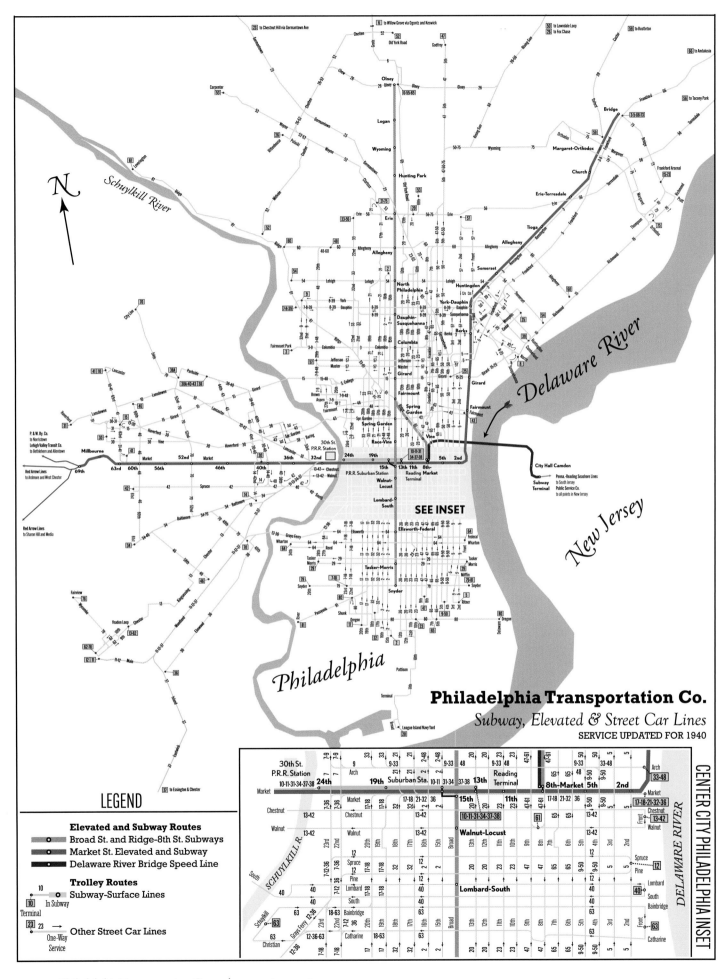

15.3 Philadelphia Transportation Co. and connecting services, 1940.

southeastern pennsylvania transportation authority
rapid transit 1974

Cheltenham Av

Bethlehem Pike

Norristown

Bridgeport
King Manor
Hughes Park
Gulph Mills
Conshohocken Rd
County Line
Radnor
Villanova
Stadium
Garrett Hill
Rosemont
Bryn Mawr
Haverford
Ardmore Av
Ardmore Jct

Carpenter Lane

Fern Rock
Olney
Logan
Wyoming
Hunting Park
Erie

Olney Av
Bridge-Pratt
Cottman A
Margaret-Orthodox
Church
Erie-Torresdale

10th & Luzerne
5th-6th & Erie

Tioga
Allegheny

Champlost Pl

Wynnewood Rd
Beechwood-Brookline
Penfield
W Overbrook
Parkview

East Falls Loop
Venango Loop
Allegheny Av
Sedgley Av

Allegheny
N Philadelphia
Susquehanna–Dauphin
Columbia

5th-6th & Allegheny
Somerset
Huntingdon
York-Dauphin
Berks

Westmoreland Loop

63rd & Malvern

Girard & Lancaster

Girard Av
Girard
Fairmount
Spring Garden
Race-Vine

Girard & 11th-12th
Girard & 4th-5th

Spring Garden
Vine

Girard
Fairmount

63rd & Girard

33rd 22nd 19th 15th 13th

11th 8th 5th 2nd

City Hall
Broadway

69th St
Millbourne
63rd 60th 56th 52nd 46th 40th 34th
30th
Sansom
37th
40th St Portal

City Hall
15th-16th
Walnut-Locust
Lombard-South
Ellsworth-Federal
Tasker-Morris
Snyder
Oregon

12th-13th
9th-10th

6th-7th & Oregon

Ferry Av
Collingswood
Westmont
Haddonfield
Ashland
Lindenwold

Media
61st & Baltimore
Garrett Rd
Baltimore Av

Yeadon Loop
Darby
Main St

Sharon Hill

88th & Eastwick

11th-12th & Bigler
Pattison

Schuylkill River
Delaware River

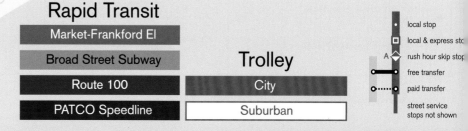

Rapid Transit
Market-Frankford El
Broad Street Subway
Route 100
PATCO Speedline

Trolley
City
Suburban

local stop
local & express sto
rush hour skip stop
free transfer
paid transfer
street service
stops not shown

15.4 SEPTA trolleys and rapid transit, 1974.

The proposed solution was simple and elegant: connect the dead-end Pennsy lines with the dead-end Reading lines and run trains through Center City instead of to Center City. Suburban Station (the old Pennsy terminal) would be converted into a through station. A new underground station, Market East, would go beneath the old Reading Terminal. Ex-Reading services would be rerouted to use Market East. Train travel between suburbs on opposite sides of the metropolis would be practical for the first time. Most importantly, this new main line could provide fast, frequent, subway-like service because all regional rail branch lines would run the length of Center City.

Workers began to dig the commuter tunnel during the chaotic late 1970s. While construction was underway, SEPTA hired civil engineering professor Vukan Vuchic, of the University of Pennsylvania, to draft a service plan for the commuter tunnel based on German and Austrian examples. The plan would bring SEPTA's regional rail operational procedures in line with international standards (fig. 15.5). Vuchic's proposed reforms consisted of most of the best practices listed previously: automated fare collection, staffing reforms, simplified transfers, and level boarding, thus enabling high-frequency service through the commuter tunnel.[6]

The Vuchic reforms were dead on arrival. Labor issues and service cuts once again interfered. In March 1983, the year before the commuter tunnel opened, SEPTA's regional rail unions went on a three-month strike over wages. The strike permanently reduced the regional rail ridership base.[7] SEPTA had 32.2 million annual regional rail passengers in 1980, and the 1983 strike reduced

that number to 12.9 million passengers. Even the opening of the commuter tunnel was insufficient to bring the lost riders back. SEPTA would not exceed its 1980 ridership level until 2008 (fig. 15.6 depicts SEPTA's present system).

SEPTA's labor strife is due to three major factors. First, unlike most states, Pennsylvania law explicitly allows public employees to strike. SEPTA unions can legally use the threat of a strike to extract concessions.[8] Second, since the SEPTA takeover, Philadelphia elected officials—including Philadelphia's mayors—have a history of supporting the unions in their disputes with SEPTA.[9] Third, Philadelphia transit unions were born out of a series of knock-down, drag-out fights with the old private transit operators, creating a combative union culture that persists to this day.

The old PTC had stridently opposed unionization. PTC recognized the largest transit union only in 1944, after seven years and the intervention of President Franklin D. Roosevelt.[10] Relations between PTC and its unions continued to be poor after the war. Between the end of World War II and the SEPTA takeover at the end of the 1960s, workers officially struck in 1946, 1949, 1961, 1963, and 1967. There were also at least nine wildcat strikes taken without official union sanction between 1955 and 1960.[11] The unions' willingness to strike has continued into the SEPTA era.[12] Willie Brown, president of the largest SEPTA union from 2008 to 2021, called himself "the most hated man in Philadelphia" for his aggressive style and his willingness to lead strikes for higher wages and benefits.[13]

The often-acrimonious relationship has encouraged SEPTA's leadership to retain its conservative

163

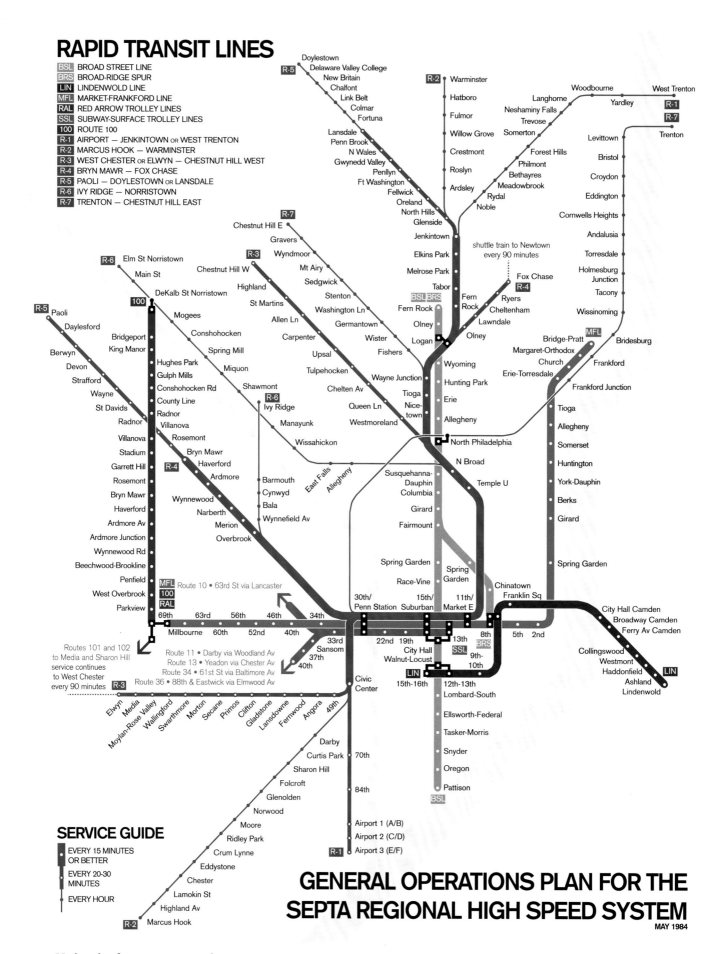

15.5 Vuchic plan for commuter tunnel service, 1984.

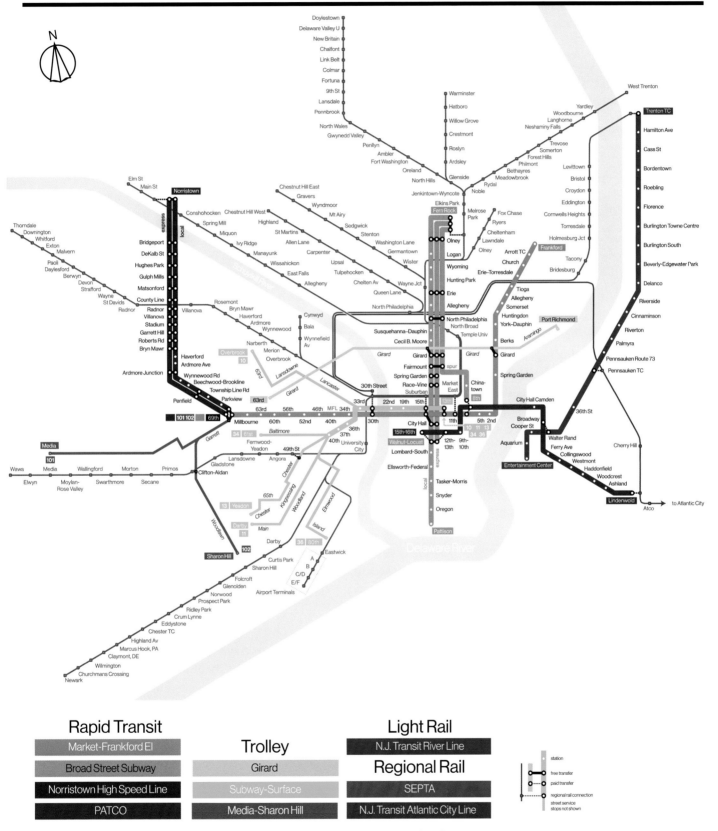

Greater Philadelphia
rapid transit and regional rail

Rapid Transit
- Market-Frankford El
- Broad Street Subway
- Norristown High Speed Line
- PATCO

Trolley
- Girard
- Subway-Surface
- Media-Sharon Hill

Light Rail
- N.J. Transit River Line

Regional Rail
- SEPTA
- N.J. Transit Atlantic City Line

- station
- free transfer
- paid transfer
- regional rail connection
- street service
- stops not shown

GUIDE TO SERVICES

15.6 SEPTA rapid transit and regional rail, 2022. In the 2010s, SEPTA temporarily sold the naming rights of Market East station to Thomas Jefferson University Hospital. SEPTA also sold the naming rights to Pattison station on the Broad Street Subway to electric utility NRG. I use the traditional names on this map due to the potential variability of naming rights contracts.

management practices and work rules. Thus, many regional rail stations even today have no ticket machines, and conductors check every ticket manually. Trains still use oversized crews. Conductors lower stairs manually at stations so passengers can board. Labor is usually the single largest operating cost for any transit authority, and this labor-intensive mode of operation makes it too expensive for SEPTA to run frequent regional rail service outside rush hour. In addition, SEPTA's old-fashioned operating procedures limit the four-track commuter tunnel to 25 trains per hour in each direction. Comparable European tunnels with only two tracks can handle 30 trains per hour.[14]

Government by Committee

There is no good way for voters, or even their elected officials, to force SEPTA to adapt. This is in part due to SEPTA's institutional governance structure. Like most transit agencies, SEPTA answers to a board of directors, and the board's makeup is a recipe for poor political accountability. No fewer than 10 entities appoint representatives to the SEPTA board. The City of Philadelphia has two votes; the four suburban counties have two votes per county; the state governor has one vote; the state house majority leader, state house minority leader, state senate majority leader, and state senate minority leader have one vote each. With no clear lines of political responsibility, it takes an overwhelming consensus at all levels of government to force SEPTA to make major reforms. Such a consensus is hard to come by, especially in the polarized political environment of the 21st century. (Pennsylvania is evenly split between Democrats and Republicans.) In earlier eras of Philadelphia transit, the city government was directly responsible for overseeing the transit network, so it was easier for the voters to throw the bums out.

Frustratingly, SEPTA's and the City of Philadelphia's planning departments both recognize that the regional rail system can be better. Both entities have put forth modernization proposals that follow the outlines of Vuchic's 1984 plan. With SEPTA's introduction of smartphone tickets in 2022, most of the physical and electronic infrastructure is in place to transform regional rail into a German-style rapid transit system. (The one major investment still needed is retrofitting existing platforms, so that train crew don't have to lower stairs.) However, that type of organizational transformation does not appear to be a high priority for SEPTA. The City of Philadelphia's planning document envisions a modernized Vuchic plan as a long-term regional rail goal for the year 2045.[15]

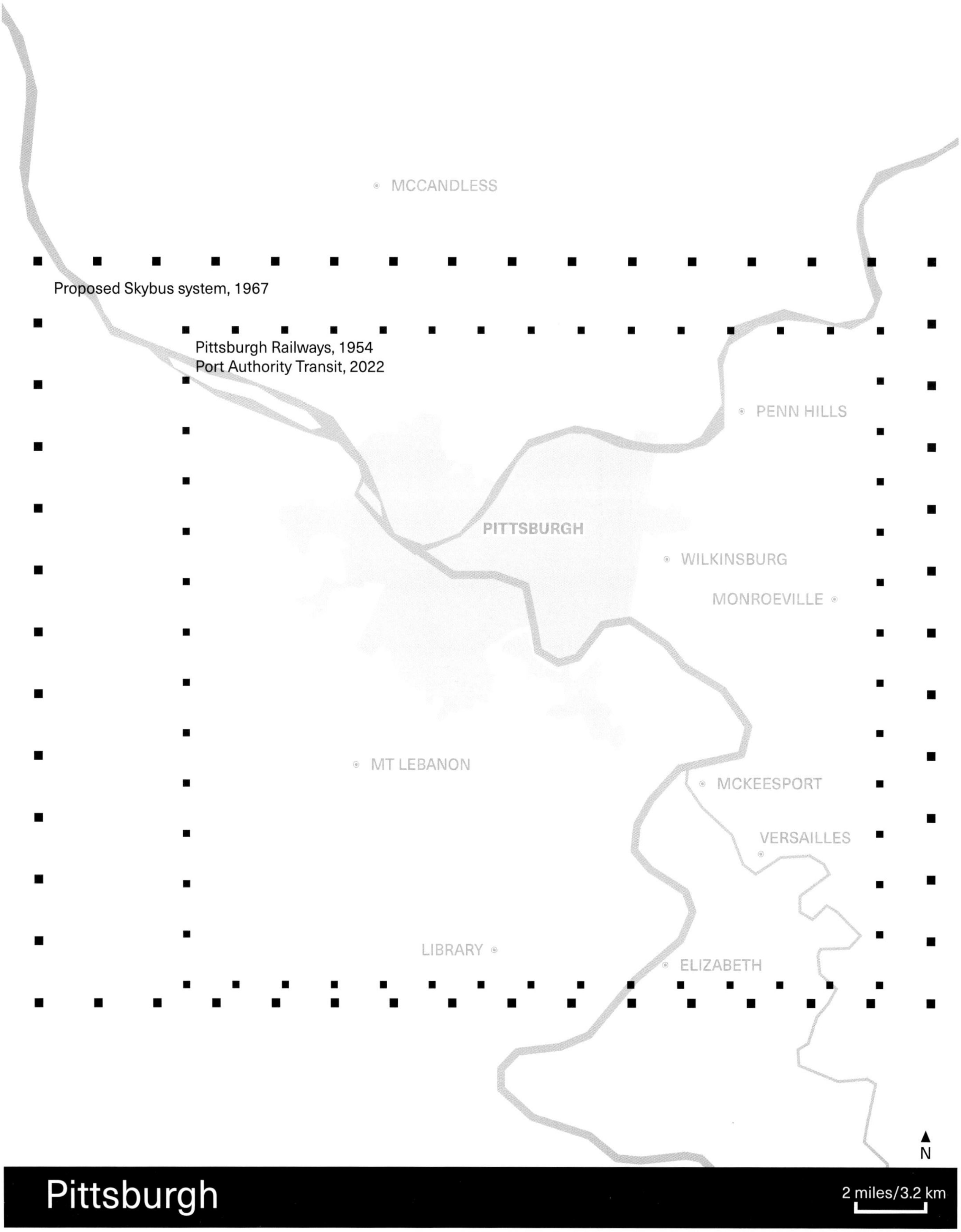

Proposed Skybus system, 1967

Pittsburgh Railways, 1954
Port Authority Transit, 2022

◎ MCCANDLESS

◎ PENN HILLS

PITTSBURGH

◎ WILKINSBURG

MONROEVILLE ◎

◎ MT LEBANON

◎ MCKEESPORT

VERSAILLES ◎

LIBRARY ◎

◎ ELIZABETH

N

Pittsburgh

2 miles/3.2 km

16

PITTSBURGH

How to Make Buses Work

16.1 Map of greater Pittsburgh.

Ideally, mass transit should do four things. One, it should be frequent; two, it should be fast; three, it should be reliable; and four, it should go where people want to go. The choice of technology matters somewhat, but only to the extent that one picks an appropriate tool for the job. Thus, Philadelphia's Regional Rail is fast, reliable, and serves the right places, but because service isn't frequent, it has never reached its potential. Miami's Metrorail is frequent, fast and reliable, but it doesn't go where people want to go. Pittsburgh's busway system, in contrast, satisfies all four criteria. Pittsburgh (fig. 16.1) is an excellent example of how a midsized city has made buses work well.

The Trolley Era

Like most of the cities in this book, Pittsburgh's transit history begins with the electric trolley. Trolley service began in the late 19th century. By the early 20th century, all trolley lines in greater Pittsburgh had been merged into one privately owned transit monopoly. That monopoly, Pittsburgh Railways, lost its strength in the face of car and bus competition in the 1920s and 1930s. By 1938, Pittsburgh Railways was bankrupt and entering a 13-year period of receivership. While in receivership, Pittsburgh Railways chose to modernize its trolley system instead of introducing buses. Over 600 state-of-the-art trolleys operated on Pittsburgh's streets at this time (fig. 16.2 shows the trolley system in 1954).

169

PITTSBURGH RAILWAYS

Trolley service updated for 1954

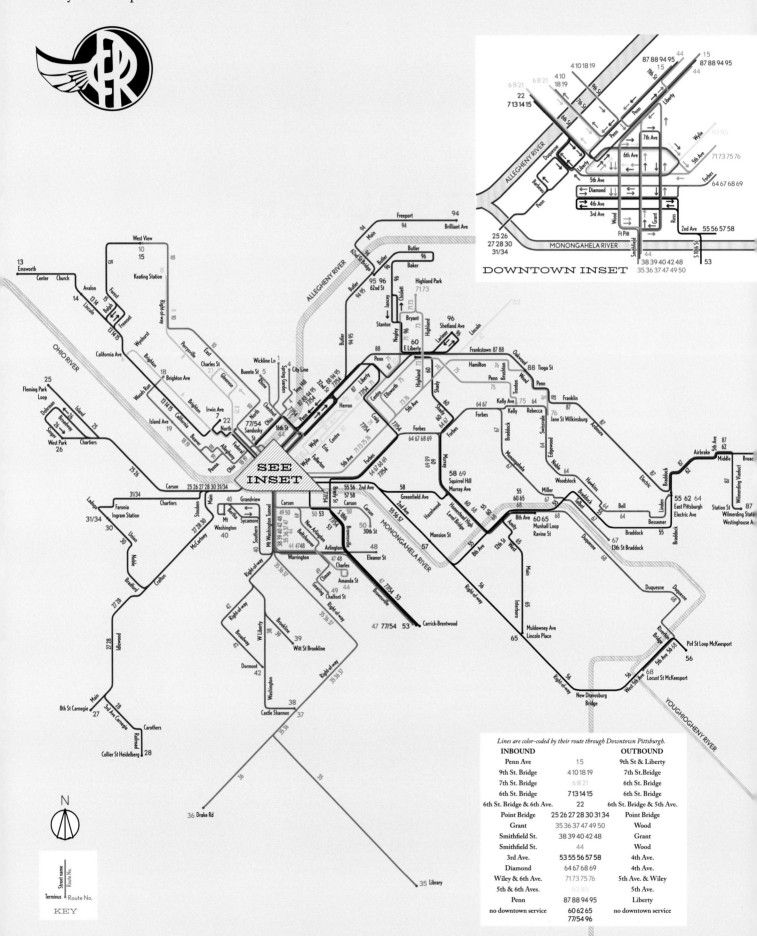

DOWNTOWN INSET

Lines are color-coded by their route through Downtown Pittsburgh.

INBOUND		OUTBOUND
Penn Ave	15	9th St & Liberty
9th St. Bridge	4 10 18 19	7th St. Bridge
7th St. Bridge	6 8 21	6th St. Bridge
6th St. Bridge	7 13 14 15	6th St. Bridge
6th St. Bridge & 6th Ave.	22	6th St. Bridge & 5th Ave.
Point Bridge	25 26 27 28 30 31 34	Point Bridge
Grant	35 36 37 47 49 50	Wood
Smithfield St.	38 39 40 42 48	Grant
Smithfield St.	44	Wood
3rd Ave.	53 55 56 57 58	4th Ave.
Diamond	64 67 68 69	4th Ave.
Wiley & 6th Ave.	71 73 75 76	5th Ave. & Wiley
5th & 6th Aves.	82 85	5th Ave.
Penn	87 88 94 95	Liberty
no downtown service	60 62 65 77/54 96	no downtown service

KEY

Street name / Route No. / Terminus / Route No.

16.2 Pittsburgh Railways trolleys, 1954.

This attempt to preserve the trolley system ended up being in vain. Pittsburgh Railways' trolley lines relied heavily on bridges to reach outlying neighborhoods. The state and local governments replaced many of these bridges during this period. Often, the new bridges had no trolley tracks.[1] Pittsburgh Railways, perpetually short on money, was happy to use these changes to replace trolley lines with buses through the 1950s and early 1960s. In 1964, the government-run Port Authority Transit (PAT, which is today Pittsburgh Regional Transit) took over Pittsburgh Railways. Nearly all the remaining trolley lines were converted to bus.[2] The few surviving trolley lines used tunnels to reach the neighborhoods south of downtown, similar to contemporary networks in Boston and Philadelphia.

But even after the trolley system's demise, it was clear to Pittsburgh's leaders in the 1960s that high-capacity rapid transit was needed. Downtown Pittsburgh had an unusually high concentration of jobs, a development pattern that persists to this day.[3] To illustrate, 21 percent of greater Pittsburgh's jobs were within three miles of the city center in the 2010s. Equivalent figures for comparable midsized metropolitan areas are Cleveland, 14 percent; St. Louis, 10 percent; and Cincinnati, 16 percent.[4] Pittsburgh's geography is responsible for much of this job density. Pittsburgh, county seat of Allegheny County, is in the Appalachian Mountains. The city grew out of a settlement established on the slivers of flat land where the Allegheny and Monongahela Rivers flow into the Ohio. Twenty-seven percent of Allegheny County is either underwater or too steep to build on.[5] Like

Seattle and San Francisco, Pittsburgh's terrain limits expressway construction, making rapid transit essential.

Pittsburgh's busway system was originally meant to be a quick and cheap interim solution while the region decided what technology to use for its permanent rail system. At the time, Pittsburgh had a vast number of lightly used industrial-age railways suitable for conversion into busways. The East, South, and West Busways all resulted from this process. Ironically, thanks to a political system that got bogged down in technological minutiae, the comprehensive rail system was never finished. The interim solution ended up becoming the permanent one.

Skybus

Through the 1960s and 1970s, Pittsburgh's politicians couldn't decide on a rapid transit system. Normally, the technology choice is not particularly complicated, as the use case dictates the technology. (For example, it would be foolhardy to build an underground metro in New Orleans, because New Orleans sits below sea level.) But in Pittsburgh, the choice of technology had major political ramifications, because a homegrown conglomerate wanted to use its hometown as a guinea pig.

That conglomerate was Westinghouse Electric, which at the time manufactured everything from washing machines to jet engines to elevators. In its Pittsburgh labs, Westinghouse had developed a fully automatic rapid transit system called Skybus (fig. 16.3). (Its little-used official name was Transit Expressway Revenue Line.) Like many modern

171

people movers, Skybus was a centrally controlled system, using lightweight, driverless rubber-tired train cars running along a concrete track. Unlike modern people movers, which generally have a carrying capacity between 1,000 and 5,000 passengers per hour in each direction, Skybus was designed for full-scale urban use, with a capacity of 19,800 passengers per hour in each direction.[6] That is, Skybus was designed to have equivalent capacity to today's Vancouver SkyTrain.[7] Westinghouse started small, first deploying Skybus in 1965 on a 1.75 mile test track at the Allegheny County Fair south of downtown. Skybus was ahead of its time—arguably, very far ahead. Feeder busways were meant to supplement the full Skybus system.

The other major alternative was to upgrade the surviving trolley lines to modern light rail standards, like the process then underway in Boston and San Francisco. Once refurbishment was complete, the upgraded network would be expanded. A heavy equipment manufacturer called WABCO presented the light rail alternative to the Allegheny County Commission in July 1969. Light rail, WABCO argued, would provide equal carrying capacity to Skybus, but at half the cost of Skybus and its supplemental busways. WABCO played hardball from the start. Hoping to gain the public's support, WABCO released the details of its proposal to the press after its presentation to the County Commission.[8]

A political standoff ensued, as light rail backers squared off against Skybus backers. Wily politicians seized on it as an electoral issue, including Pittsburgh city councilman and mayoral candidate Pete Flaherty. At public hearings in August

1969, WABCO and Flaherty charged that PAT and the county commissioners were making a shady backroom deal. Flaherty told the press that the Skybus plan was being pushed through because of "an emotional commitment by the majority of the PAT Board and the Pittsburgh industrial establishment."[9] But in September 1969, the Allegheny County Commission approved funding for Skybus anyway.

Unluckily for Skybus's backers, Flaherty won the mayoral election. He immediately went to court to halt Skybus in its tracks.[10] Seven years of ugly litigation and political stalemate ensued between Mayor Flaherty and the county. During the litigation, PAT began construction on the South Busway, even with the fate of Skybus unsettled.[11] Under threat of losing federal funds, the parties finally settled and hired neutral consultants to determine a way forward. The solution that the consultants came up with in 1976 was relatively conservative: modernize the surviving trolley lines to light rail standards, underground the upgraded rail lines through downtown, and build out the busway system to serve the suburbs, paving over disused industrial-era railways to carry buses.[12] The parties agreed that the consultants' solution was basically sound. In 1979, with the technological issue finally settled, the federal government allocated $265 million ($1.1 billion, inflation-adjusted) to PAT to begin construction.

Elegant Simplicity

In 1984, after a five-year rebuild, PAT's light rail reopened for business, running through the new

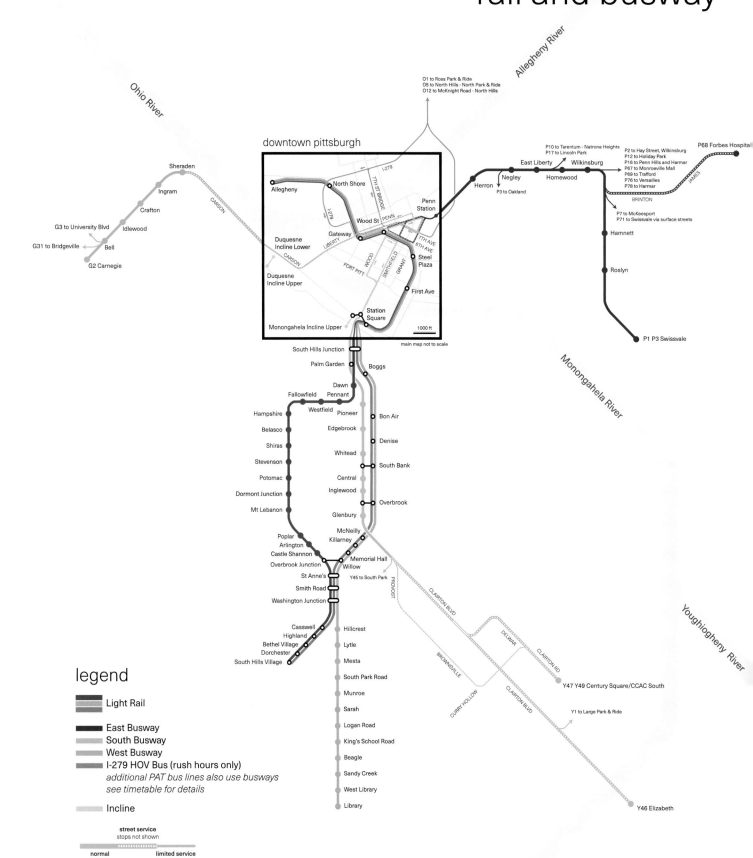

port authority transit
rail and busway

N

downtown pittsburgh

1000 ft

main map not to scale

O1 to Ross Park & Ride
O5 to North Hills - North Park & Ride
O12 to McKnight Road - North Hills

P10 to Tarentum - Natrona Heights
P17 to Lincoln Park

P2 to Hay Street, Wilkinsburg
P12 to Holiday Park
P16 to Penn Hills and Harmar
P67 to Monroeville Mall
P69 to Trafford
P76 to Versailles
P78 to Harmar

P68 Forbes Hospital

East Liberty
Wilkinsburg
Homewood
Herron
Negley
P3 to Oakland

BRINTON
JAMES

P7 to McKeesport
P71 to Swissvale via surface streets

Hamnett
Roslyn

P1 P3 Swissvale

Allegheny
North Shore
7TH ST BRIDGE
I-279
Penn Station
Wood St
PENN
Gateway
LIBERTY
7TH AVE
6TH AVE
Duquesne Incline Lower
CARSON
FORT PITT
WOOD
SMITHFIELD
GRANT
Steel Plaza
Duquesne Incline Upper
First Ave
Station Square
Monongahela Incline Upper

Sheraden
Ingram
Crafton
Idlewood
Bell
G3 to University Blvd
G31 to Bridgeville
G2 Carnegie
CARSON

Allegheny River
Ohio River
Monongahela River
Youghiogheny River

South Hills Junction
Palm Garden
Boggs
Dawn
Fallowfield
Pennant
Westfield
Pioneer
Hampshire
Bon Air
Belasco
Edgebrook
Shiras
Denise
Stevenson
Whitead
Potomac
South Bank
Dormont Junction
Central
Inglewood
Mt Lebanon
Overbrook
Glenbury
McNeilly
Poplar
Killarney
Arlington
Castle Shannon
Memorial Hall
Overbrook Junction
Willow
St Anne's
Y45 to South Park
Smith Road
Washington Junction

PROVOST
CLAIRTON BLVD
DELWAR
CLAIRTON RD
BROWNSVILLE
CURRY HOLLOW
CLAIRTON BLVD

Casswell
Hillcrest
Highland
Bethel Village
Lytle
Dorchester
South Hills Village
Mesta
South Park Road
Munroe
Sarah
Logan Road
King's School Road
Beagle
Sandy Creek
West Library
Library

Y47 Y49 Century Square/CCAC South
Y1 to Large Park & Ride
Y46 Elizabeth

legend

Light Rail

East Busway
South Busway
West Busway
I-279 HOV Bus (rush hours only)
additional PAT bus lines also use busways
see timetable for details

Incline

street service
stops not shown

normal service
limited service
see timetable

downtown light rail tunnel. The South, East, and West Busways were completed in 1977, 1983, and 2000, respectively (fig. 16.4). The East Busway is the single busiest transit corridor in greater Pittsburgh, carrying about 24,000 passengers per weekday in January 2020, about the same as Pittsburgh's three light rail lines combined.[13]

To understand why the busways work, it's necessary to reexamine the East Busway using the four criteria mentioned at the beginning of this chapter: frequency, speed, reliability, and area coverage. The lines using the East Busway run frequently, with a bus every 12 minutes or less during the daytime. The busway lines don't share road space with car traffic. Thus, they're time-competitive with driving on the expressway and more reliable than the traditional city bus. The buses also take passengers where they need to go. The main busway route,

Route P1/P2, is busy all day. But PAT also uses the inherent flexibility of buses to provide an enormous amount of supplemental express service to outlying areas which otherwise might not be suitable for mass transit. The busways also serve as a force multiplier for the local bus network, because local buses can use busway segments to bypass especially congested roads.

The system works well, judging from the ridership figures. Pre-pandemic, 18 percent of Pittsburghers commuted to work by mass transit. This is one and a half times the rate of Buffalo and two and a half times the rate of Detroit. After getting bogged down for a decade and a half on the details of whiz-bang transit technology, it's ironic that Pittsburgh ended up building a transit system that relies on the intelligent use of conventional buses and light rail.

MECHANICSVILLE

SCOTT'S
ADDITION

Early streetcar system, 1891

DOWNTOWN

RICHMOND

CARYTOWN

OLD MANCHESTER

BON AIR

N

Richmond

1 mile/1.6 km

17

RICHMOND
The First Streetcar System

17.1 Map of Richmond.

In 1888, inventor Frank J. Sprague opened the first full-scale electric streetcar system in Richmond, Virginia (fig. 17.1), on the Richmond Union Passenger Railway. Every operating light rail and streetcar system in the world today is a lineal descendant of Sprague's work. Sprague's electric streetcar system marked a revolution in transportation. Before then, electrically powered trains were considered unsafe, and cities relied on a variety of less-than-satisfactory solutions. Steam engines were noisy, smoky fire hazards. Many cities banned them outright. Manual cable cars, like the ones that still exist in San Francisco, were slow, expensive to build, and prone to mechanical failure. Horses and mules left excrement in the streets and had the unfortunate tendency to drop dead from overwork. The average horse pulling a horsecar had a life expectancy of two years.[1]

The Sprague system solved all these problems. Richmond's terrain is hilly, with grades of up to 10 percent. This made the city a good test case, and the experimental system worked superbly (fig. 17.2). Electric streetcars had better hill-climbing capacity than steam engines, horses, or mules, and they were cheaper than the mechanically complex cable cars. Within 15 years the entire Richmond horsecar network had been converted to use electricity. The electric streetcar became the universal means of transportation for the last decade of the 19th century and the first three decades of the 20th. Because of its ubiquity, the streetcar system provided a stage for the great social dramas of the period to play out.

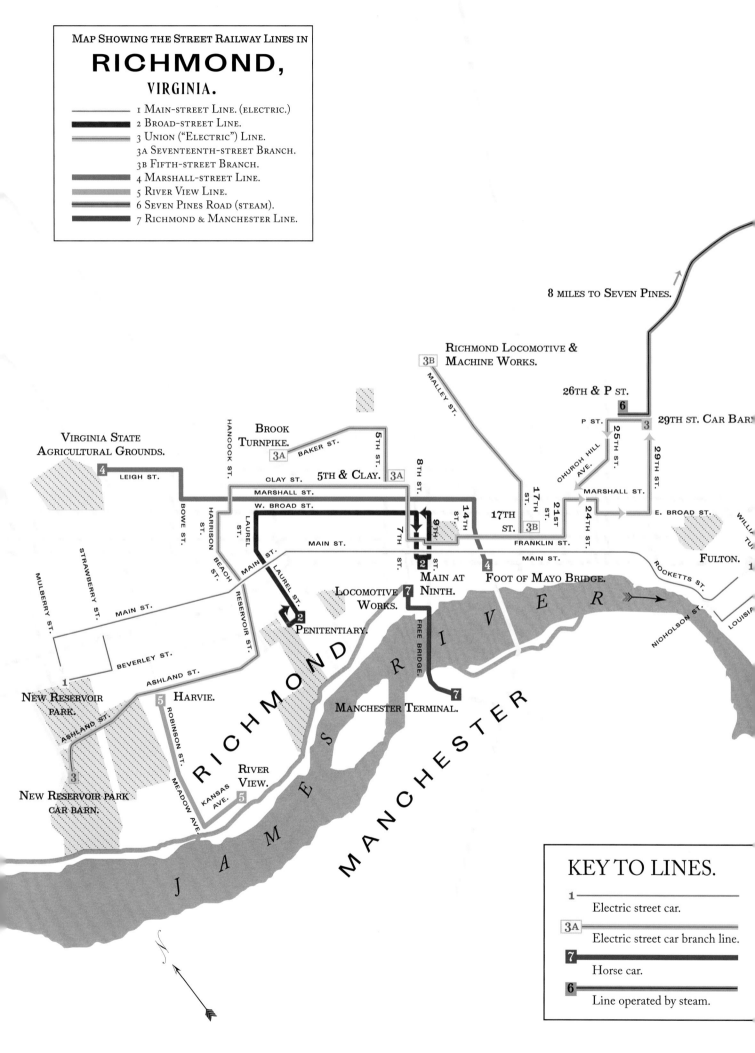

17.2 Streetcars, horsecars, and urban steam
      railroads of Richmond, 1891.

Notably, when the Virginia state legislature began to debate mandatory racial segregation on public transit in 1901, an unlikely alliance of white streetcar company owners and black activists fought the proposals tooth and nail.[2] While black opposition was a no-brainer, the white streetcar company owners initially opposed segregation for nakedly commercial reasons. The companies' logic was that enforcing Jim Crow laws was expensive. Streetcar operators feared that the loss of black ridership could be financially crippling. All the same, when the state government decided to allow (but not mandate) segregation on transit in 1904, the alliance between blacks and streetcar operators rapidly fell apart. The Virginia Passenger and Power Company, successor to the Union Passenger Railway, decided to force blacks to sit at the back.

To enforce segregation, the company armed its (white) conductors and authorized them to move passengers on the streetcars as they saw fit. Virginia Passenger and Power claimed that such measures were necessary to prevent the state legislature from mandating all-black and all-white streetcars. Richmond's black population responded with a six-month-long streetcar boycott beginning in April of that year, prefiguring the Montgomery bus boycott of the 1950s. The loss of black ridership was enough to force Virginia Passenger and Power into bankruptcy by July 1904. (Virginia Passenger and Power was already in weak financial shape due to a conductors' strike.) Unlike the Montgomery bus boycott, the Richmond streetcar boycott was in vain. While the boycotters had successfully bankrupted Virginia Passenger and Power, they could not pre-vent the state legislature from passing a mandatory streetcar segregation law two years later, in 1906.[3]

Outside Richmond, the Sprague system was revolutionary, and electrification spread like wildfire. Within a year, 110 companies were building or planning to build electric railways based on Sprague's technology. By 1896, Boston had put its electric streetcars underground, creating North America's first electric rapid transit line in the process. By 1910, the horsecar was nearly extinct. The overwhelming majority of the old horsecars were replaced with electric streetcars. The last horsecar line in North America, located in Sulphur Rock, Arkansas, population 277, closed in 1926, bringing the horsecar era to a close.

As with most of the other cities in this book, Richmond's streetcar system fell victim to competition from motor vehicles, shrinking during the Great Depression and shutting down after World War II. Sixty-one years after the first streetcar ran on the Richmond Union Passenger Railway, the last Richmond streetcars were replaced with buses on November 25, 1949. The surviving streetcars were burned to allow the metal parts to be recovered and scrapped. Today, there is no urban rail transit in Richmond, although there is one busway and long-distance Amtrak service. As for Sprague himself, he continued his distinguished career as an electrical engineer, inventing the electric elevator, designing train cars for the early Chicago Elevated, and working on the electrification of the tunnels that lead into Grand Central Terminal in New York City.

LAKE ONTARIO

GREECE

IRONDEQUOIT

ROCHESTER

BRIGHTON

Rochester Subway, 1929

N

Rochester

2 miles/3.2 km

ROCHESTER

The Only City to Open a Subway, Then Close It

18.1 Map of Rochester.

Rochester, New York—population 211,000 as of 2020—is the smallest city in North America to ever build a subway system. Rochester (fig. 18.1) also has the unhappy distinction of being the only city in the world to build and operate a full-blown subway system, then abandon it entirely. (Cincinnati partially built its subway but never ran any trains.) By accident of history, the route of the Rochester Subway has evolved to follow the progress of transport technology: first used as a canal in the 19th century, the route was converted to a subway in the 20th century. In the 21st century much of it is used as an expressway.

Rochester grew up around the Erie Canal, the first great public works project of the early American republic. The canal, running from Buffalo on Lake Erie to Albany on the Hudson River, opened in 1825 and provided the first all-water route from the Atlantic to the Great Lakes. Transportation costs dropped by 90 percent, securing New York City's position as the premier Atlantic commercial port. Rochester grew up where the Erie Canal met the waterfalls of the Genesee River, making the city a logical point to mill Midwestern grain into flour.

By the early 20th century, the canal had served its purpose and was obsolete, due to railroad competition and the opening of the larger New York State Barge Canal in 1918. The old canal bed was primed for redevelopment. Some suggested that the canal be converted to an expressway, but the idea seemed like "impractical or wild dreaming" at the time. Fewer than 3,000 autos existed in the entire county.[1] The final decision was to

181

retrofit the canal for rapid transit. It was thought that the subway conversion would raise property values, strengthen the city's commercial core, and take interurbans and freight trains off city streets.[2]

The Rochester Subway took six years to build and cost $12 million ($193 million, adjusted for inflation), double the original estimate. At its completion in 1927, the streetcar era was already beginning to wane nationwide (fig. 18.2). The subway entered a decline after an initial period of optimism. Aside from a short extension to Rochester's General Motors plant in 1938, the system's decay continued apace through the Great Depression. Although the subway operated profitably during World War II, it became a money loser after the war. No great measures were taken to save it when the expressway era arrived.[3] Subway ridership peaked at 5.1 million per year in 1947, or about 20,000 average weekday riders—about the same as the ridership in 2019 of the Tren Urbano in San Juan, Puerto Rico.[4]

The expected economic benefits of the subway never really materialized, either. This is chiefly because Rochester never changed its land use patterns to encourage real estate development around the subway stations during the 1920s and 1930s. As

discussed in greater detail in the chapter on Dallas, transport and land use decisions are two sides of the same coin. Transit should serve the places where lots of people live and work to be effective. Even at the time, it was considered a major fault that the subway did not serve camera giant Eastman Kodak's industrial complex, Kodak Park.[5] (For a brief period in the 1920s a few rush-hour subway trains served Kodak Park using preexisting streetcar trackage on the surface, but the routing was short-lived.) Outside of downtown, suburban-style single-family homes surrounded the stations, rather than businesses and apartment buildings. As the *Rochester Times-Union* editorialized in 1949, "The subway's fault is that it starts nowhere and goes nowhere."[6]

By 1952, ridership had dropped so quickly that Sunday and holiday service was eliminated. Passenger service on the Rochester Subway ended June 30, 1956. Freight deliveries on the western half would continue into the 1990s. The eastern half was demolished, and the route was repurposed to carry Interstates 490 and 590, completing the corridor's evolution from waterway, to railway, to expressway.

LAKE ONTARIO

IRONDEQUOIT BAY

BEACH AVE.
CHARLOTTE

WINDSOR BEACH

GENESEE RIVER

DURAND EASTMAN PARK

SEA BREEZE PARK
NATATORIUM

SENECA PARK

MAPLEWOOD PARK

3 LEWISTON AVE.
DEWEY AVE.

RIDGE RD.
N. CLINTON AVE.

NORTON ST.
JOSEPH AVE.

NORTON ST.
HUDSON AVE.

NORTON ST.
N. GOODMAN ST.

GLEN HAVEN 13

13
SCHALFER'S STOP

CULVER RD.
CLIFFORD AVE.

During rush hour, some Subway trains run
from **Lewiston Ave.** via the **No. 3** surface line.

DRIVING PARK AVE.

EMERSON ST.

EDGER-TON PK.

EDGERTON PARK
Municipal Museum · Exposition Grounds

6 LYELL AVE.
GLIDE ST.

LYELL AVE.

9 MAPLE ST.
OREN ST.

CULVER RD.
BAY ST.

CULVER RD.
PARSELLS AVE.

for N.Y.C.R.R. Depot:
use lines 7, 10 or 12
for Erie R.R. Depot:
use line 1

CITY HALL
City Hall Interurban Terminal
BR. & P. and Penna. R.R. depots

ELLISON PARK

BLOSSOM RD.
CITY LINE

Main St.

10 13

Court St.
Lehigh Valley R. R. depot

4 CULVER RD.
UNIVERSITY AVE.

EAST AVE.
PROBERT ST.

2

2 LINCOLN AVE.
WEST AVE.

Monroe Ave. **Culver Rd.**

3 **Winton Rd.**

Colby St.

COBBS HILL PARK

GENESEE RIVER

JEFFERSON AVE.
COTTAGE ST.

12

4 THURSTON RD.
BROOKS AVE.

Meigs St.
Goodman St. S.

7 FIELD ST.
S. CLINTON AVE.

HIGHLAND BOTANICAL PARK

Highland Ave.

Elmwood Ave.

8 GENESEE VALLEY PARK
ELMWOOD AVE.

GENESEE VALLEY PK.

5 CRITTENDEN PARK

1 **Rowlands**

N

ROCHESTER SUBWAY & STREETCAR LINES

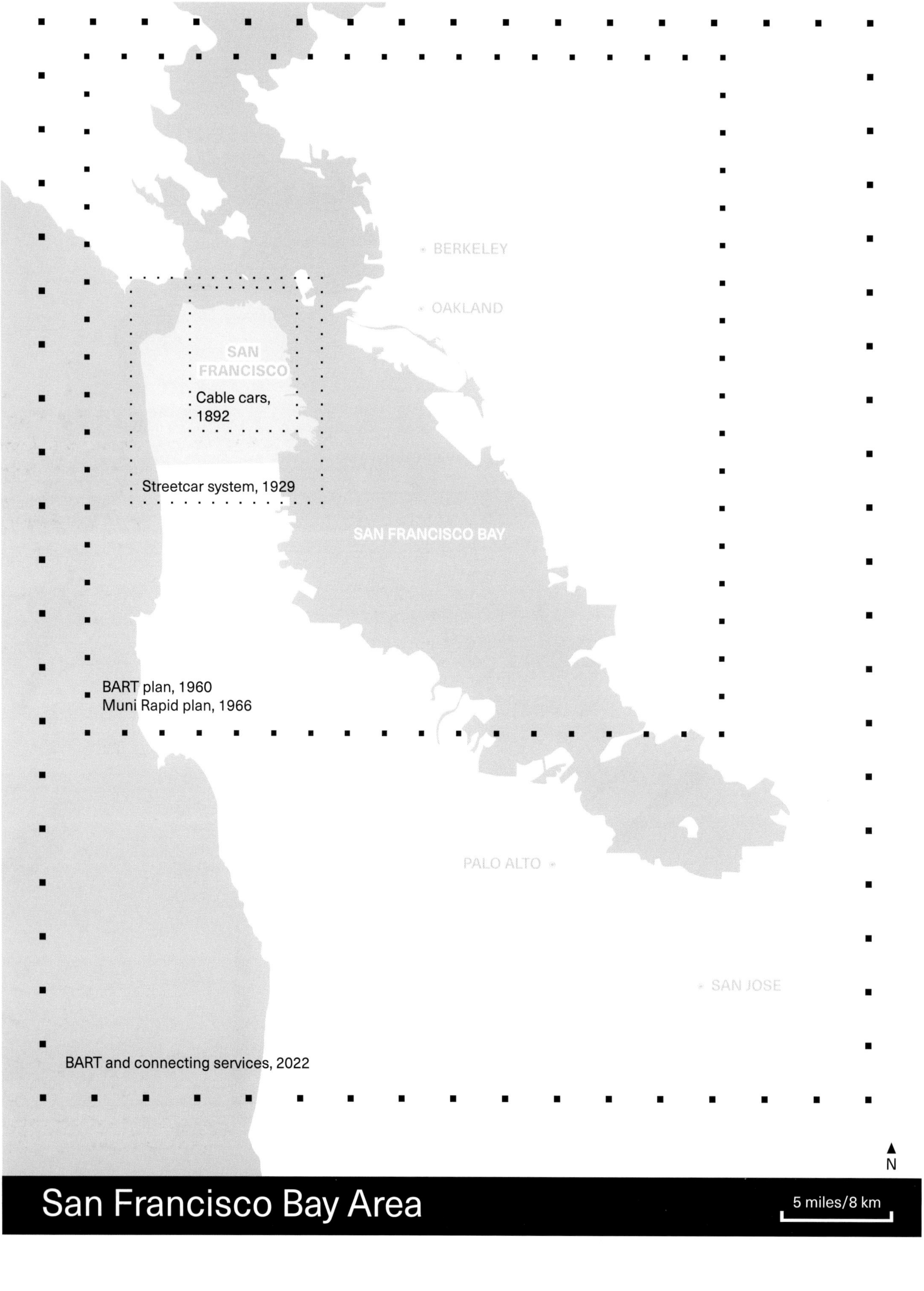

BERKELEY

OAKLAND

SAN
FRANCISCO

Cable cars,
1892

Streetcar system, 1929

SAN FRANCISCO BAY

BART plan, 1960
Muni Rapid plan, 1966

PALO ALTO

SAN JOSE

BART and connecting services, 2022

N

San Francisco Bay Area

5 miles/8 km

19

SAN FRANCISCO

The View from Geary Street

19.1 Map of the San Francisco Bay Area.

The San Francisco Bay Area is the world's technological mecca, home to Apple, Google, Facebook, and countless other household names. Per capita, the Bay Area (fig. 19.1) is the wealthiest major metropolitan area in North America.[1] But it doesn't feel like the metropolis of the future if you visit. Parks, underpasses, and sidewalks are filled with the homeless. The housing shortage is so bad that the average Bay Area home sold for $1.25 million in August 2022, nearly triple the national average.[2] Public infrastructure, much of which dates to the post–World War II decades, is showing its age. New transit and housing projects are liable to get nitpicked to death, making things worse.

It wasn't always like this. The Bay Area once was an innovator, a place where it really was possible to get things done. San Francisco's famously hilly terrain was home to the first successful large-scale deployment of the mechanical cable car in the late 19th century. Voters established the San Francisco Municipal Railway (Muni) in 1912, North America's first publicly owned transit agency, to challenge the city's detested private transit monopoly, the Market Street Railway.[3] The Bay Area fought urban freeways in the 1950s, a battle known as the Freeway Revolt, and instead built the Bay Area Rapid Transit (BART) subway system. The Bay Area's takeaway from the Freeway Revolt was that the status quo should be carefully protected, and that any urban change was highly suspect.

This approach has left the Bay Area reeling from its prosperity.

19.2 Cable car system, 1892. The Geary cable line is in purple.

This chapter chronicles this evolution by taking a long look at the Geary corridor in the city of San Francisco.[4] Geary, named after a 19th-century mayor, is San Francisco's major east–west thoroughfare, running six miles from downtown San Francisco to the Pacific Ocean. Before World War I, Geary hosted, in turn, a steam railroad, a cable car line and Muni's first streetcar line. After World War II, the Geary corridor became the target of urban renewal, and streetcars were replaced with buses. Since then, these overcrowded buses have been the sole transit option. Multiple plans to upgrade the corridor to rail have failed.

Today, Geary still has no high-quality rapid transit. Because of the post–Freeway Revolt reforms, San Francisco has been unable to adequately upgrade the Geary bus line. New apartment buildings along the Geary corridor, as in the rest of San Francisco, go through the same painful, drawn-out process. As a result, transit service stays bad, housing remains costly, and the homelessness crisis gets worse.

Origin of the Species

Geary has been a major transit corridor since the 19th century. The first transit line on Geary opened in 1880, pulled by steam locomotives. The line was financially successful, and the Market Street Railway bought it in 1887. In 1892, cable cars replaced the steam locomotives. The cable car network near its peak is depicted in figure 19.2. The line's profitability made Geary ground zero for the City of San Francisco's decades-long battle with the Market Street Railway.

The Market Street Railway held a near-monopoly over San Francisco transit in the early 20th century. It was thoroughly corrupt, had atrociously bad labor relations, and had no qualms about paying bribes to get its way. Countless officials were on the take, all the way up to Tirey Ford, attorney general of California.[5] Ford resigned as attorney general to become head of the Market Street Railway's legal department. He was later indicted in 1907 for offering bribes on the company's behalf.

Opposition to the Market Street Railway was strongest in San Francisco's Union Labor Party. One of the Union Labor Party's chief goals was city ownership of all public utilities, including the public transport network. Its candidate, a musician named Eugene Schmitz, won the 1901 mayoral election. Shockingly, the party also took control of the city council, known as the Board of Supervisors. In 1902, Mayor Schmitz and his allies asked city voters to approve bonds to take over the Geary line. The bond measure didn't get the necessary supermajority, but supporters of a municipal railway were not deterred. The issue went before the electorate multiple times over the course of the decade. In 1909, the Muni referendum finally got the necessary support.

Muni's first action was to take over the Geary line. The last privately owned Geary cable car ran a little after midnight, on May 6, 1912. Construction immediately began to convert it into a streetcar line. Seven months later, at 12:30 p.m., on December 29, 1912, North America's first city-owned streetcar left the corner of Geary and Market, outbound to Golden Gate Park.

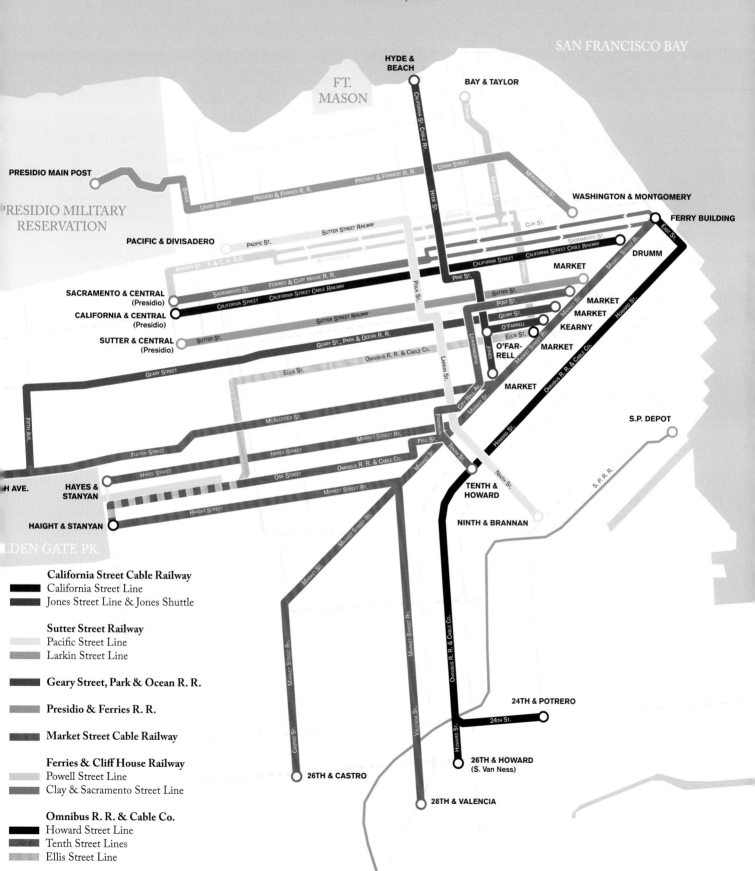

Map of the SAN FRANCISCO
CABLE CAR SYSTEM
REV. DECEMBER 30, 1892.

SAN FRANCISCO BAY

HYDE & BEACH

BAY & TAYLOR

FT. MASON

PRESIDIO MAIN POST

PRESIDIO MILITARY RESERVATION

WASHINGTON & MONTGOMERY

FERRY BUILDING

PACIFIC & DIVISADERO

SUTTER STREET RAILWAY

DRUMM

MARKET

SACRAMENTO & CENTRAL (Presidio)

MARKET

CALIFORNIA & CENTRAL (Presidio)

MARKET

KEARNY

SUTTER & CENTRAL (Presidio)

O'FAR-RELL

MARKET

O'FARRELL

MARKET

S.P. DEPOT

HAYES & STANYAN

TENTH & HOWARD

HAIGHT & STANYAN

NINTH & BRANNAN

GOLDEN GATE PK.

California Street Cable Railway
California Street Line
Jones Street Line & Jones Shuttle

Sutter Street Railway
Pacific Street Line
Larkin Street Line

Geary Street, Park & Ocean R. R.

Presidio & Ferries R. R.

Market Street Cable Railway

Ferries & Cliff House Railway
Powell Street Line
Clay & Sacramento Street Line

Omnibus R. R. & Cable Co.
Howard Street Line
Tenth Street Lines
Ellis Street Line

24TH & POTRERO

26TH & CASTRO

26TH & HOWARD (S. Van Ness)

28TH & VALENCIA

The Roar of the Four

Muni was young, hungry, and full of ambition, and it expanded dramatically in its first 20 years. Muni built three major tunnels to speed streetcar service, allowing development in the sand dunes that made up the western half of San Francisco. For decades, San Francisco's main drag, Market Street, carried four streetcar tracks, two each for Muni and the Market Street Railway. The din of the competing streetcars was such that the situation was dubbed "the roar of the four" (fig. 19.3).[6]

By the 1930s, the combination of the Great Depression, competition from Muni, and the automobile had put the Market Street Railway on the ropes. Muni was flush enough with cash that it was able to buy state-of-the-art steel streetcars, at a time when the Market Street Railway's equipment was largely made of wood.[7] Even in the middle of the Great Depression, Muni felt comfortable enough with its position that it planned to put its busiest streetcar lines underground through downtown.[8] In the mid-1930s, the Market Street Railway still carried 70 percent of San Francisco's transit riders. By 1944, Muni had drawn even.[9] That year, San Francisco's electorate voted to buy out the Market Street Railway. Muni had won.

Make No Small Plans

During this era, the two great bridges which connect San Francisco to the rest of the region were completed. The Golden Gate Bridge finally provided a fixed link to the Marin County suburbs to the north. The Bay Bridge across San Francisco Bay did the same for commuters from Oakland and the East Bay suburbs. The Golden Gate Bridge carried only motor vehicles, while the Bay Bridge carried both motor traffic and interurbans.

The bridges were immediately swamped with traffic. After the end of World War II, the Bay Area was rapidly growing, the East Bay interurbans were in terminal decline, and Muni only operated within the city proper. Something had to be done. In 1947, a joint army-navy board had concluded that it would be necessary to build a rail tunnel under San Francisco Bay. A state report in 1957 agreed, suggesting that a regional subway system would be the most cost-effective way to deal with the problem. This led to the formation of the BART district. BART's plans included a line under Geary, shown in figure 19.4.

The Bay Area's leaders were thinking big, with changes planned for the entire metropolis. In addition to the BART subway, a massive urban freeway network would link the Bay Area together. San Francisco's iconic but obsolete cable car lines were slated for closure.[10] Muni's aging streetcars would be replaced with brand-new buses. San Francisco's leaders prepared to convert portions of Geary into an expressway and to demolish nearby minority neighborhoods in the name of urban renewal.

There was only one problem with all these grand plans: the notoriously stubborn residents of the Bay Area weren't all onboard. The citizenry rebelled. Much like the dynamic which would play out in Washington, DC, 10 years later, most of the grand plans had to be greatly scaled back. The Freeway Revolt forced the Board of Supervisors to cancel seven of the ten freeways planned through

CITY OF SAN FRANCISCO

STREET CAR AND CABLE SERVICE
May, 1929.

MUNICIPAL STREETCAR — **MARKET STREET RAILWAY STREETCAR** — **MARKET ST. RY. CABLE**

A	Geary-Park
B	Geary-Ocean
C	Geary-California
D	Geary-Van Ness
J	Church
K	Market-Ingleside
L	Market-Taraval
N	Judah
E	Union
F	Stockton
H	Potrero
M	Ocean View

1	Sutter-California
2	Sutter-Clement
3	Sutter-Jackson
4	Turk-Eddy
5	McAllister-Fulton
6	Haight-Masonic
7	Haight-Ocean
8	Market
9	Valencia
17	Haight-Ingleside
21	Hayes
32	Hayes-Oak
SH	Divisadero Shuttle

10	Guerrero
11	Mission-24th
12	Mission-Ingleside
14	Mission-Daly City
18	Mission-Daly City
26	Guerrero-Daly City
40	San Mateo
15	Kearny-No. Beach
16	3rd-Kearny
29	3rd-Kearny-B'way
SH	Visitacion Shuttle
19	Polk

20	4th-Ellis-O'Farrell
22	Fillmore-16th
23	Fillmore-Valencia
24	Mission-Richmond
SH	Hill Shuttle
25	San Bruno
27	Bryant
30	8th-Army
28	Harrison-Depots
35	Howard
36	Folsom
33	18th-Park

WJ	Wash.-Jackson
PM	Powell-Mason
SC	Sacramento-Clay
CS	Castro
PA	Pacific

CAL. ST. CABLE R.R.

CA	California
HY	Hyde-O'Farrell
SH	Jones Shuttle

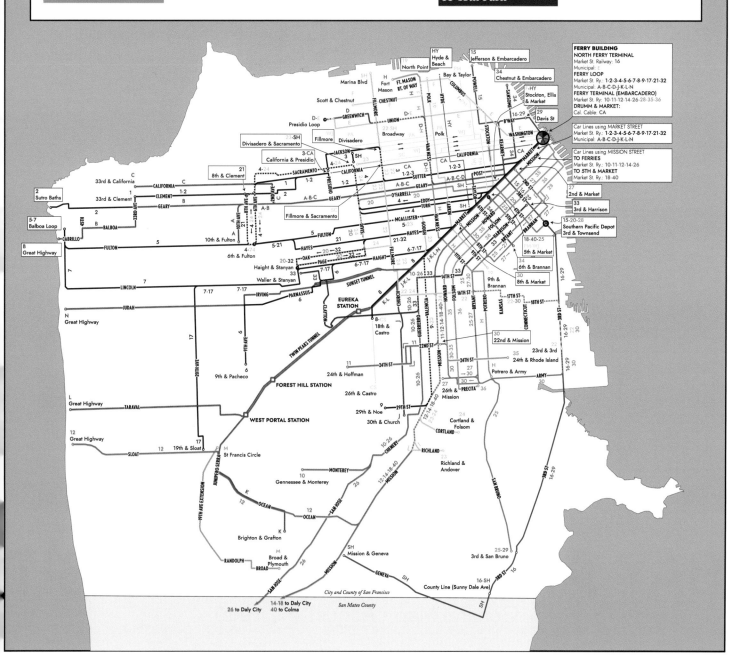

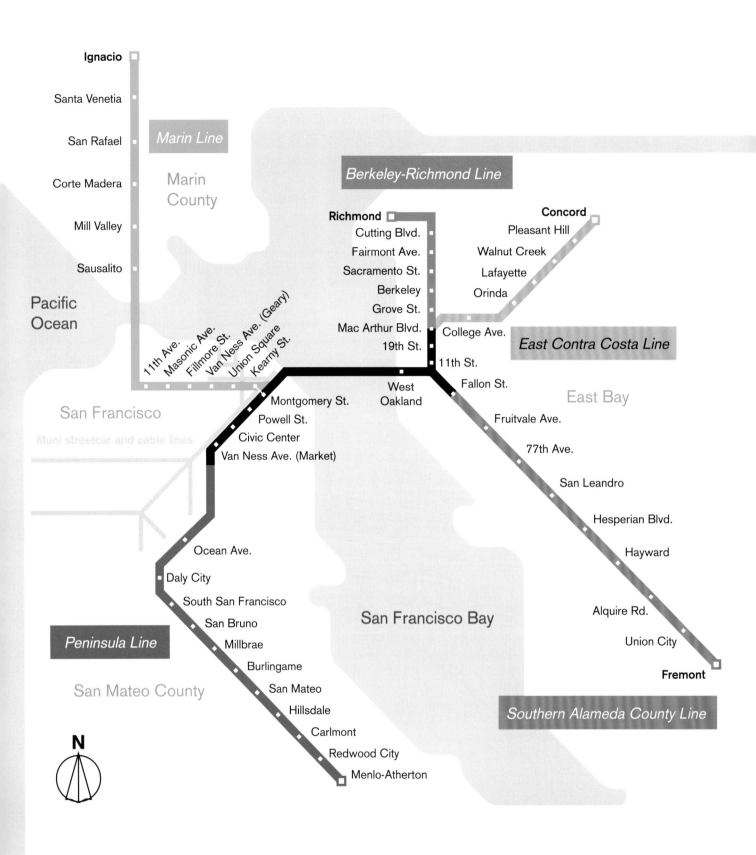

BAY AREA RAPID TRANSIT
General Map - August 11, 1960

19.4 Early BART system plans, 1960.
   The Geary line is in light blue.

the city in 1959. A society maven led a successful referendum campaign to force Muni to keep the cable cars running.[11]

Geary got the worst of both worlds. It got poorly designed urban renewal, without the accompanying modern transit. As planned, the city transformed much of Geary into an expressway. The Asian and black neighborhoods respectively known as Japantown and the Fillmore were razed. Seventy city blocks were torn down and over 10,000 residents were displaced. Vibrant working-class neighborhoods were replaced with sterile apartment blocks and a Japanese-themed shopping mall. Muni's B-Geary streetcar was eliminated in 1956 and replaced with buses on the assumption that a BART line was on the way.

The Geary BART line never happened. A supermajority of voters across the Bay Area had to approve the BART plan in 1962, which meant that one county's intransigence could torpedo the whole thing. San Mateo County, immediately to the south of San Francisco, was the pinch point. San Mateo's suburban merchants feared they would lose shoppers to downtown San Francisco if BART came down the peninsula.[12] Led by a mall developer named David Bohannon, the merchants objected to the cost, and whipped up fears that BART would bring minorities and "blight" to lily-white San Mateo County.[13] San Mateo pulled out, and without San Mateo's tax base there wasn't enough money for the Geary line.[14] Voters approved the reduced BART system without a Geary line. Four years later, in 1966, the voters of San Francisco rejected a ballot measure that would have put a Muni subway line under Geary and up-

graded most of Muni's surviving streetcar lines to subway operation. (This iteration of the Geary line is the Richmond Rapid in fig. 19.5.) BART opened in 1972.

San Francisco was vindicated in its decision to oppose urban freeways and build a subway. (The city later took its anti-freeway stance even further, tearing down two existing freeways in the late 20th century.[15]) Freeways through cities devastated the neighborhoods around them with smog, pollution, and noise. Large-scale slum clearance and urban renewal eliminated the traditional, organic rhythms of city life. The Bay Area avoided the worst of these excesses because it zigged when others zagged. At a time when everybody else was building freeways, the Bay Area approved a state-of-the-art subway and pledged tax dollars to do it. But the Bay Area learned the lessons of the Freeway Revolt too well. The Bay Area soon adopted a posture that *any* changes to the urban fabric were presumptively bad, and that exhaustive study of any such changes would be necessary.

Procedure Is the Last Refuge of the Scoundrel

The centerpiece of the post–Freeway Revolt reforms was a state law called the California Environmental Quality Act (CEQA, pronounced *seek-wha*). Signed by future president Ronald Reagan in 1970, CEQA requires state and local governments to study and mitigate potential environmental impacts of government decisions.[16] In theory, CEQA is a look-before-you-leap law. In practice, it allows anybody with enough free time to bring billion-

191

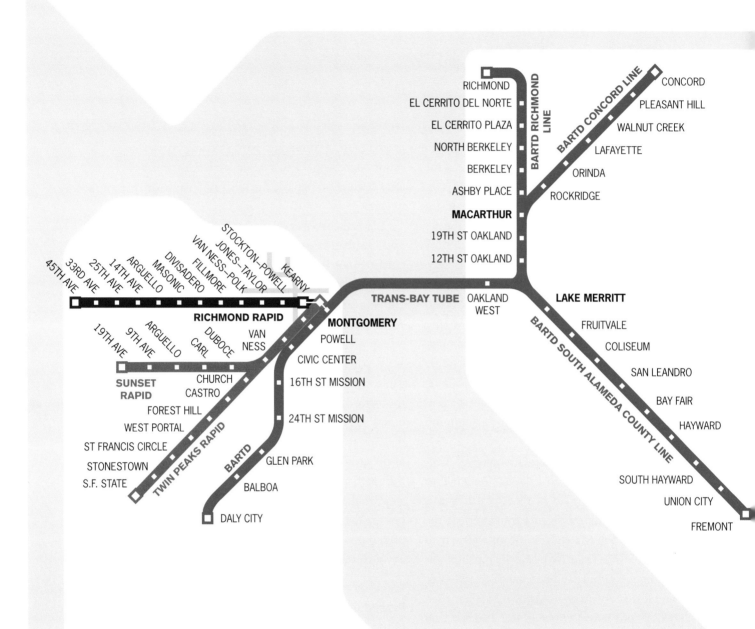

RICHMOND
EL CERRITO DEL NORTE
EL CERRITO PLAZA
NORTH BERKELEY
BERKELEY
ASHBY PLACE
MACARTHUR
19TH ST OAKLAND
12TH ST OAKLAND

BARTD RICHMOND LINE

BARTD CONCORD LINE

CONCORD
PLEASANT HILL
WALNUT CREEK
LAFAYETTE
ORINDA
ROCKRIDGE

STOCKTON–POWELL
JONES–TAYLOR
VAN NESS–POLK
FILLMORE
DIVISADERO
MASONIC
ARGUELLO
14TH AVE
25TH AVE
33RD AVE
45TH AVE

KEARNY

RICHMOND RAPID

VAN NESS

MONTGOMERY

POWELL
CIVIC CENTER
16TH ST MISSION
24TH ST MISSION

TRANS-BAY TUBE   OAKLAND WEST

OAKLAND WEST

LAKE MERRITT

FRUITVALE
COLISEUM
SAN LEANDRO
BAY FAIR
HAYWARD

BARTD SOUTH ALAMEDA COUNTY LINE

19TH AVE
9TH AVE
ARGUELLO
CARL
DUBOCE
CHURCH
CASTRO
FOREST HILL
WEST PORTAL
ST FRANCIS CIRCLE
STONESTOWN
S.F. STATE

SUNSET RAPID

TWIN PEAKS RAPID

BARTD

GLEN PARK
BALBOA
DALY CITY

SOUTH HAYWARD
UNION CITY
FREMONT

SAN FRANCISCO MUNICIPAL RAILWAY
PROPOSED RAPID TRANSIT ROUTES
AND APPROVED BAY AREA RAPID TRANSIT
APRIL 1966

19.5 Muni Rapid proposal, 1966. The Geary line is the Richmond Rapid, while the approved BART system uses its contemporary acronym, BARTD.

dollar public works projects and multimillion-dollar apartment buildings to a halt, by claiming that the government hasn't studied a particular issue enough. Anyone can bring a CEQA lawsuit and there are no consequences for bringing frivolous CEQA suits.[17] Unsurprisingly, abuse of the process is rife. In 2020, nearly half of all new housing in the state of California was the target of CEQA litigation.[18] As my old professor Troy McKenzie used to say, "procedure is the last refuge of the scoundrel."

At the local level, the Bay Area's governments expanded the discretionary powers granted to city councils and planning boards. This effectively empowered aggrieved property owners to sandbag even minor governmental decisions. The group, known as NIMBYs (an acronym for "not in my backyard"), was disproportionately older, whiter, and richer than the population at large.[19] The Bay Area's NIMBYs found allies in the ascendant Bay Area hard left, which was born of the hippie movement and opposition to the Vietnam War.[20]

Only the Lawyers Win

The reforms of the 1960s and 1970s had immediate effect. By 1980, California's average housing cost was 80 percent above the national average.[21] San Francisco's great political battles of the 1980s were fought over the city's alleged Manhattanization, a fight that the NIMBY–hard left alliance won. From the 1980s onward, San Francisco built less than a quarter of the housing necessary to meet demand.[22] This result satisfied both sides of the alliance. The NIMBYs got high property values. The

hard left got to stick it to real estate developers. While land values went up, NIMBY homeowners were insulated from rising property taxes by the 1978 property tax law called Proposition 13.[*] This was a problem, because the Bay Area's research labs were busy inventing the future. The great computing revolution of the late 20th century was underway, and the Bay Area was at its center. Money and talent began to flood in.

The Bay Area's local governments were not set up to capitalize on this newfound prosperity. Quite the opposite. The post–Freeway Revolt reforms meant that new infrastructure and housing proposals invariably led to brutal local political battles. In retrospect, the opening of BART in 1972 marked the end of the era when the Bay Area could build great public works projects at reasonable speed and cost.[23] By the early 21st century, no matter the size or importance of the change to the urban fabric, the Bay Area's NIMBYs would show up in force to gum up the works.[24]

The Geary corridor spotlights how this dysfunction has stymied the construction of badly needed transport and housing. On the transport front, the 38-Geary bus, which replaced the old B-Geary streetcar, is the Bay Area's busiest bus line. It has held that title for over 50 years. Upgrading the 38-Geary to light rail or subway is an obvious improvement, but this has been impossible to carry out. In 1989, San Francisco voters raised taxes to make major improvements on four corridors,

---

* Because of Proposition 13, landowners pay taxes based on a property's initial purchase price, not its current market value. See chapter 9, on Los Angeles.

193

19.6 BART and connecting services, 2022. The Central Subway is the portion of the T-Third light rail line north of 4th & King.

including Geary. The improvements plan eventually released in 1995 included a surface light rail line on the western, residential half of Geary and a tunnel through downtown San Francisco.[25] But nothing got done, as the project got bogged down in procedural delays.

San Francisco voters renewed the tax in 2003, but by then costs had risen so much that the light rail line had been downgraded to a busway. From a technical perspective, a busway is not complicated to build. All a transit agency has to do is paint bus lanes, put in ticket machines, retime stoplights to prioritize buses, and build a few extra curbs so the buses can run in the center lanes without interference from other traffic. Little heavy construction is needed, compared to a light rail or subway line. Nonetheless, the Geary busway nearly suffered death by bureaucracy. After the 2003 tax vote, the city carried out 15 years of studies, public outreach, and environmental review, and fought off a CEQA lawsuit alleging that the city hadn't adequately studied the effects of better bus service on the environment.[26] Construction didn't begin until 2018, and the busway wasn't completed until late 2022. The process took 33 years, start to finish.

Even so, to get the busway built, Muni made major compromises that significantly watered down its effectiveness. First, the bus lanes don't extend the full length of Geary, so buses still have to contend with other traffic for much of the route. Second, Muni was forced to switch the bus lanes from the median to the curb lanes.[27] In the curb lanes, buses can be blocked by right turns and illegally parked vehicles. This means worse service.

These long delays and bureaucratic holdups for new transit are normal in San Francisco. In another example, voters approved the 1.7 mile Central Subway extension of the T-Third light rail in 2003 (fig. 19.6). It didn't open until late 2022. Thankfully, the State of California has realized that process streamlining is necessary. The state finally exempted transit, bicycle, and pedestrian improvements from CEQA in 2020. Nevertheless, San Francisco's labyrinthine local planning laws, which exist separately from CEQA, remain in place.[28]

San Francisco's housing bureaucracy is no more efficient than its transit-planning bureaucracy. For example, on Geary's western half, it took from 2000 to 2011 to replace the bankrupt Coronet Theater with rent-controlled senior apartments.[29] This timescale is not unusual. The median San Francisco apartment building has to survive two years and three months of planning bureaucracy before a single shovel is put into the ground. But longer timelines are common. Planning approval *alone* took 11 years and two months in one case.[30] (In comparison, Sacramento, two hours inland, gives planning approval for most new housing within 90 days.) All the while, San Francisco keeps adding more jobs. A "balanced" jobs-to-housing ratio is considered between 1.4:1 and 1.8:1.[31] Between 2009 and 2019, the city of San Francisco added 7.1 new jobs for every new unit of housing.[32] Unsurprisingly, San Francisco rents are astronomical in large part because of the housing shortage, and homelessness is out of control.[33]

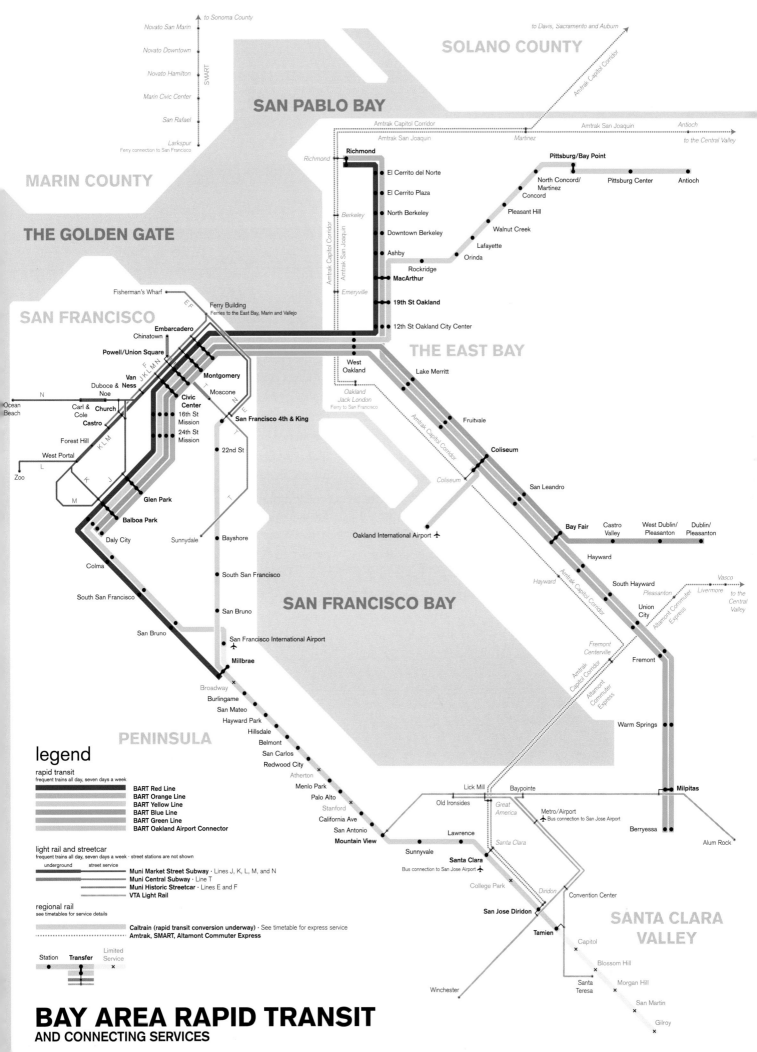

BAY AREA RAPID TRANSIT
AND CONNECTING SERVICES

Dysfunction and Red Tape

The Bay Area used to be able to undertake projects at a grand scale. After all, when San Francisco's electorate voted to build a city-owned Geary streetcar, Muni built the line within six months of taking ownership. By setting up Muni, San Francisco's voters took a chance that the city could eventually beat the local transit monopoly at its own game. No city had ever tried this before. San Francisco chose to build a subway and cancel most of its freeways at a time when nearly every other metropolis had freeway fever. The city was a pioneer in urban freeway removal. All these gambles paid off.

But having been vindicated in its decision to avoid freeways, San Francisco has hamstrung its ability to confront the problems of the 21st cen-tury, like its inadequate transit infrastructure and its housing shortage. Even the simplest changes require an army of lawyers and lobbyists, whether for bus lanes or apartment buildings. This combination of dysfunction and red tape is the norm in the Bay Area. It's comparatively easy to pave over farmland for exurban subdivisions, or to extend rail service ever-farther into the suburbs. But putting people near transit, or transit near people, is hard. San Francisco's city government, like the bulk of the Bay Area, has continued to keep these restrictive practices in place, even at the risk of a state crackdown.[34] While the desire to prevent a repeat of the freeway era is understandable, the cure was worse than the disease.

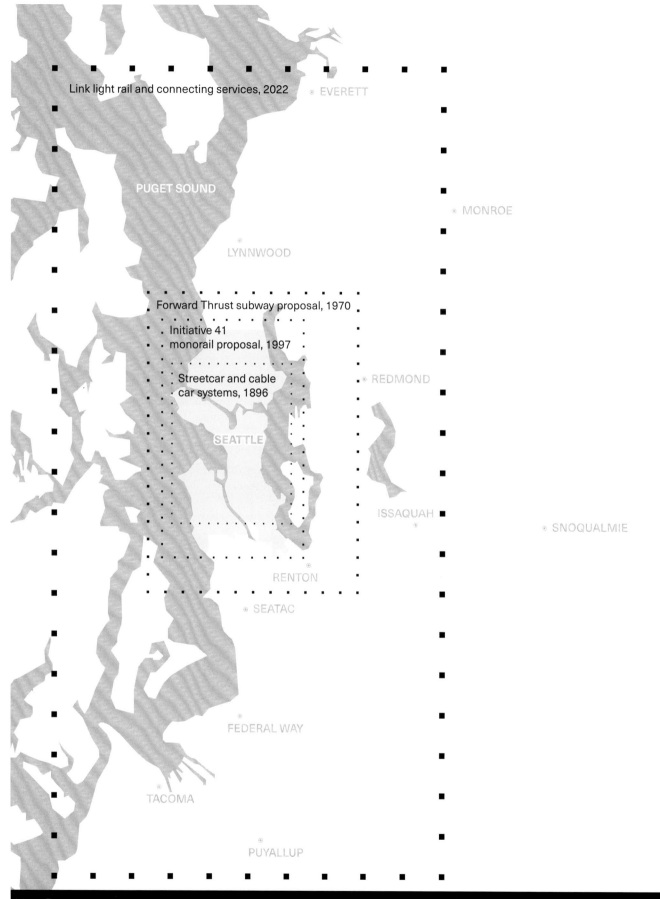

Link light rail and connecting services, 2022

PUGET SOUND

EVERETT

MONROE

LYNNWOOD

Forward Thrust subway proposal, 1970

Initiative 41
monorail proposal, 1997

Streetcar and cable
car systems, 1896

REDMOND

SEATTLE

ISSAQUAH

SNOQUALMIE

RENTON

SEATAC

FEDERAL WAY

TACOMA

PUYALLUP

N

Seattle

2 miles/3.2 km

SEATTLE

Consensus through Exhaustion

20.1 Map of greater Seattle. Seattle is on an isthmus, surrounded by water and mountains (fig. 20.1). It is ideal terrain for mass transit, because its geography places major limitations on where freeways can be built. Rapid transit has been a long time coming. After the demise of Seattle's prewar streetcar and cable car system, plans to build a rapid transit network stalled for half a century. In part, these plans fell through because Seattle's political culture puts a premium on consensus, letting the perfect become the enemy of the good. This interminably slow method of public deliberation even has its own name: the Seattle Process. The Seattle Process leads to the same high costs and long delays as in San Francisco, but the public deliberations are different in style. (Seattle has countless hearings and consultations; San Francisco has pitchforks and torches.) The Seattle Process was on full display in the late 1990s and early 2000s, when Seattle was choosing between a light rail and a monorail.

Streetcars, Cable Cars, and Forward Thrust

Seattle's transit history began in relatively standard fashion, with a dozen competing streetcar and cable car companies in the late 19th century (fig. 20.2). Many of Seattle's hills were too steep for electric streetcars to climb. Thus, a few of the mechanical cable car lines depicted were never electrified. The City of Seattle agreed to buy the privately owned transit system in 1918 and put it under municipal ownership, but the city overpaid. With the city budget in financially straitened circumstances, necessary upgrades were simply too expensive. As a result, the city's streetcar and cable car system would slowly fall apart over the following two decades. All cable cars and streetcars were replaced with motor buses and trolleybuses by 1941.

Attempts to build large-scale rapid transit systems had been mooted since the 1910s, but none ever got much traction. Seattle came closest between 1960 and 1970, during an era of local politics when the city was receptive to ambitious changes pushed from above. The first rapid transit built was the Seattle Center Monorail, installed as a demonstrator for the 1962 world's fair. Even today, the Seattle Center Monorail shuttles commuters and tourists from downtown to the Space Needle, about a mile away.

The other major push for rapid transit during this period was spearheaded by a group of Seattle's prominent citizens known as the Forward Thrust Committee. In 1968, the committee put a series of large public works measures on the ballot to fund sewers, parks, firehouses, and a rapid transit system. Most of this so-called Forward Thrust package was popular and passed easily. The rapid transit system was not so popular, and it failed to pass. The committee was undeterred. Two years later, the committee put rapid transit on the ballot again. This time, the federal government had agreed to pay for 80 percent of construction costs. It made no difference. Seattle was rapidly sliding into a recession due to Boeing layoffs. Voters were leery of the price tag, and Forward Thrust 1970 didn't get the necessary supermajority from the electorate (fig. 20.3). After this second failure, the federal government pulled its funding commitment. The money earmarked for Seattle would go to Atlanta. Atlanta used the money to build the MARTA subway.

Looking back, Forward Thrust's failure is understandable. The committee was criticized at the time as white-collar downtown professionals shoving their schemes down the throats of blue-collar aviation plant employees. The critics had a point. The Forward Thrust Committee was lined with lawyers, bankers, and businesspeople. Its plans were born in the halls of the Chamber of Commerce, the Rotary Club, and the Washington Athletic Club.[1] Factory workers in neighborhoods like South Seattle, the Rainier Valley, and Renton simply didn't find the committee's pro-rapid transit arguments convincing. These voters opposed Forward Thrust by nearly a 2–1 margin.

Forward Thrust was the last time that Seattle would ever try to plan anything from above. As in San Francisco, Seattle's local governments changed the laws to allow individual citizens direct input into the hard decisions of governing, through public hearings, commissions, and the like. In previous eras, such decisions would be made quickly, either by elected officials or within the bureaucracy. But by 1983, endless deliberations and public consultations had become so well established that the *Seattle Weekly* coined a name for it: "the usual Seattle process of seeking consensus through exhaustion."[2]

Half Measures

Consensus through exhaustion often resulted in

20

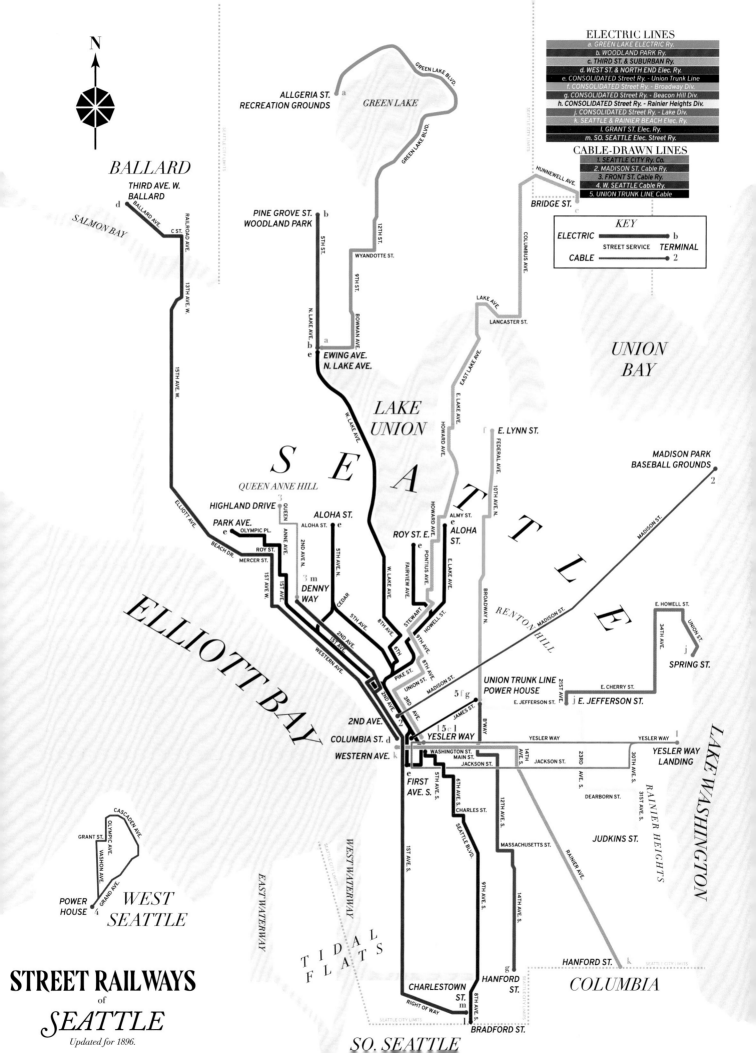

N

BALLARD
THIRD AVE. W.
BALLARD
d
BALLARD AVE.
C ST.

SALMON BAY

RAILROAD AVE.

13TH AVE. W.

15TH AVE. W.

ELLIOTT AVE.

ALLGERIA ST.
RECREATION GROUNDS

GREEN LAKE

GREEN LAKE BLVD.

GREEN LAKE BLVD.

a

12TH ST.

PINE GROVE ST.
WOODLAND PARK
b

5TH ST.

WYANDOTTE ST.

9TH ST.

BOWMAN AVE.

N. LAKE AVE.

a
b
e
EWING AVE.
N. LAKE AVE.

W. LAKE AVE.

*LAKE
UNION*

ELECTRIC LINES
a. GREEN LAKE ELECTRIC Ry.
b. WOODLAND PARK Ry.
c. THIRD ST. & SUBURBAN Ry.
d. WEST ST. & NORTH END Elec. Ry.
e. CONSOLIDATED Street Ry. - Union Trunk Line
f. CONSOLIDATED Street Ry. - Broadway Div.
g. CONSOLIDATED Street Ry. - Beacon Hill Div.
h. CONSOLIDATED Street Ry. - Rainier Heights Div.
j. CONSOLIDATED Street Ry. - Lake Div.
k. SEATTLE & RAINIER BEACH Elec. Ry.
l. GRANT ST. Elec. Ry.
m. SO. SEATTLE Elec. Street Ry.

CABLE-DRAWN LINES
1. SEATTLE CITY Ry. Co.
2. MADISON ST. Cable Ry.
3. FRONT ST. Cable Ry.
4. W. SEATTLE Cable Ry.
5. UNION TRUNK LINE Cable

HUNNEWELL AVE.

BRIDGE ST.
c

LAKE AVE.

LANCASTER ST.

COLUMBUS AVE.

EAST LAKE AVE.

E. LAKE AVE.

HOWELL AVE.

KEY
ELECTRIC ──────── b
STREET SERVICE   TERMINAL
CABLE ──────── 2

*UNION
BAY*

f E. LYNN ST.

FEDERAL AVE.

10TH AVE. N.

MADISON PARK
BASEBALL GROUNDS
2

S E A T T L E

QUEEN ANNE HILL

HIGHLAND DRIVE
3
Queen
PARK AVE.
e OLYMPIC PL.
ROY ST.
BEACH DR.
MERCER ST.

ANNE AVE.

2ND AVE. N.

1ST AVE.

1ST AVE. W.

3 m
DENNY
WAY

ALOHA ST.
ALOHA ST. e

5TH AVE. N.

W. LAKE AVE.

6TH AVE.

5TH AVE.

2ND AVE.

1ST AVE.

WESTERN AVE.

CEDAR

6TH

ROY ST. E.
e
FAIRVIEW AVE.
PONTIUS AVE.

STEWART

HOWELL ST.

9TH AVE.

8TH AVE.

PIKE ST.

UNION ST.

MADISON ST.

ALMY ST.
e
ALOHA
ST.

HOWARD AVE.

E. LAKE AVE.

BROADWAY N.

RENTON HILL

MADISON ST.

MADISON ST.

UNION TRUNK LINE
POWER HOUSE

E. CHERRY ST.

21ST AVE.

E. JEFFERSON ST.
j E. JEFFERSON ST.

E. HOWELL ST.

34TH AVE.

UNION ST.

j
SPRING ST.

ELLIOTT BAY

2ND AVE.
2
COLUMBIA ST. d
WESTERN AVE. k

3RD AVE.

JAMES ST.

5 f g

B'WAY

15 c l
YESLER WAY

YESLER WAY

YESLER WAY

l

YESLER WAY
LANDING

LAKE WASHINGTON

WASHINGTON ST.
MAIN ST.
JACKSON ST.

e
FIRST
AVE. S.

5TH AVE. S.

6TH AVE. S.

CHARLES ST.

JACKSON ST.

14TH AVE. S.

23RD
AVE. S.

30TH AVE. S.

RAINIER HEIGHTS

CASCADEN AVE.

OLYMPIC AVE.

GRANT ST.

VASHON AVE.

GRAND AVE.

POWER
HOUSE
4

*WEST
SEATTLE*

EAST WATERWAY

WEST WATERWAY

*T I D A L
F L A T S*

1ST AVE. S.

SEATTLE BLVD.

12TH AVE. S.

14TH AVE. S.

DEARBORN ST.

MASSACHUSETTS ST.

RAINIER AVE.

JUDKINS ST.

31ST AVE. S.

CHARLESTOWN
ST. m

RIGHT OF WAY

8TH AVE. S.

9TH AVE. S.

HANFORD
ST.
g

HANFORD ST.
k

COLUMBIA

BRADFORD ST.
l

SO. SEATTLE

STREET RAILWAYS
of
SEATTLE
Updated for 1896.

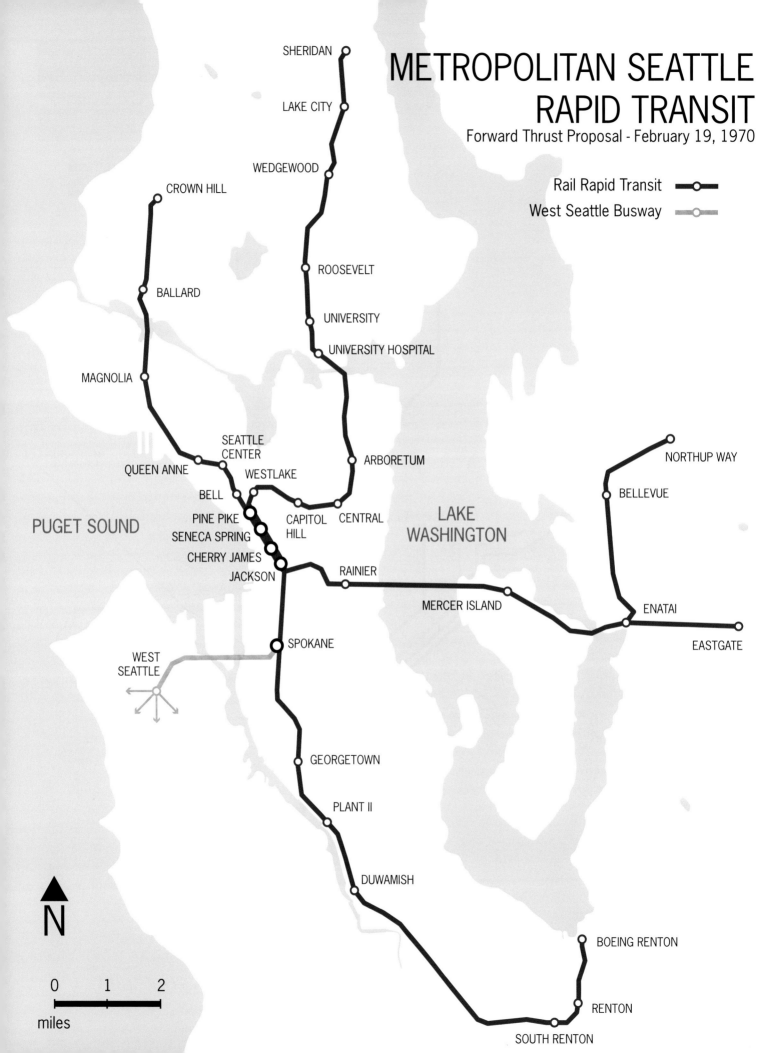

20.3 The subway proposal of
Forward Thrust 1970.

compromises that left few happy. For example, in the late 1980s, Seattle built a downtown transit tunnel which was neither fish nor fowl. When the tunnel opened in 1990, it had electric wires, train tracks, and full-scale underground stations, but no trains.[3] The tunnel wasn't designed to vent diesel exhaust, so ordinary motor buses couldn't use it either. For the first 25 years of its existence, the only vehicles using the tunnel were expensive custom-built "tunnel buses" with both trolleybus equipment for tunnel use and diesel engines for the surface. The only proper rapid transit during this period was the aging two-stop monorail left over from the world's fair.

All the while, the regional transit authority, Sound Transit, was slowly, painfully building support to convert the tunnel to carry trains. After two tries, Sound Transit got voter approval in November 1996 to convert the downtown tunnel and build a $1.7 billion ($3.1 billion, inflation-adjusted), 25-mile light rail system called Link. Sound Transit board member Maggi Fimia told the *Seattle Times*, "Now we stop going to meetings and start building it."[4]

The meetings did not, in fact, stop. Rather, for the following nine years, there would be two dueling rapid transit proposals. The Link light rail plan, conceived by professionals, would run through a gauntlet of micromanagement. Costs ballooned. But there was also a monorail expansion plan run by enthusiastic amateurs in over their heads. Thanks to the Seattle Process and its overemphasis on consensus, the inmates very nearly took over the asylum.

Enthusiastic Amateurs

The monorail expansion project came from a most unlikely place: a populist cab driver. This cabbie was, by his own admission, a gadfly who had been on the fringes of Seattle politics for decades. His name was Dick Falkenbury. Falkenbury, in his memoir, wrote: "I had started this monorail political movement with very few resources and the hope that it might allow me to build something huge and become famous. If I became rich in the bargain, that would be good, too."[5] Falkenbury's monorail project was possible only because citizens' initiatives are legal in Seattle. That is, if a group collects enough voter signatures for a proposed law, the entire electorate has to decide whether to enact that law at the next election, full stop. This meant that Falkenbury and his allies were able to get the monorail proposal onto the 1997 ballot, no matter what Seattle's political class or engineering experts thought. Voters approved the plan mapped in figure 20.4, formally known as Initiative 41, by 53 percent to 47 percent.

There was only one problem with Initiative 41: no money. Initiative 41 assumed that the monorail could be funded by private investors and passenger fares, with little or no public investment.[6] Private investors didn't bite. A slow-motion train wreck ensued because the monorail's backers had so underestimated the difficulty of building it out. Elected officials were mostly unwilling to challenge monorail advocates' unrealistic assumptions. Over the following eight years, the citizens of Seattle were asked to vote four more times on whether the

203

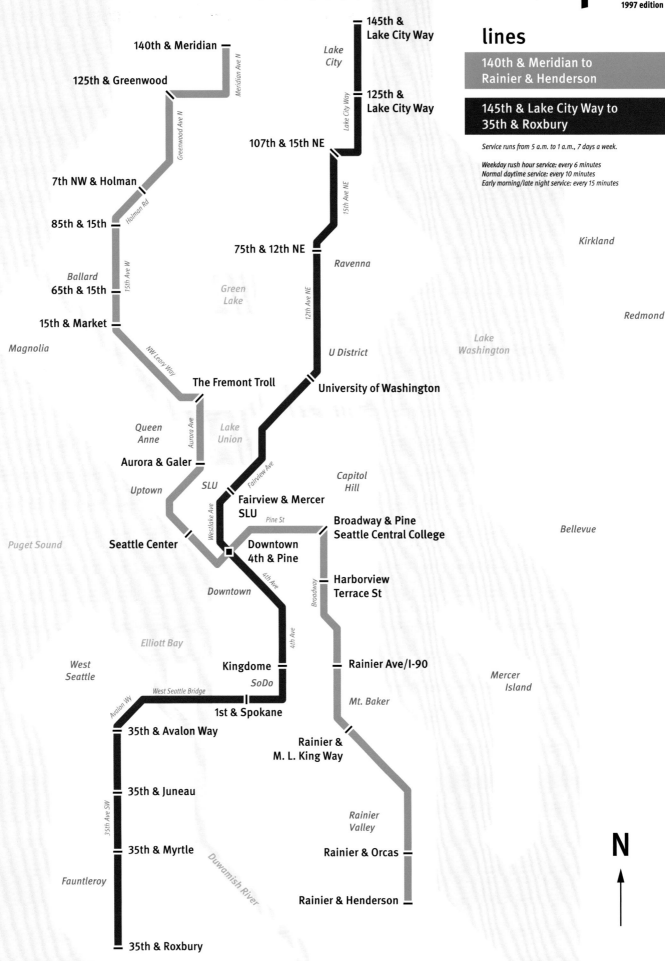

seattle monorail plan

1997 edition

lines

140th & Meridian to Rainier & Henderson

145th & Lake City Way to 35th & Roxbury

Service runs from 5 a.m. to 1 a.m., 7 days a week.

Weekday rush hour service: every 6 minutes
Normal daytime service: every 10 minutes
Early morning/late night service: every 15 minutes

140th & Meridian

125th & Greenwood

7th NW & Holman

85th & 15th

65th & 15th

15th & Market

The Fremont Troll

Aurora & Galer

Seattle Center

Fairview & Mercer SLU

Downtown 4th & Pine

Kingdome

1st & Spokane

35th & Avalon Way

35th & Juneau

35th & Myrtle

35th & Roxbury

145th & Lake City Way

125th & Lake City Way

107th & 15th NE

75th & 12th NE

University of Washington

Broadway & Pine
Seattle Central College

Harborview
Terrace St

Rainier Ave/I-90

Rainier & M. L. King Way

Rainier & Orcas

Rainier & Henderson

Lake City

Ravenna

Green Lake

U District

Lake Union

Capitol Hill

Queen Anne

Uptown

SLU

Pine St

Downtown

SoDo

West Seattle Bridge

West Seattle

Elliott Bay

Puget Sound

Magnolia

Ballard

Fauntleroy

White Center

Duwamish River

Rainier Valley

Mt. Baker

Mercer Island

Bellevue

Lake Washington

Kirkland

Redmond

Meridian Ave N

Greenwood Ave N

Holman Rd

15th Ave W

NW Leary Way

Aurora Ave

Westlake Ave

4th Ave

Fairview Ave

12th Ave NE

15th Ave NE

Lake City Way

Broadway

Avalon Wy

35th Ave SW

N

20.4 Initiative 41's proposed monorail, 1997.

monorail should go forth. Little by little, the price tag crept up, and the scope shrunk.

Initiative 41 proposed a two-line, 54-mile system, funded by fares and private money, and needing little public investment. (The monorail authority was granted $200,000 in onetime start-up funding from the city budget.) With private investors not forthcoming, the city council repealed Initiative 41 in the summer of 2000. Falkenbury and his allies were undeterred, winning a second ballot referendum in the November 2000 election. The second referendum authorized the City to spend $6 million of public money to design the system.[7] The next year, unwilling to alienate pro-monorail voters, all three mayoral candidates supported monorail expansion, including eventual winner Greg Nickels.[8]

In 2002, the monorail promoters went back to the Seattle public a third time. This time, they had political cover from Mayor Nickels and the backing of one of the city's two major newspapers. The pro-monorail side asked Seattle voters to fund a one-line, 14-mile monorail with $1.75 billion ($2.76 billion in 2022 dollars) in new taxes. Voters narrowly approved the tax, but revenues were stubbornly below projections. The two major bidders to build the monorail withdrew in 2004. After this withdrawal, anti-monorail forces got their own referendum on the ballot attempting to cancel the monorail, without success. But that was before the public knew the true cost. In June 2005, the monorail authority finally released the price tag to finance and build a single 14-mile monorail line: $11 billion ($15 billion, adjusted for inflation), all in public money.[9] After a hailstorm of criticism, May-

or Nickels withdrew his support. Monorail backers tried to salvage the project by shrinking the project even further. This fifth referendum asked voters to approve an even shorter, 10.5-mile system. Voters rejected this system in the November 2005 election by a 2–1 margin, killing the monorail for good. Eight years and five referendums later, $125 million in public money had been spent with nothing to show for it.[10]

The monorail fiasco represents the excesses of the Seattle Process in action. Seattle was unwilling or unable to have a frank discussion of the benefits and costs of Falkenbury's monorail upfront, so good money was thrown after bad. When presented with an unrealistic vision, Seattle's political institutions were paralyzed.

Micromanaged Professionals

In the meantime, Link had the same problem, but from the opposite direction. Link was a basically sound project that was nearly derailed by the need to study and discuss every detail ad nauseam. After voter approval in 1996, Sound Transit embarked on two years of environmental review. The proposed starter system was a single line, running north–south from Northgate to the airport, which is today Line 1 (fig. 20.5).

But nothing is ever simple in Seattle. Shortly after voter approval, interest groups began to come out of the woodwork. The University of Washington was unhappy that Sound Transit planned to tunnel underneath the university. The neighborhoods of South Seattle demanded a tunnel rather than a surface line. Capitol Hill business interests

205

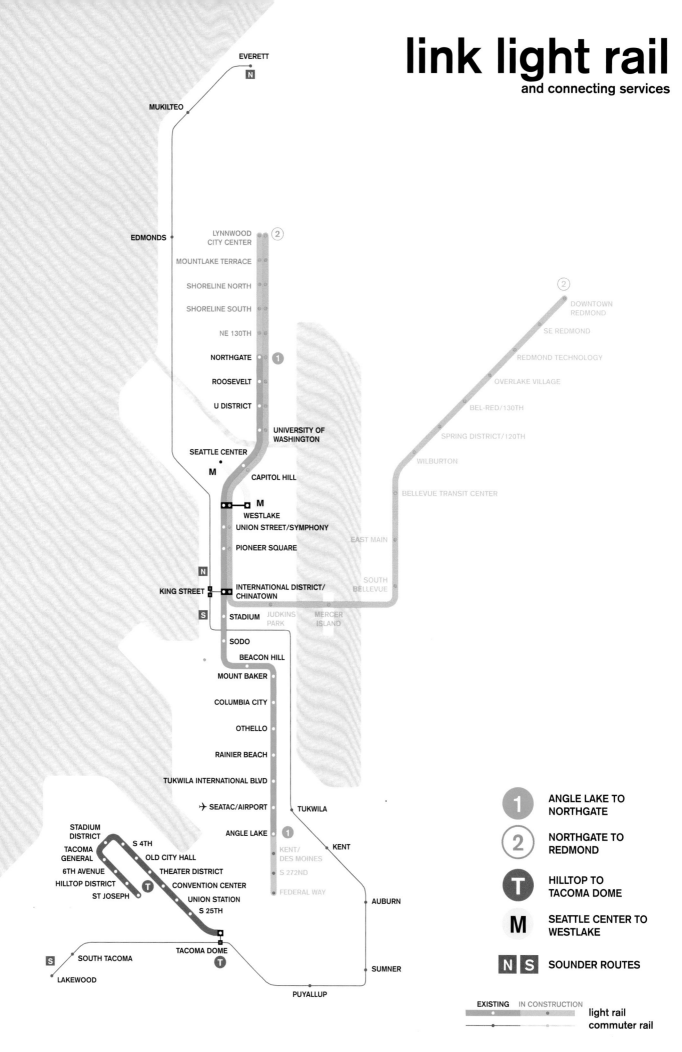

20.5 Link light rail, 2022. The Seattle
      Center Monorail, built for the World's
      Fair, is also shown as line M.

east of downtown objected to the disruptions caused by tunnel construction. The suburban city of Tukwila demanded routing changes to run the line via the city's main street instead of the freeway. These changes were expensive and time-consuming. By 2000, the cost of the initial line had doubled from the initial estimates. Instead of a 25-mile line opening in 2006, Seattle got a 14-mile line that opened in 2009.[11] The rest of the starter system was delayed for 15 years, until 2021. In the meantime, Seattle voters raised their own taxes two more times to expand Link to cover the whole region, in 2008 and 2016.

On paper, this is a nation-leading expansion of transit, but its sheer slowness is telling. There's little reason to believe that timelines or budgets will be met, given the Seattle Process and the region's history of persistent planning delays. To compare like with like, the first Forward Thrust proposal had a projected 17-year implementation timeline. Had Forward Thrust 1968 been approved, full build-out would be complete by 1985.[12] The initial 1996-approved Link system plus its 2008-approved expansion is roughly the same size as Forward Thrust. These projects have a projected completion date of 2024, 28 years from start to finish. Completion of the 2016-approved expansion, called Sound Transit 3, was originally expected in 2041.[13] As of this writing, predicted completion of Sound Transit 3 has already been pushed back three years, to 2044.[14]

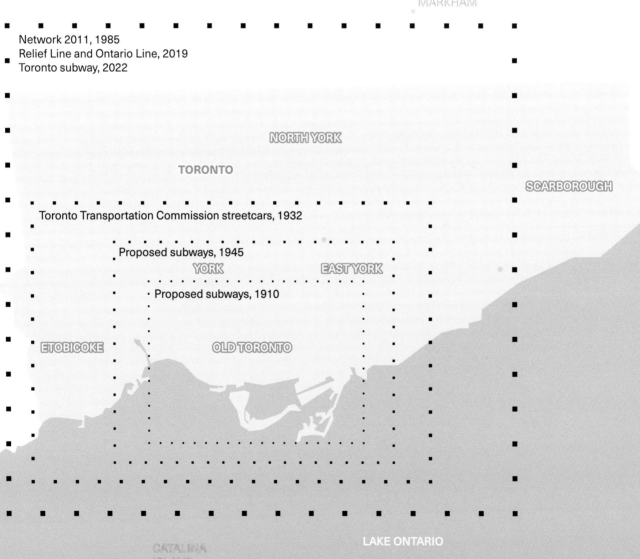

RICHMOND HILL

VAUGHAN

MARKHAM

Network 2011, 1985
Relief Line and Ontario Line, 2019
Toronto subway, 2022

NORTH YORK

TORONTO

SCARBOROUGH

Toronto Transportation Commission streetcars, 1932

Proposed subways, 1945

YORK   EAST YORK

Proposed subways, 1910

ETOBICOKE   OLD TORONTO

LAKE ONTARIO

CATALINA
ISLAND

N

Toronto

2 miles/3.2 km

21

TORONTO

Subway Line as Political Football

21.1 Map of the Greater Toronto Area.

Toronto surpassed Montreal as Canada's largest and wealthiest city after World War II. Unlike its neighbors south of the border, Toronto (fig. 21.1) relied mostly on subways rather than expressways to handle a growing population after the war. Moreover, when it built a subway, Toronto chose to put businesses and homes near transit, unlike most of the American cities in this book. It's common in the Greater Toronto Area to see clusters of towers surrounding subway stations even in the suburbs, so from the air Toronto looks rather like an archipelago. In some ways, the Toronto Subway has fallen victim to its own success. The Yonge Street subway (*Yonge* is pronounced like *young*), now numbered Line 1, has been overcrowded for nearly half a century. Toronto's transit operator, the Toronto Transit Commission (TTC), keeps extending the line deeper into the suburbs. This makes the crowding problems worse at the downtown end. There has been broad consensus for decades that a second downtown subway is needed to relieve the Yonge line. But this second downtown subway has not been built. For four decades, the governments of the Greater Toronto Area have treated the line as a political football.

Tubes for the People

Toronto followed a relatively standard North American development pattern in the first half of the 20th century. It should thus come as no surprise that Toronto suffered from severe streetcar congestion. Toronto city con-

PLAN OF A RAPID TRANSIT SYSTEM
prepared for the Mayor and Council of the city of
TORONTO

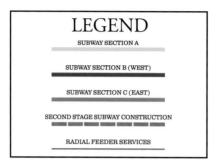

LEGEND

SUBWAY SECTION A

SUBWAY SECTION B (WEST)

SUBWAY SECTION C (EAST)

SECOND STAGE SUBWAY CONSTRUCTION

RADIAL FEEDER SERVICES

21.2 Jacobs and Davies subway proposal,
       1910. Section C is the eventual second
       downtown subway corridor.

troller Horatio Hocken made congestion his signature issue when he ran for mayor in 1910. Hocken pushed a publicly funded subway as an alternative to the privately owned and much-reviled Toronto Railway Company. "Tubes for the People" was his slogan. Hocken lost the election, but the proposal's popularity convinced the Toronto city council to hire engineers to design a subway. The engineers, Charles M. Jacobs and J. Vipond Davies, had respectively served as chief engineer and tunnel designer on the Hudson Tubes, between New Jersey and New York City (fig. 21.2).[1] A proposal to fund the Jacobs-Davies subway went on the ballot in 1912, but the measure failed. Ironically, that same year, Hocken was elevated to the mayoralty, but he was in no position to revive Tubes for the People after its defeat at the ballot box.

The streetcars (fig. 21.3), taken over by the TTC in 1921, continued to carry the bulk of Toronto's transit passengers. The King lines and Queen lines were extremely busy, just as they are today. Subways would not gain political traction again until the 1940s, when World War II juiced transit ridership. Toronto politicians realized that that the coming postwar influx of automobiles would choke the city in car traffic. (Unusually, Toronto kept its streetcar network after World War II, because the TTC decided to buy lightly used streetcars secondhand from cities that were closing their systems.)

While World War II prevented major civilian construction from happening, the subway project returned to the front burner at war's end. The TTC put the 1945 subway plan (fig. 21.4) to the voters on New Year's Day 1946. Unlike Tubes for the People,

the Queen Street line was meant to be a streetcar tunnel through downtown, like the venerable Green Line in Boston. The subway vote passed, although the Queen Street line was eventually cut for lack of federal government funding.

Network 2011

After it opened in 1954, the subway was an immediate success. But after an initial period of rapid expansion, subway growth stalled in the late 1970s and early 1980s. Money was short, and the densest parts of the Greater Toronto Area were adequately served by the extended Yonge subway and the Bloor-Danforth subway (Line 2), which opened in the 1960s. For a time, the TTC and longtime Ontario premier Bill Davis dabbled with upgrading the commuter rail system to a maglev—a hovering, high-speed train propelled by powerful magnets— to relieve Yonge subway congestion. (The maglev technology never quite worked out, because it didn't function well in Canada's snow and ice.) With maglev a nonstarter, Davis decided to support an expansion of traditional subway service. His government released the plan, called *Network 2011*, in 1985. The plan was revised multiple times as to station locations and routings; figure 21.5 is based on the assumptions used in the final *Network 2011* proposal.[2]

Here, the game of political hot potato began. *Network 2011* was intimately tied politically to Davis and his center-right Progressive Conservative Party, known popularly as the Tories. Tories had controlled Ontario politics uninterrupted since 1943. Under the Tories, the TTC subway system

211

TORONTO TRANSPORTATION COMMISSION

NORMAL STREETCAR ROUTING IN EFFECT MAY 1, 1932

Additional rush hour services are not shown

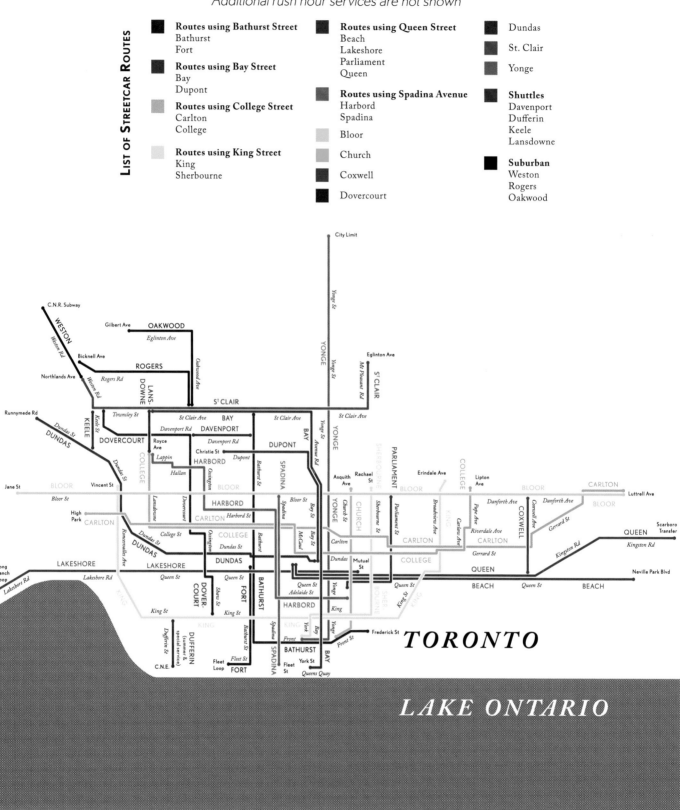

21.3 TTC streetcars, 1932. I use the period-correct name, the Toronto Transportation Commission, not its modern name, the Toronto Transit Commission.

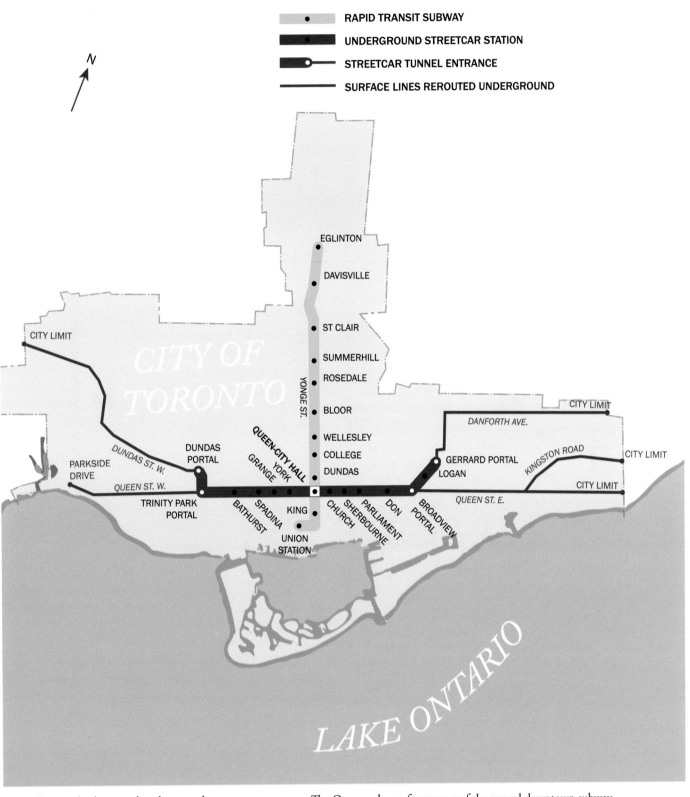

RAPID TRANSIT FOR TORONTO

A Statement of Policy
Toronto Transit Commission, 1945

● RAPID TRANSIT SUBWAY

● UNDERGROUND STREETCAR STATION

○ STREETCAR TUNNEL ENTRANCE

SURFACE LINES REROUTED UNDERGROUND

N

CITY OF TORONTO

CITY LIMIT

EGLINTON

DAVISVILLE

ST CLAIR

SUMMERHILL

ROSEDALE

YONGE ST.

BLOOR

WELLESLEY

COLLEGE

DUNDAS

QUEEN-CITY HALL

CITY LIMIT

DANFORTH AVE.

KINGSTON ROAD

CITY LIMIT

CITY LIMIT

GERRARD PORTAL
LOGAN

DUNDAS ST. W.

DUNDAS PORTAL

PARKSIDE DRIVE

QUEEN ST. W.

TRINITY PARK PORTAL

GRANGE

BATHURST

SPADINA

KING

YORK

UNION STATION

CHURCH

SHERBOURNE

PARLIAMENT

DON

BROADVIEW PORTAL

QUEEN ST. E.

CITY LIMIT

LAKE ONTARIO

21.4 Proposed subway and underground streetcar routes, 1945. The Queen subway, forerunner of the second downtown subway, is colored red.

N

North York Centre

Finch

Bayview

Bessarion

Leslie

Don Mills

Consumers

Victoria Park

Warden

Kennedy N

Downsview

Faywood

Bathurst N

Senlac

Sheppard–Yonge

Agincourt

Scarborough Centre

Wilson

York Mills

Ellesmere

Midland

McC

Yorkdale

Lawrence W

Lawrence

Lawrence E

Glencairn

Eglinton W–Allen

York City Centre

Caledonia

Keele N

Dufferin N

Scarlett

Jane N

Eglinton

Davisville

Kennedy

Renforth

Martin Grove

Kipling N

Islington N

Royal York N

St Clair

Summerhill

Warden

Dupont

St Clair W

Rosedale

Victoria Park

Bathurst

St George

Bloor–Yonge

Christie

Ossington

Spadina

Greenwood

Woodbine

Main St

Lansdowne

Dufferin

Dundas W

Keele

High Park

Runnymede

Jane

Old Mill

Royal York

Islington

Kipling

Museum

Queen's Park

St Patrick

Osgoode

St Andrew

Bay

Wellesley

College

Dundas

Queen

King

Chester

Pape

Donlands

Coxwell

Broadview

Castle Frank

Sherbourne

Gerrard

Queen East

Union

Lower Spadina

Convention Centre

St Lawrence

Atiritari

TORONTO TRANSIT COMMISSION

Network 2011 Final Report

System 3

21.5 Network 2011, as proposed by Bill Davis's
    Tories, 1985. The Relief Line is in pink.

had been built out and supported with large-scale investment. *Network 2011* was no exception. The plan's centerpiece was a full-scale second downtown subway called the Relief Line, designed to relieve the pressure on the Yonge line. The Relief Line was the largest single line item in *Network 2011*.

The Tories couldn't stay in office long enough to enact the plan. In the 1985 provincial election, the rival center-left Liberal Party took control for the first time in four decades. The new Liberal premier, David Peterson, junked *Network 2011* and funded suburban expressways instead. Peterson promised to put together a totally separate package of transit improvements. Five years later, in April 1990, Peterson's package came out, shortly before the Liberals lost the election. Peterson's successor, Bob Rae of the left-wing New Democratic Party, cut back most of the Liberals' proposed improvements. In turn, Rae's Tory successor, Mike Harris, slashed things further in 1995. Harris's platform, called the Common Sense Revolution, called for budget austerity, reduced social spending, and lower taxes, much like the contemporary Contract with America backed by the U.S. Republican Party.[3] There was little appetite for large-scale subway expansion in such a political environment.

Subways for the Suburbs

The second downtown subway ended up on hold until the early 21st century, when the regional transportation authority, Metrolinx, called for the revival of the Relief Line in 2008. Downtown Toronto was undergoing a building boom at the time. The overcrowding issues on the Yonge subway, which had never really gone away, had become serious again. Nonetheless, Toronto wasn't able to get the second downtown subway built this time, either. This time around, the line fell victim to the squabbles between Toronto's city center and suburbs.

Some local context for Toronto's politics is necessary, because the local city-suburb split is quite different from what non-Torontonians might expect. Most of Toronto's inner-ring suburbs merged with the old city of Toronto in 1998. The amalgamated city covers 243 square miles—about the same land area as Chicago. The pre-1998 city limits, called Old Toronto, make up about one-sixth of Toronto's land area and account for about one-third of its population. Old Toronto is denser, richer, whiter, and more left-wing than the rest of the city. In contrast, suburban parts of Toronto lean toward the Tories. Immigrants to Canada tend to live in the suburbs, and they are unusually willing to vote for right-wing populists, unlike most places in the developed world.[4] Because Torontonians take transit at high rates, it's generally not a debate whether better mass transit is needed. Rather, it's about who should get it first.

In the 2010 mayoral election, this political dynamic was on clear display. Mayor David Miller had received provincial funding and approval for a project called Transit City. Transit City was meant to blanket Toronto's suburban neighborhoods with light rail, busways, and bike lanes. The new trains and buses would run in the medians of Toronto's grid of main streets. (Miller's plan did not include the second downtown subway.) Miller had chosen not to run for a third term, leaving Transit City's fate in the hands of future politicians.[5] In the race to succeed Miller, there were three competing visions. Ontario's Deputy Premier George Smitherman, running on the Liberal ticket, supported the continuation of Transit City. Deputy Mayor Joe Pantalone, running on the New Democratic ticket, supported an expanded version of Transit City, plus more bike lanes.[6] Then there was the Tory, Rob Ford.

Ford was a loudmouthed, alcoholic, crack-smoking Toronto city councilman who ran on a right-wing populist platform.[7] He was an unapologetic creature of the suburbs, hammering gays, feminists, creatives, and "the elitist downtown latte-sipping media and socialist hordes" of Old Toronto.[8] This reactionary populism resonated with suburban, socially conservative immigrants from the Indian subcontinent, the Caribbean, and the Middle East who made up key parts of Ford's electoral coalition. His message also resonated with traditional conservative blocs, like affluent white Protestant suburbanites.[9] The second downtown subway was unsurprisingly a low priority for Ford. "The downtown people have enough subways already," he later told the press.[10]

On the transportation front, it meant that Ford opposed bike lanes—"war on the car," he said. He also vowed to cancel Transit City's light rail lines in favor of subways for the suburbs.[11] Ford won the election, sweeping the suburbs and losing Old Toronto. He then followed through and killed off the parts of Transit City that hadn't been funded already. But he couldn't get most of his suburban subways through the Toronto City Council. Nor could Mayor Ford stop the second incarnation of the Relief Line. TTC chief Andy Byford—who would later run the New York MTA—publicly called for the revival of the Relief Line in 2012. Planning continued apace as well. Metrolinx voted to add the updated Relief Line to Toronto's regional transportation plan in 2013.[12] The next year, Mayor Ford's unstable behavior and substance abuse issues eventually led the City Council to strip him of his powers. A cancer diagnosis and continued scandals meant that he was unable to run for reelection in 2014. He died in March 2016. Two weeks after his death, the Toronto City Council gave its approval to build the Relief Line.

Three years later, Mayor Ford's elder brother Doug barged in and scuttled the whole thing.

Still Waiting

Doug Ford was also a Tory politician, and he shared his brother's politics and suburban base of support. He was elected premier of Ontario in 2018. Doug Ford was a more competent politician than his brother, more disciplined and less prone to unstable outbursts. Premier Ford canceled the Relief Line outright in 2019, and moved to seize control of the TTC from local authorities.[13] Instead, he proposed an automated, lower-capacity, partially elevated line called the Ontario Line. The Ontario Line would follow a different route from the manned, all-underground route of the Relief Line (fig. 21.6). To get the city of Toronto's support, Premier Ford cut a deal. The city agreed to switch its support to the Ontario Line, and in return, Ford agreed to drop his efforts to take over

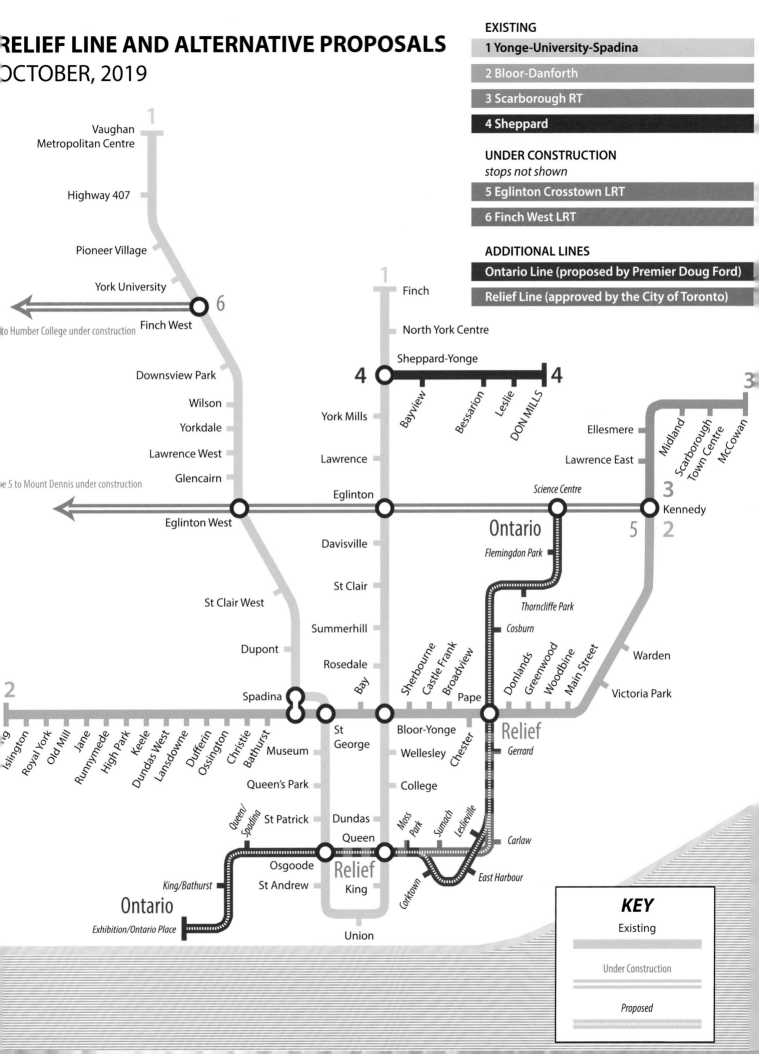

TORONTO SUBWAY

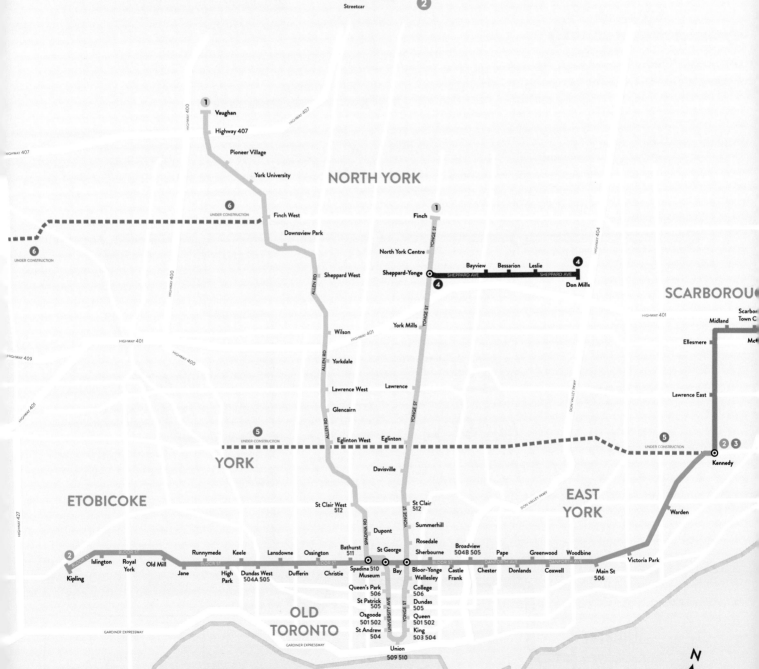

1 Yonge-University   **4** Sheppard

2 Bloor-Danforth   **5** Eglinton

3 Scarborough   **6** Finch West

Terminal

Station   Transfer
Streetcar
2

1 Vaughan

Highway 407

Pioneer Village

York University

6 UNDER CONSTRUCTION   Finch West

Downsview Park

NORTH YORK

6 UNDER CONSTRUCTION

Sheppard West

1 Finch

North York Centre

Bayview   Bessarion   Leslie   **4**

Sheppard-Yonge   SHEPPARD AVE   SHEPPARD AVE   Don Mills
4

SCARBOROU

Wilson

York Mills

Midland   Scarbor Town C

Yorkdale

Lawrence

Ellesmere   Mc

Lawrence West

Glencairn

Lawrence East

5 UNDER CONSTRUCTION   Eglinton West   Eglinton   **5** UNDER CONSTRUCTION   **2 3**

YORK

Davisville

Kennedy

ETOBICOKE

St Clair West
512

St Clair
512

EAST
YORK

Warden

Dupont

Summerhill

Rosedale

Bathurst
511

St George

Broadview
504B 505

Pape   Greenwood   Woodbine

Victoria Park

2 Runnymede   Keele   Lansdowne   Ossington

Sherbourne

Islington   Royal
York   Old Mill   Jane   High
Park   Dundas West
504A 505   Dufferin   Christie   Spadina 510
Museum   Bay   Bloor-Yonge   Castle
Wellesley   Frank   Chester   Donlands   Coxwell   Main St
506

Kipling

Queen's Park
506

College
506

St Patrick
505

Dundas
505

Osgoode
501 502

Queen
501 502

St Andrew
504

King
503 504

OLD
TORONTO

Union
509 510

LAKE ONTARIO

N

0   1   2   3
kilometres

21.7 Toronto Subway, 2022.

the TTC. At its cancellation, about one-sixth of the design work for the Relief Line had been carried out. Millions of dollars had been spent. Due to the differences between the two plans, little of the design work could be reused.[14]

This aggressive politicking has delayed construction even further. The Toronto City Council's 2016 iteration of the Relief Line was projected to open in 2028. The squabbling between Premier Ford and the City of Toronto, and the wholesale switch to the Ontario Line, delayed the project by another three years. As of this writing, the Ontario Line is not projected to open until 2031. The existing subway (fig. 21.7) is still overcrowded. Given the second downtown subway line's troubled history, the only thing that seems certain is that completing the Ontario Line will require navigating difficult political shoals. These hazards should not be underestimated. After all, those hazards already wracked Relief Line 2.0, *Network 2011*, and Tubes for the People.

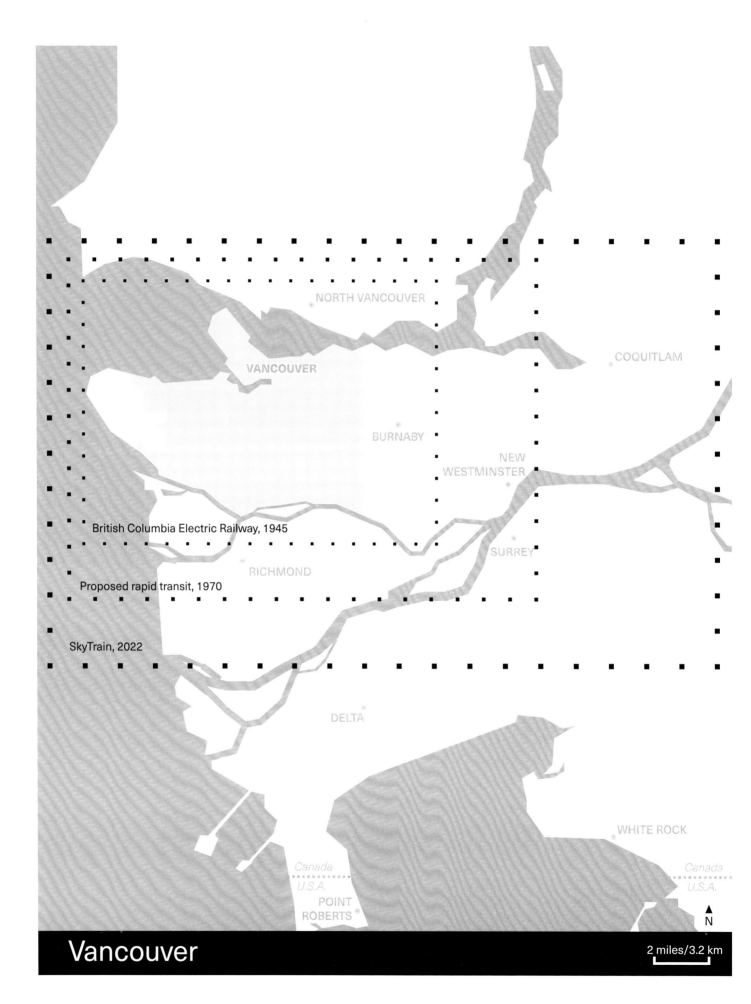

NORTH VANCOUVER

VANCOUVER

COQUITLAM

BURNABY

NEW
WESTMINSTER

British Columbia Electric Railway, 1945

RICHMOND

SURREY

Proposed rapid transit, 1970

SkyTrain, 2022

DELTA

WHITE ROCK

Canada
U.S.A.

Canada
U.S.A.

POINT
ROBERTS

N

Vancouver

2 miles/3.2 km

22

VANCOUVER

An Exceptional Elevated

22.1 Map of greater Vancouver.

Barring unusual circumstances, like a preexisting railway corridor that can be upgraded, the cheapest and simplest way to build a metro is to build an elevated, full stop. Compared to a subway, an elevated can be built faster, its engineering is far simpler, and it can go over nearly any street or alleyway. This reduces the need to requisition private property. For all these reasons, it was a natural choice to build elevateds before World War I, when speed and construction cost were paramount concerns. But despite their utility, primitive elevated lines were coal-burning public nuisances. Even after technical advancements eliminated most of the problems with elevated structures, elevateds had such a bad reputation that cities spent huge sums to replace otherwise functional elevated lines in the mid-20th century. From the 1920s onward, large-scale elevated construction fell out of favor in North America and never really returned, except where required by reasons of geology or economy.

Vancouver's SkyTrain, the Western world's longest automated metro, is the exception to this rule.[1] SkyTrain, which opened in 1985, was meant to be all elevated from the start. The only underground segment in the initial system was a converted freight tunnel from 1933. Vancouver's choice is even more unusual because light rail and underground metro projects were in vogue at the time SkyTrain was built. It took a unique combination of circumstances for Vancouver (fig. 22.1) to build an elevated: the British Columbia provincial government's desire to build a showcase project for the 1986 World's Fair, the Canadian federal government's wish to show off Canadian-developed rapid transit technology, and the need to prop up a failing rail car manufacturer owned by the Ontario provincial government.

The Greatest Litigation the World Has Ever Witnessed

To understand why elevated lines were so out of fashion at the time, some background is necessary. North America's first rapid transit line was the steam-powered Ninth Avenue Elevated in Manhattan, which opened in 1868. The Ninth Avenue Elevated was a major success and fabulously profitable for its owners. Four decades of widespread, privately funded elevated construction followed in New York and elsewhere. The elevated lines were a godsend for commuters accustomed to being packed into overcrowded, slow horsecars and carriages.

The elevateds were also a godsend for lawyers.[2] The trains ran 19 hours a day, seven days a week. They belched coal smoke into the air, and rained sparks, grease, and occasionally parts onto the streets below. By 1893, Manhattan's elevateds had caused so many lawsuits that New York attorney Frank Mackintosh wrote in the *Yale Law Journal*: "The litigation which has been in progress for the past 16 years between the various elevated railroad companies . . . and the owners of real estate abutting upon the streets and avenues of New York City through which the lines of such railroads run, is probably, quantitively speaking, the greatest which the world has ever witnessed. Some idea of its mag-

nitude can be had when one is informed that more than two thousand cases are constantly pending, and that it costs the defendant companies upward of a quarter of a million [$7.9 million in 2022 dollars] annually to carry on the litigation, entirely aside from judgments, costs and allowances paid, or voluntary settlements entered into by them."[3] Mackintosh estimated that Manhattan's elevated companies were liable for at least $20 million ($661 million, inflation-adjusted) in damages. A quarter of that sum was paid in voluntary settlements, and the balance was court judgments won by plaintiffs.[4] (For comparison, when the litigation from the Flint, Michigan, water crisis settled in 2021, the defendants agreed to pay $626 million.)[5]

The pace of elevated construction in city cores dropped off after underground lines became widespread, even though electrification had eliminated the pollution and much of the noise. The big shift from elevated to subway happened in the 1920s. Thus, Philadelphia's Market-Frankford Line, which opened in 1905, was mostly elevated, with a relatively short center-city tunnel. Philadelphia's next rapid transit line was the all-underground Broad Street Line, which opened in 1928. In New York, the Interborough and BMT rapid transit lines that opened between 1904 and 1920 were a mix of elevated and subway. A decade later, the Independent Subway of the 1930s was nearly all underground. Boston stopped building new elevated lines in 1919, and all future lines ran on the surface or in tunnels. Los Angeles voters rejected an elevated system in 1926.

Existing elevateds were routinely slated for demolition. In the 1950s, 14 percent of the Market-

Frankford Line was shuttered and replaced with tunnels. Boston began tearing down its elevated lines in 1938, a process largely complete by 1987. New York closed Manhattan's elevated lines by 1955 and continued closing elevateds in the Bronx, Queens and Brooklyn through the 1980s. Chicago's venerable downtown elevated Loop also faced the same fate well into the 1970s. The Loop survived demolition only because Chicago's politicians couldn't agree on how to replace it.

New urban elevated lines were generally only built where no good alternative existed, especially after World War II. In the 1960s, the Bay Area Rapid Transit (BART) system in greater San Francisco included long elevated sections because of cost concerns. In other cases, elevated construction was required by geology. The entire Miami Metrorail system built in the 1980s was elevated because Miami's high water table makes underground lines impractical. But where not required by cost or terrain, most postwar metro systems made heavy use of underground lines. All of the Montreal Metro is underground. So are both of Los Angeles's subway lines. In Washington, DC, where the Washington Metro goes deep into the suburbs, 40 percent of the system's mileage is underground, including nearly all the routes through Washington's core. Vancouver is the notable exception to this trend. Only 16 percent of the Vancouver SkyTrain's trackage is in tunnels.[6]

The Right Tool for the Job

Vancouver's path to an elevated was a winding one. Its electric utility, the British Columbia Electric

Railway, built out the region's original pre–World War II streetcar and interurban system. This system closed in phases between 1947 and 1958 (fig. 22.2 shows the 1945 streetcar and interurban system immediately before the shutdown began). An American-style freeway network was proposed for Vancouver in the decades after World War II. Nothing ever came of it. Vancouver started its freeway planning relatively late, and the drawbacks of freeways had become apparent by the late 1960s and early 1970s. Freeway opponents in Vancouver wielded the specter of Los Angeles's smoggy, congested suburbia like a cudgel against pro-freeway politicians. This debate culminated in a 1972 provincial election that put an end to any large-scale freeway efforts.[7] With freeways a political nonstarter, and urban growth and congestion getting worse, the only practical option was better transit. The question was, what form should that transit take?

At the time, the transit industry was in a relative state of flux due to the large-scale technological advancements of the age. Postwar transit systems offered real improvements over the systems of the past. BART pioneered automated trains. Montreal introduced the rubber-tired metro to North America. Boston updated its legacy streetcar system to light rail standards, buying modern, high-capacity, air-conditioned trains. At the time, there was also interest in adapting these technological tools to use in smaller cities with smaller budgets. In both the United States and Canada, this took the form of investments in revived street-level rail networks (i.e., light rail), and in automated small-scale metro technology.

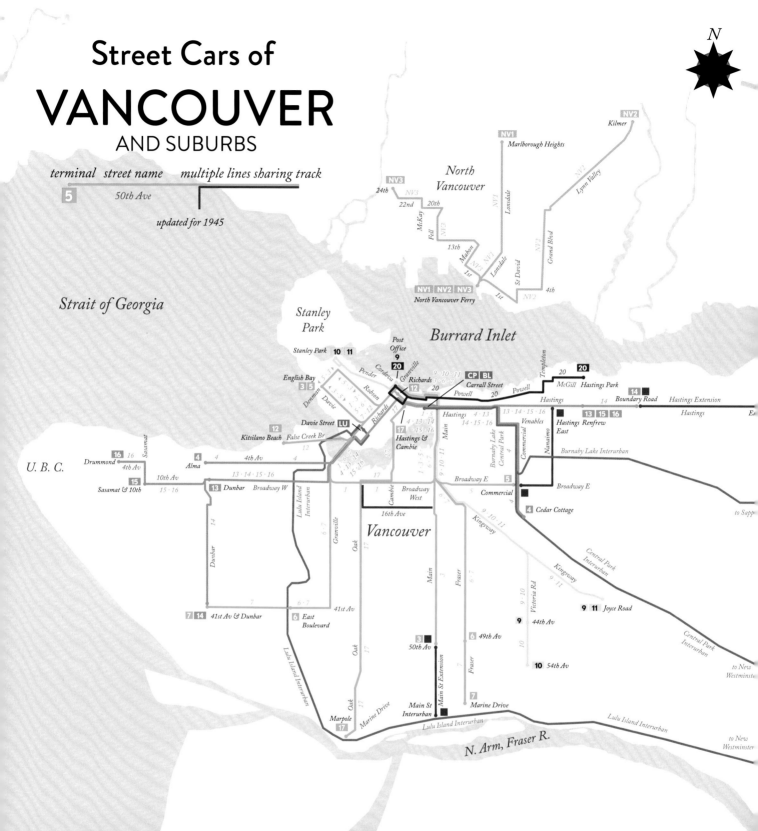

Street Cars of
VANCOUVER
AND SUBURBS

terminal   street name   multiple lines sharing track

5   *50th Ave*

updated for 1945

N

Strait of Georgia

North Vancouver

NV3 *24th*
NV3
22nd
McKay *20th*
Fell
13th
Mabon
1st *NV1*
NV3 *Lonsdale*
NV1 *Marlborough Heights*
NV2 *Kilmer*
NV1 *Lonsdale*
St David
NV2 *Grand Blvd*
Lynn Valley

NV1 NV2 NV3
North Vancouver Ferry
1st *NV2* *4th*

Stanley Park

Burrard Inlet

Templeton

Stanley Park **10** **11**

Post Office
9
20

English Bay
3 **5**
Denman *Pender* *Cordova* *Granville*
Robson *Richards* *20*
Davie
9·10·11 *Richards*
30·65·7 CP BL
Carrall Street
20 *Powell*
McGill *20* **20** *Hastings Park*

12 *Kitsilano Beach* *Davie Street* LU *False Creek Br*

Richards *17*
17 **17**
Hastings & Cambie
4·13·14
15·16
Hastings *4·13*
14·15·16
13·14·15·16
Venables
Hastings East **13** **15** **16**
Renfrew
14 *Boundary Road* *Hastings Extension*
Hastings *E*

16 *Drummond* *Sasamat*
4th Av *4*
4 *Alma* *4* *4th Av*
U.B.C.
15 *Sasamat & 10th* *10th Av* *13·14·15·16* *15·16*
13 *Dunbar* *Broadway W*
1·3·5 *6·7*
1·3·5·6·7
Hastings *4·13·14·15·16*
Main
9·10·11
Burnaby Lake Central Park
Commercial
5 *Broadway E*
Broadway E

Dunbar *14* *Lulu Island Interurban*
13·14·15·16
Broadway West
16th Ave
Cambie *1*
Vancouver *1*
Granville *6·7*
Oak *17*
9·10·11 *Broadway E*
Kingsway
Commercial **4** *Cedar Cottage*
Central Park Interurban
to Sapp

Dunbar *14*
41st Av *East Boulevard* **6**
7 **7** **14** *41st Av & Dunbar*
Oak *17*
6·7
9·10·11 *Kingsway*
Victoria Rd
9 *44th Av*
9 **11** *Joyce Road*
Central Park Interurban

Lulu Island Interurban
Oak *17*
3 *50th Av*
Fraser *7*
6 *49th Av*
10 **10** *54th Av*
to New Westminster

Marpole **17**
Marine Drive
Main St Interurban
Main St Extension
7 *Marine Drive*
N. Arm, Fraser R.
Lulu Island Interurban
Lulu Island Interurban
to New Westminster

STREET CAR SERVICES

Cambie	17		Powell	1·3·5·6·7		No. Vancouver	NV1·NV2·NV3
Granville	1·3·5·6·7·12		Richards	4·13·14·15·16		Interurban	Lulu Island · Central Park · Burnaby Lake
Pender	10·11		Shuttle	Main St. Extension · Nanaimo St. · 16th Ave. · Hastings St. Extension			

22.2 British Columbia Electric Railway
      streetcar and interurban network, 1945.

Fancy New Technology

To cash in on the expected small-scale metro market, the provincial government of Ontario established a manufacturer called the Urban Transportation Development Corporation. The corporation's task was to develop medium-capacity transit systems for cities too small to justify a traditional metro but too large to rely solely on buses. They spent most of the 1970s developing a fully automated small metro system called the Intermediate Capacity Transit System (ICTS).[8]

Broadly speaking, ICTS was an evolutionary improvement of the technologies used with BART, scaled down to fit the needs of smaller cities. For example, while BART's trains were run by a central control computer, BART still required a driver to open and close train doors. ICTS, developed a decade later, was sophisticated enough to not require a driver at all. While BART trains max out at 80 miles per hour, ICTS trains had a speed limit of 50 miles per hour. BART was built for 710-foot trains carrying 2,000 people per train; ICTS's trains were 250 feet, capable of carrying 480 passengers per train. With shorter, lighter trains, the footprint of the elevated structure could be reduced. This was a principal design aim of the ICTS team. In its own words, "The system is being designed for maximum community acceptability at-grade or on elevated structures to avoid the high cost of underground construction to the greatest extent possible."[9] Despite these benefits, ICTS had large drawbacks. In particular, ICTS permanently locked the purchaser into one vendor due to its nonstandard, proprietary components.

Light rail, the other common option of the era, was a safe, mature technology. Light rail trains used human drivers and standard-issue railway signals. Many different suppliers sold light rail equipment. Light rail trains from different makers can serve the same line with no problems, provided that physical dimensions are compatible. The Green Line in Boston, for example, has used five distinct types of light rail cars over its history. Each model of train car came from a different manufacturer.

Light rail was the standard late-1970s choice for midsized cities on both sides of the border. In Canada, Edmonton and Calgary built light rail. In the United States, San Diego, Buffalo, and Portland were doing the same. Most of the early plans for Vancouver's rapid transit system assumed that light rail would be the chosen technology. But not all. Some plans were agnostic as to the actual choice of technology and considered monorail, traditional rail, and busway all to be viable options (fig. 22.3).[10] Under ordinary circumstances, light rail would have made perfect sense for a Canadian metropolis with a population of 1 million. But Vancouver's circumstances were anything but ordinary, because Vancouver was going to host a world's fair.

Expo 86

Vancouver put in a bid to host a world's fair in 1979. In the summer of 1980, the city was awarded the 1986 World Exposition on Transportation and Communication, or simply, Expo 86. Expo 86 was a US$1.1 billion ($3.75 billion, adjusted for inflation to 2022) coming-out party for Vancouver.[11] The theme of Expo 86 was "World in

225

GREATER VANCOUVER AREA RAPID TRANSIT STUDY

SELECTED RAPID TRANSIT NETWORK
SEPTEMBER 1970

PREPARED FOR THE JOINT TRANSPORTATION COMMITTEE
GREATER VANCOUVER REGIONAL DISTRICT &
BRITISH COLUMBIA HYDRO & POWER AUTHORITY

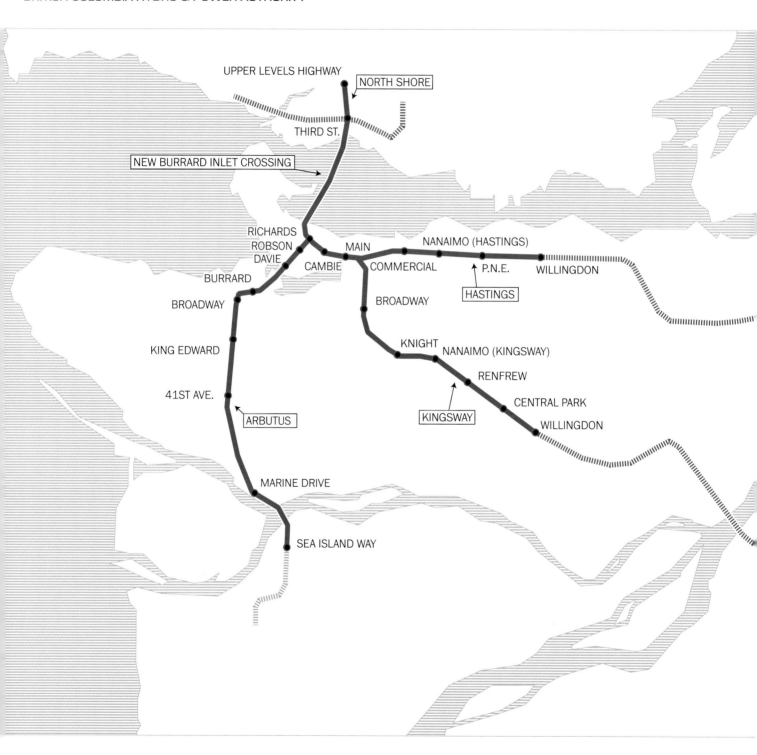

UPPER LEVELS HIGHWAY

NORTH SHORE

THIRD ST.

NEW BURRARD INLET CROSSING

RICHARDS
ROBSON
DAVIE
BURRARD
BROADWAY

KING EDWARD

41ST AVE.

ARBUTUS

MARINE DRIVE

SEA ISLAND WAY

CAMBIE
MAIN
COMMERCIAL
BROADWAY

NANAIMO (HASTINGS)
P.N.E.
HASTINGS
WILLINGDON

KNIGHT
NANAIMO (KINGSWAY)
RENFREW
KINGSWAY
CENTRAL PARK
WILLINGDON

PROPOSED RAPID TRANSIT SYSTEM FOR BUILD-OUT BY 1990

PROPOSED EXTENSIONS - FURTHER STUDY NECESSARY

KEY

22.3 Proposed rapid transit system, 1970.

Motion—World in Touch." For obvious reasons, one couldn't hold such an expo in a city without a rapid transit system. The *Vancouver Sun* snarked in 1980: "Actually it makes a lot of sense to stage an international fair about transportation in our little old timber town in the West Coast rain forest. It would, after all, be too obvious to hold it in cities like Montreal, Toronto, or Edmonton, all of which have efficient transportation systems. . . . But Vancouver is a transportation nightmare and no doubt will still be in 1986."[12] The provincial government of British Columbia was also aware of the problem. British Columbia's premier, Bill Bennett of the conservative Social Credit Party, had to find a transportation system fast.

Like Montreal's Jean Drapeau in the 1960s, Bennett had to find a system that was exceptional. The eyes of the world were going to be on Vancouver, after all. Conveniently, there was a homegrown transit system that could satisfy both the Canadian federal government's need to showcase Canadian technology and the technical requirements of a midsized city: ICTS. The provincial government of Ontario pitched Bennett hard on both the political and technological aspects of ICTS. "Implementation of a high quality, Canadian rapid transit system to link the central [Expo] 86 sites with other downtown attractions would demonstrate the Canadian capability in advanced urban transit technology, and would be an exhibit in motion which provides a reliable and necessary transportation service during the exposition," they wrote.[13]

The government of Ontario was desperate to sell ICTS because nobody wanted to buy it. Millions had been spent with little to show for it. Not even Toronto wanted to use the system, even though the technology was invented in Ontario. The Toronto Transit Commission initially rejected ICTS for the Toronto Subway's Line 3 Scarborough. The provincial government had to take heavy-handed measures to force the Toronto Transit Commission to adopt the technology.[14]

Luckily for the province of Ontario, Bennett bought ICTS for Vancouver. And once Bennett bought it, the federal and provincial governments moved to bring the doubters in line. Publicly, the federal government supplied financial guarantees to make sure that ICTS would be ready in time for Expo 86. Behind the scenes, Bennett and his allies made it crystal clear that building an ICTS-based elevated was nonnegotiable. Mike Harcourt, mayor of Vancouver at the time, said in retrospect, "One of the shakedowns of it all was that the federal and provincial governments wanted a Canadian system to demonstrate . . . and we ended up with the [ICTS] system."[15]

All's Well That Ends Well

But unlike so many other experimental projects, Vancouver's ICTS, dubbed "SkyTrain" in 1985, did the job remarkably well. Expo 86 went off without a hitch. Even naysayers like Mayor Harcourt were sold. Harcourt was later elected premier of British Columbia in the 1990s. During his premiership from 1991 to 1996, the SkyTrain system continued to expand. ICTS ended up as a commercial success, and it is now in use in a half dozen other

227

Vancouver SkyTrain
and connecting regional transit

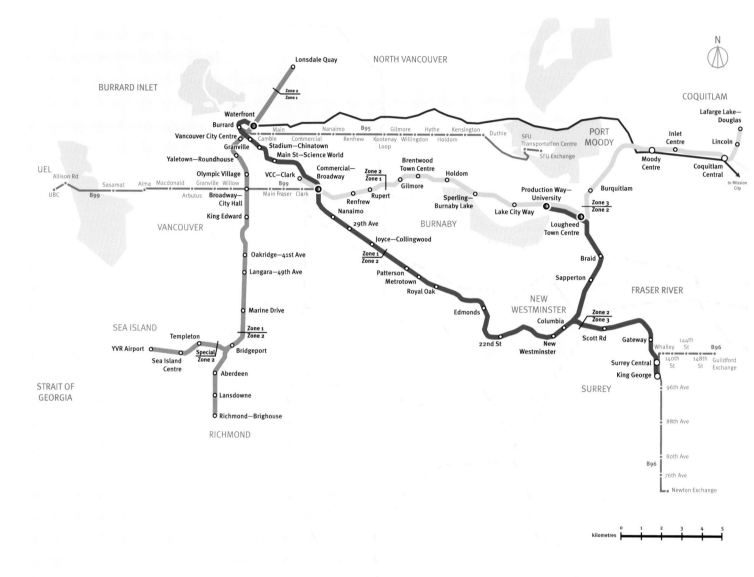

N

BURRARD INLET

NORTH VANCOUVER

COQUITLAM

Lonsdale Quay

Zone 2
Zone 1

Waterfront
Burrard
Vancouver City Centre
Granville

Lafarge Lake—
Douglas

Inlet
Centre
Lincoln

PORT
MOODY

Moody
Centre

Coquitlam
Central

to Mission
City

UEL

Allison Rd
UBC

Sasamat
B99

Alma   Macdonald
Granville Willow
Arbutus

Main   Nanaimo   B95   Gilmore   Hythe   Kensington   Duthie
Cambie   Commercial   Renfrew   Kootenay Willingdon Holdom   SFU
         Loop   Transportation Centre
                                                                    SFU Exchange

Stadium—Chinatown
Main St—Science World

VCC—Clark
B99
Main Fraser   Clark

Yaletown—Roundhouse

Olympic Village

Commercial—
Broadway

Zone 2
Zone 1

Brentwood
Town Centre

Holdom

Burquitlam

Production Way—
University

Zone 3
Zone 2

VANCOUVER

Broadway—
City Hall

King Edward

Rupert
Renfrew
Nanaimo

Gilmore

Sperling—
Burnaby Lake

Lake City Way

BURNABY

Lougheed
Town Centre

29th Ave

Joyce—Collingwood

Oakridge—41st Ave

Langara—49th Ave

Zone 1
Zone 2

Braid

Sapperton

FRASER RIVER

Patterson
Metrotown

Royal Oak

Marine Drive

Edmonds

NEW
WESTMINSTER

Columbia

Zone 2
Zone 3

SEA ISLAND

Templeton

YVR Airport

Sea Island
Centre

Special
Zone 2

Bridgeport

Aberdeen

22nd St

New
Westminster

Scott Rd

Gateway

Whalley

144th
St

B96

Surrey Central

King George

140th
St

148th
St

Guildford
Exchange

STRAIT OF
GEORGIA

Lansdowne

Richmond—Brighouse

RICHMOND

SURREY

96th Ave

88th Ave

80th Ave

B96

76th Ave

Newton Exchange

kilometres   0   1   2   3   4   5

Rail	Ferry	Express Bus	Commuter Rail
Canada Line	SeaBus	B-Line	West Coast Express
Expo Line			
Millennium Line			

22.4 SkyTrain and connecting
     services, 2022.

places outside Canada. ICTS still suffers for being proprietary technology, but the system is basically sound.

Since its inception, SkyTrain has expanded steadily over the past three and a half decades, and it is very popular among Vancouverites (fig. 22.4 shows the system in 2022). SkyTrain is practically a unicorn among North American metro systems, because it's an elevated that nearly everybody likes. It was possible to get to that point only because of the unusual confluence of political circumstances that virtually dictated building an elevated. After all, not every city that wants to build an elevated metro can count on having a world's fair as an excuse.

ROCKVILLE

Maryland

Virginia

Capital Transit Co., 1942
Proposed underground trolley lines, 1944

SILVER SPRING

HYATTSVILLE

RESTON

TYSONS

WASHINGTON D.C.

ARLINGTON

FAIRFAX CITY

ALEXANDRIA

Proposed subway system, 1962

Washington Metro, 2022

N

Washington D.C.

5 miles/8 km

WASHINGTON, DC

The Freeway Revolt and the Creation of Metro

23.1 Map of greater Washington, DC.

The United States embraced the freeway like nowhere else in the industrialized world. In the decades after World War II, the United States spent immense amounts of money bulldozing its cities to optimize them for the car. This happened because of a complex combination of circumstances seen elsewhere in this book: the United States' tremendous postwar prosperity, subsidized suburban growth, the bad reputations of privately owned transit companies, the strength of the automobile lobby, racial strife, and the rush to rid cities of "blight" that characterized the spirit of the age. No matter what the specific pressures were in a city, the end result was usually the same. American cities shredded their cores so that suburbanites could drive downtown as quickly as possible.

Ironically, the national capital was mostly spared this fate, even though Washington, DC (fig. 23.1), came of age during the freeway era. Now, in the 2020s, the District of Columbia has less freeway mileage within its city limits than any other major city in the Northeast. The city also has the second-busiest metro system in the United States, the Washington Metro. Unusually for an American city, even the suburban metro stations are surrounded by dense commercial and residential development. This atypical situation came to pass because of one of North America's most successful freeway revolts. It happened in spite of the objections of stubborn, obstructive members of Congress.

The Growing Government Colossus

The initial development of Washington, DC, followed the patterns seen elsewhere in this book, with 19th-century steam railways and horsecars giving way to the electric streetcar around 1900. During this period, the federal bureaucracy was still relatively compact, and the nation's capital was of less commercial importance than nearby Baltimore. (Baltimore had a deepwater port and better inland transportation connections than Washington.) Eventually, the streetcar system was unified under the umbrella of the privately owned Capital Transit Company.

Unusually, during the early streetcar era, the federal government built a small subway system to connect the various congressional buildings, the Capitol Subway. The half-mile Capitol Subway was for the exclusive use of congresspeople, senators, and their staff, shuttling them between the various Senate and House office buildings in downtown Washington. There were calls as early as 1909 for the system to be extended to cover most of downtown. In that year the *Washington Post* pushed a subway under the headline "Why Not a Real Subway System for Washington?" The *Post* argued that an extension of the Capitol Subway would help "lift the nation's Capital City out of the ruck of mediocrity."[1] Nothing came of it. Money was not forthcoming, and the transit system remained surface-only (fig. 23.2).

During World War II, the federal government began to look again into building downtown transit tunnels. The district's population had greatly swelled during the war, with the influx of war workers causing a housing shortage. The growth of the federal bureaucracy had made traffic congestion ever worse. Moreover, rubber and gasoline rationing meant that its streetcar system once again became the city's the primary means of transportation after decades of slow decline. The result was a plan in 1944 to build streetcar tunnels through downtown. Streetcars would continue on the surface in outer areas (fig. 23.3). In other words, it would create a system similar to the subway-surface trolley lines in Philadelphia and what is now the Green Line in Boston. The 1944 plan never gained sufficient political traction and was ultimately shelved. A 1959 plan by Capital Transit's owner to build a 116-mile monorail met a similar fate.

In the years immediately following the war, the automobile was in clear ascendancy. Federal politicians and urban planners looked to a dramatically different tool to alleviate the city's traffic congestion: the freeway. Unusually, the battle over Washington's freeway infrastructure played out in the halls of Congress, as opposed to the state, local, and provincial halls of power that take up so much of this book. This is because Washington was not self-governing between 1874 and 1973. During this period, the federal government directly administered the District of Columbia. Residents were unable to elect their own mayor or city councilors.[2] Direct federal control meant that congressmembers and even the president of the United States regularly involved themselves in the day-to-day minutiae of local government. It also meant that

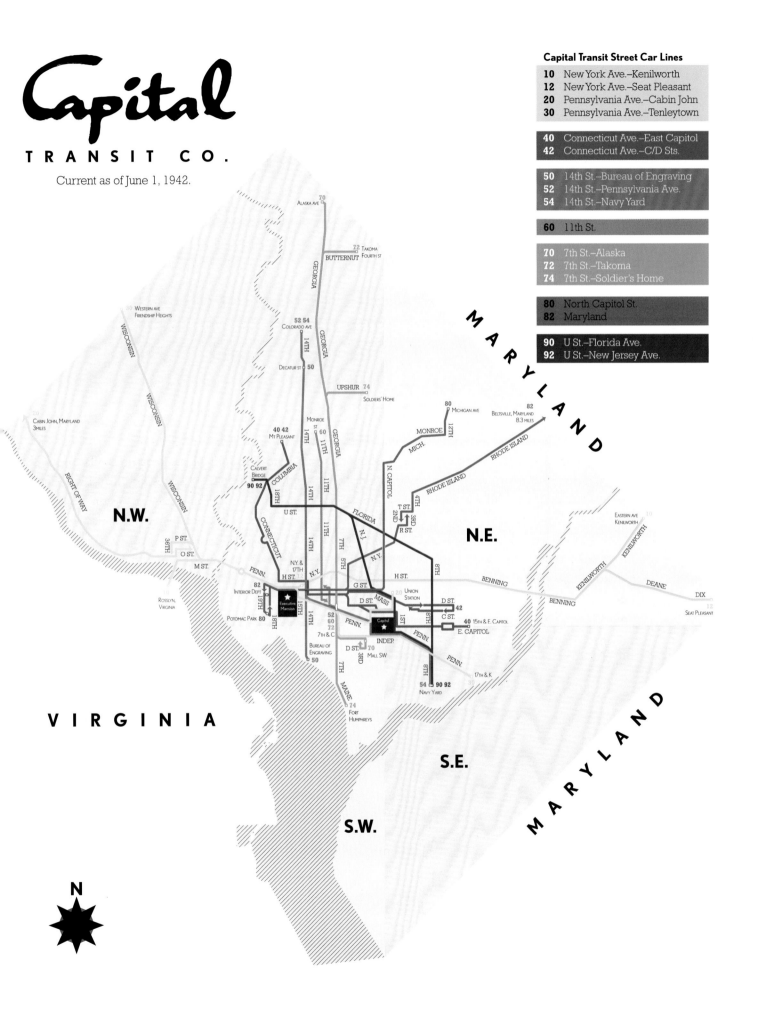

WASHINGTON D.C. SUBWAY & STREETCAR ROUTES
FINAL PROPOSAL - OCTOBER 1944

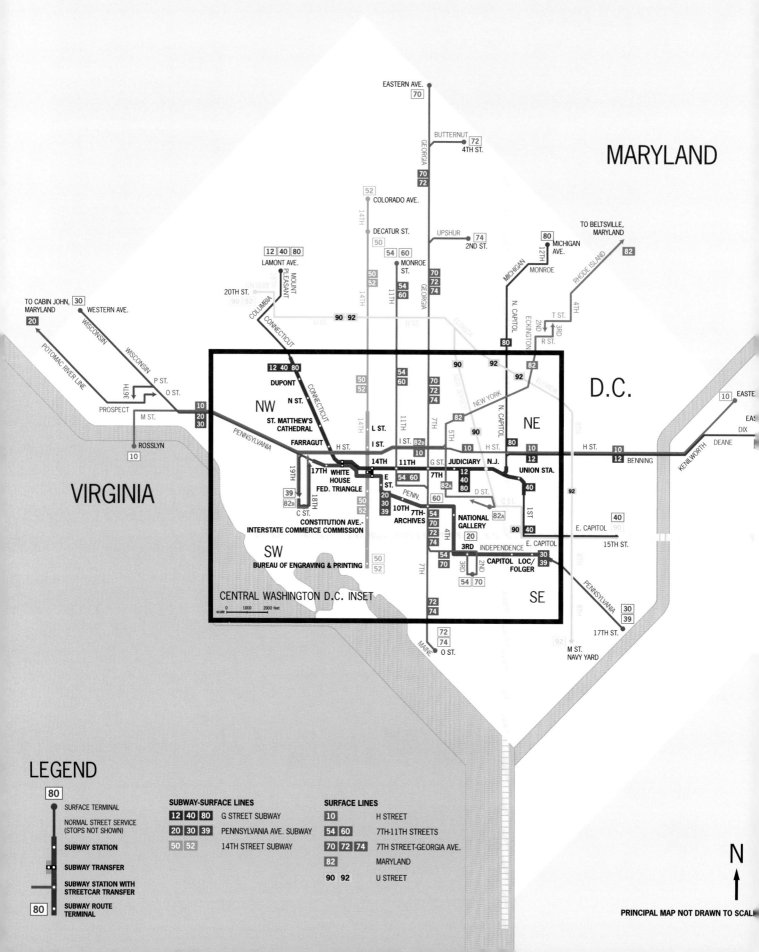

MARYLAND

D.C.

VIRGINIA

EASTERN AVE.
70

BUTTERNUT
72
4TH ST.

GEORGIA

70
72

52

COLORADO AVE.

14TH

DECATUR ST.
50

UPSHUR
74
2ND ST.

TO BELTSVILLE,
MARYLAND

MONROE
ST.
54 60

70
72
74

MICHIGAN
AVE.
80

MICHIGAN

MONROE

RHODE ISLAND

4TH

82

12 40 80
LAMONT AVE.

MOUNT PLEASANT

CALVERT

50
52

54
60

GEORGIA

11TH

14TH

80
N. CAPITOL

ECKINGTON

2ND

3RD

T ST.
R ST.

TO CABIN JOHN,
MARYLAND
30
WESTERN AVE.

20

WISCONSIN

WISCONSIN

20TH ST.
90 92

COLUMBIA

CONNECTICUT

U ST.

U ST.

90 92

FLORIDA

90

92

92

82

D.C.

10
EASTE

POTOMAC RIVER LINE

36TH

P ST.
O ST.

PROSPECT

M ST.

10
20
30

12 40 80
DUPONT

N ST.

ST. MATTHEW'S
CATHEDRAL

FARRAGUT

CONNECTICUT

14TH

50
52

L ST.

I ST.

H ST.

54
60

11TH

7TH

I ST. 82B

5TH

82

NEW YORK

NEW JERSEY

90

NE

N. CAPITOL

FLORIDA

82

10

80

10

H ST.

12

10

BENNING

EAS

DIX

DEANE

KENILWORTH

ROSSLYN
10

NW

PENNSYLVANIA

19TH

39
82B

18TH

17TH
WHITE
HOUSE
FED. TRIANGLE

C ST.

14TH

E
ST.

50
52

11TH

54 60

G ST.

70
72
74

10

PENN.

JUDICIARY N.J.

12
40
80

82A

D ST.

C ST.

UNION STA.

40

82A

1ST

1ST

90

90

40

40

E. CAPITOL
90

15TH ST.

H ST.

12

10

40

92

CONSTITUTION AVE.-
INTERSTATE COMMERCE COMMISSION

SW
BUREAU OF ENGRAVING & PRINTING

20
30
39

10TH
7TH-
ARCHIVES

7TH

54
70

4TH

60

NATIONAL
GALLERY

20
3RD
INDEPENDENCE

E. CAPITOL

54
70

3RD

2ND

CAPITOL LOC/
FOLGER

30
39

PENNSYLVANIA

8TH

E. CAPITOL

30
39

17TH ST.

54 70

50
52

CENTRAL WASHINGTON D.C. INSET

scale   0   1000   2000 feet

MAINE

72
74

O ST.

72
74

92

M ST.
NAVY YARD

SE

LEGEND

80

● SURFACE TERMINAL

NORMAL STREET SERVICE
(STOPS NOT SHOWN)

SUBWAY STATION

SUBWAY TRANSFER

SUBWAY STATION WITH
STREETCAR TRANSFER

80 SUBWAY ROUTE
TERMINAL

SUBWAY-SURFACE LINES

12 40 80   G STREET SUBWAY

20 30 39   PENNSYLVANIA AVE. SUBWAY

50 52   14TH STREET SUBWAY

SURFACE LINES

10   H STREET

54 60   7TH-11TH STREETS

70 72 74   7TH STREET-GEORGIA AVE.

82   MARYLAND

90 92   U STREET

N

PRINCIPAL MAP NOT DRAWN TO SCAL

23.3 Proposed downtown subways, 1944.

Inside the Beltway

By the 1950s, the freeway-building process had already begun nationwide. The Federal-Aid Highway Act of 1956 obligated the federal government to pay for 90 percent of construction costs of new freeways. State and local authorities were on the hook for the remaining 10 percent. These state and local governments wasted no time jumping on this gravy train, laying waste to entire city neighborhoods to build urban freeways. This aggressive remaking of the city was unique to the United States. In other advanced countries, governments generally built freeways around existing city cores rather than through them. For example, there is little or no freeway mileage inside of Madrid's Autopista de Circunvalación M-30, London's M-25 Motorway, or Paris's Boulevard Périphérique. In greater Washington, the plans were for an orbital beltway through the Maryland and Virginia suburbs. Radial freeways would carry motor vehicle traffic into Washington proper. A domino-shaped freeway loop would surround the White House and the Capitol.[3]

The Capital Beltway was approved and built without too much fuss, as were the radial freeways through the Maryland and Virginia suburbs. Many of these areas were relatively sparsely populated. But when freeway work began in the built-up core of the District of Columbia, the populace fought back. Bureaucratic opposition mounted and counterproposals abounded.[4] The most popular option among the alternatives was a rail rapid transit system. Feeling the political heat from Washingtonians, Congress punted and established the National Capital Transportation Agency (NCTA) in 1960. NCTA had a mandate to plan for both freeways and mass transit. Congress also placed a five-year moratorium on new freeway construction through the center of the District of Columbia. The team led by NCTA's head, Darwin Stolzenbach, spent two years assembling a system of freeways and metro lines for the region (fig. 23.4). NCTA's plans required significant lip service to the period's freeway mania.[5] Even so, Stolzenbach's plan got him into hot water with pro-freeway forces because of its emphasis on downtown metro lines, instead of downtown freeways.[6]

Support for rapid transit grew over the course of the 1960s. By 1968, President Lyndon Johnson had approved the construction of a mass transit system and created an authority to run it, the Washington Metropolitan Area Transit Authority (WMATA—pronounced *wah-mah-tah*, rhyming with *piñata*). But even so, the nascent Metro system was still held hostage to the parochial interests of pro-freeway members of Congress.

White Man's Roads through
Black Men's Homes

The bottleneck was Congressman William Natcher, of Warren County, Kentucky, contemporary population 45,491. Natcher had been appointed as

235

RAPID TRANSIT SYSTEM RECOMMENDED
FOR THE NATIONAL CAPITAL REGION

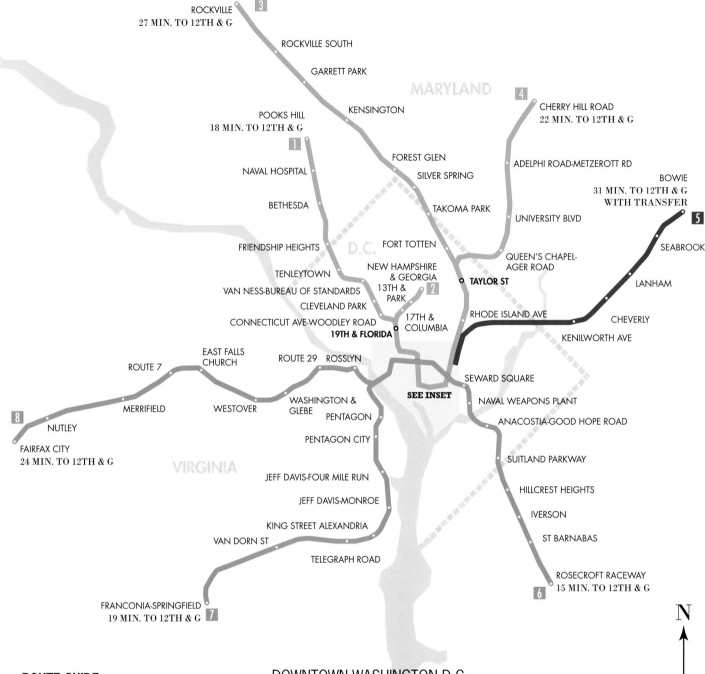

ROCKVILLE
27 MIN. TO 12TH & G

ROCKVILLE SOUTH

GARRETT PARK

MARYLAND

KENSINGTON

CHERRY HILL ROAD
22 MIN. TO 12TH & G

POOKS HILL
18 MIN. TO 12TH & G

ADELPHI ROAD-METZEROTT RD

NAVAL HOSPITAL

FOREST GLEN
SILVER SPRING

BOWIE
31 MIN. TO 12TH & G
WITH TRANSFER

BETHESDA

TAKOMA PARK

UNIVERSITY BLVD

SEABROOK

FRIENDSHIP HEIGHTS

D.C.

FORT TOTTEN

QUEEN'S CHAPEL-
AGER ROAD

LANHAM

TENLEYTOWN

NEW HAMPSHIRE
& GEORGIA
13TH &
PARK

TAYLOR ST

VAN NESS-BUREAU OF STANDARDS

CLEVELAND PARK

RHODE ISLAND AVE

CHEVERLY

CONNECTICUT AVE-WOODLEY ROAD

17TH &
COLUMBIA

KENILWORTH AVE

19TH & FLORIDA

EAST FALLS
CHURCH

ROUTE 29   ROSSLYN

ROUTE 7

SEWARD SQUARE

SEE INSET

MERRIFIELD

WESTOVER

WASHINGTON &
GLEBE
PENTAGON

NAVAL WEAPONS PLANT

NUTLEY

ANACOSTIA-GOOD HOPE ROAD

FAIRFAX CITY
24 MIN. TO 12TH & G

VIRGINIA

PENTAGON CITY

SUITLAND PARKWAY

JEFF DAVIS-FOUR MILE RUN

HILLCREST HEIGHTS

JEFF DAVIS-MONROE

IVERSON

KING STREET ALEXANDRIA

ST BARNABAS

VAN DORN ST

TELEGRAPH ROAD

ROSECROFT RACEWAY
15 MIN. TO 12TH & G

FRANCONIA-SPRINGFIELD
19 MIN. TO 12TH & G

N

ROUTE GUIDE

VIA DOWNTOWN SUBWAY "A"
6 HENSON CREEK - ANACOSTIA
7 SPRINGFIELD - ALEXANDRIA
8 ROUTE 66 - ROSLYN

VIA DOWNTOWN SUBWAY "B"
1 NORTHWEST - BETHESDA
2 COLUMBIA HEIGHTS - PETWORTH
3 ROCKVILLE - SILVER SPRING
4 ROUTE 95 - QUEEN'S CHAPEL

VIA PENNSYLVANIA R. R.
5 BOWIE - CHEVERLY

DOWNTOWN WASHINGTON D.C.

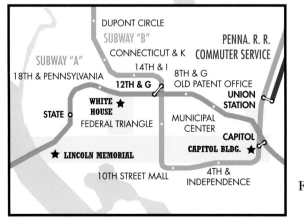

DUPONT CIRCLE

SUBWAY "B"

PENNA. R. R.
COMMUTER SERVICE

CONNECTICUT & K

SUBWAY "A"

14TH & I

18TH & PENNSYLVANIA

8TH & G
OLD PATENT OFFICE

12TH & G

UNION
STATION

STATE

WHITE
HOUSE

FEDERAL TRIANGLE

MUNICIPAL
CENTER

CAPITOL

LINCOLN MEMORIAL

CAPITOL BLDG.

10TH STREET MALL

4TH &
INDEPENDENCE

**NATIONAL CAPITAL
TRANSPORTATION AGENCY**
TRANSPORTATION IN THE
NATIONAL CAPITAL REGION:
FINANCE AND ORGANIZATION
NOVEMBER 1, 1962

23.4 NCTA's proposed subway and
      commuter rail lines, 1962.

chair of the House Appropriations Subcommittee for the District of Columbia in 1961. This position gave him a de facto veto over the District's finances. Natcher wanted to ram as many freeways through the District of Columbia as possible. In figure 23.5, a map of the proposed downtown freeways is overlaid on today's Metro.

Washingtonians largely hated the idea of more freeways through the city. They showed an uncommon degree of racial solidarity in opposition to those freeways. In many other cities in this book, support for freeways often broke down on racial lines, and the targeted demolition of minority neighborhoods proceeded apace. Like Detroit and Atlanta, whites were leaving the District of Columbia for the suburbs, and a major demographic shift was underway. Washington proper changed from being 65 percent white, 35 percent black in 1950, to 28 percent white, 71 percent black in 1970. But unlike Detroit and Atlanta, Washingtonians across political and racial lines maintained a relatively united opposition to urban freeway construction. Freeway opponents included both the white, wealthy residents of Georgetown and the working- and middle-class blacks who made up the bulk of Washington's population at the time. Famously, anti-freeway forces put up a banner in front of houses that had been condemned for the North Central Freeway reading: "White Man's Roads through Black Men's Homes."[7] (In figure 23.5, the North Central Freeway parallels the eastern branch of the modern Red Line.)

But this local opposition was of no import to Natcher. He was unmoved, and he raised the stakes even further in mid-1969. If Congress didn't approve his downtown freeway network, starting with a bridge over the Potomac River called the Three Sisters Bridge, he would block funding for Metro.[8] The city council—at the time appointed by the president—approved the Three Sisters Bridge in August 1969. All hell broke loose. The council meeting turned into a full-scale brawl. The police had to intervene, making 14 arrests.[9] Bridge construction began in September, leading to months of protests. All through September and October, Georgetown University students, backed by anti-freeway activists, tried to occupy the construction site. Police arrested over 200 demonstrators.[10] DC's two major newspapers, the *Washington Star* and the *Washington Post*, had a field day, putting pictures of the demonstrations on their front pages. The protests ended in November only because persons unknown tried to firebomb the construction site with Molotov cocktails.[11]

If at First You Don't Succeed, Try, Try Again

With direct action failing, the protesters sued in court, backed by environmental groups. A federal judge issued an injunction preventing the bridge from being built. With the Three Sisters Bridge in judicial limbo, a two-year stalemate ensued. Natcher continued to block Metro funding, but construction on his pet bridge project was also halted. President Richard Nixon got personally involved in late 1971. Nixon began to pressure Natcher's congressional colleagues, setting up a

237

WASHINGTON METRO
and unbuilt downtown freeways

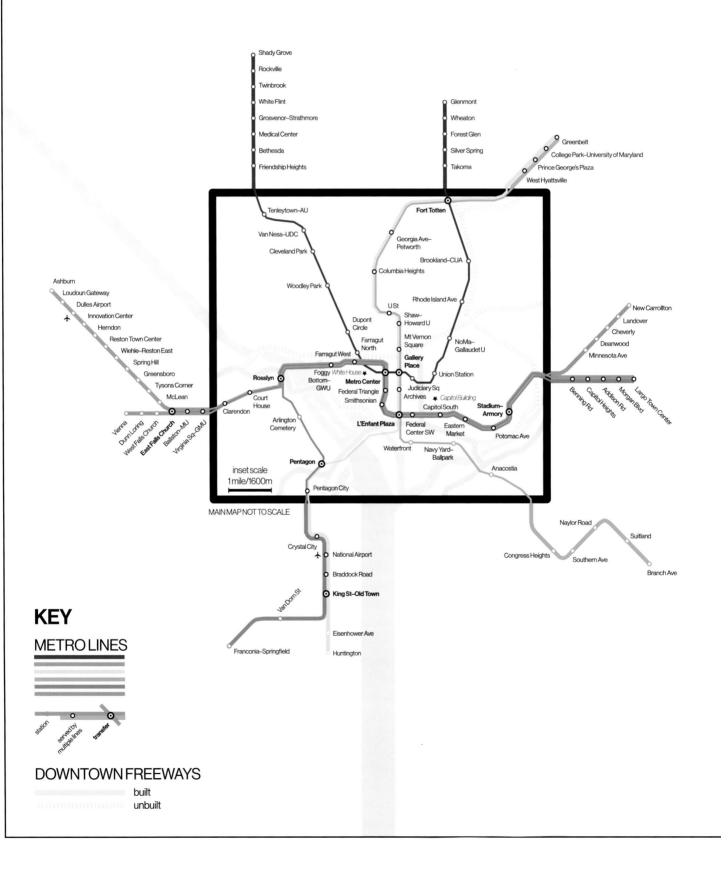

Shady Grove
Rockville
Twinbrook
White Flint
Grosvenor–Strathmore
Medical Center
Bethesda
Friendship Heights

Glenmont
Wheaton
Forest Glen
Silver Spring
Takoma

Greenbelt
College Park–University of Maryland
Prince George's Plaza
West Hyattsville

Tenleytown–AU
Van Ness–UDC
Cleveland Park
Woodley Park

Fort Totten

Georgia Ave–Petworth
Brookland–CUA
Columbia Heights
Rhode Island Ave

Ashburn
Loudoun Gateway
Dulles Airport
Innovation Center
Herndon
Reston Town Center
Wiehle–Reston East
Spring Hill
Greensboro
Tysons Corner
McLean

New Carrollton
Landover
Cheverly
Deanwood
Minnesota Ave

U St
Shaw–Howard U
Mt Vernon Square
Gallery Place
NoMa–Gallaudet U

Dupont Circle
Farragut North

Farragut West

Rosslyn

Foggy Bottom–GWU
White House
Metro Center
Federal Triangle
Smithsonian

Union Station

Judiciary Sq
Archives
Capitol Building
Capitol South

Benning Rd
Capitol Heights
Addison Rd
Morgan Blvd
Largo Town Center

Court House
Clarendon
Ballston–MU
Virginia Sq–GMU

Arlington Cemetery

L'Enfant Plaza

Federal Center SW
Eastern Market

Stadium–Armory

Vienna
Dunn Loring
West Falls Church
East Falls Church

Potomac Ave

inset scale
1 mile/1600m

Pentagon

Waterfront
Navy Yard–Ballpark

Anacostia

MAIN MAP NOT TO SCALE

Pentagon City

Crystal City

National Airport
Braddock Road
King St–Old Town

Van Dorn St

Naylor Road
Suitland
Congress Heights
Southern Ave
Branch Ave

Franconia–Springfield

Eisenhower Ave
Huntington

KEY

METRO LINES

station
served by multiple lines
transfer

DOWNTOWN FREEWAYS

built
unbuilt

23.5 Washington Metro, 2022, with
canceled downtown freeways.

showdown in the House of Representatives. Nixon's lobbying led a bipartisan coalition to narrowly overrule Natcher that December, 195–174. The Metro funds were unlocked. Natcher was humiliated on the floor of the House.[12]

After this high-stakes political drama, Metro construction began in stages. The first parts of the system opened in 1976. Stolzenbach was appointed Metro's first general manager. Construction proceeded over the decades, and the last sections of the initial system opened in 2001.

Greater Washington ended up using the Metro to anchor its postwar development, building lots of office buildings and apartments near the train stations—in marked contrast to cities like Dallas, Los Angeles, and Rochester. The Metro has been a remarkable success in this respect. However, the Metro still suffers from the same governance issues that have troubled it since William Natcher's day. Because WMATA is a joint venture between the federal government, the states of Maryland and Virginia, and the District of Columbia, it has been historically difficult to coordinate oversight and provide long-term funding stability. Congress, the legislatures of Virginia and Maryland, and the DC City Council were prone to bickering over who should pay for what. The result was chronic under-

funding and deferral of long-term repairs.

A series of fatal derailments and tunnel fires resulted. To illustrate: the Metro was originally designed for computer-controlled trains. Metro's computer control system was left to decay, leading to a 2009 crash that left nine dead and 80 injured. Since then, the trains have been operated manually. The funding problem was resolved only in 2018, when Virginia, Maryland and the District of Columbia each agreed to set aside a dedicated revenue source for WMATA. This accord has solved most of the funding issues. But significant work in the decade to come is necessary to bring the nearly 50-year-old system to a state of good repair. As of 2022, there is no timeline for the return of full computer control.[13]

As for the Three Sisters Bridge, it was never completed. Hurricane Agnes destroyed the unfinished bridge in June 1972. Construction was never restarted, putting a seemingly permanent halt to new freeways in the District of Columbia. Three Sisters Bridge funding was reallocated for Metro construction instead. Natcher continued to serve in Congress representing Warren County, Kentucky until 1994, when he died in office. Fittingly, the state of Kentucky named a freeway and a highway bridge in his memory.

CONCLUSION

Americans and Canadians who travel to Europe or East Asia are often astounded by the quality of mass transit there and delighted by the beauty and convenience of those cities' walkable, dense neighborhoods. This book is an effort to explain just how North American cities ended up with such sprawl, car dependence, and poor transit. This situation didn't arise out of thin air. Rather, it was the result of a uniquely North American combination of dysfunctional politics, the desire to reshape the metropolis for the automobile, racial strife, and the law of unintended consequences, especially in the middle decades of the 20th century. We still live with the aftereffects of those decisions.

At times it might seem an impossible problem to address, especially in light of a climate crisis which pressures society to switch to sustainable transportation as quickly as possible. But the future is hard to predict, as history teaches us. In 1915, it was unthinkable that the Pacific Electric Railway's grip over Los Angeles would be broken. But 40 years later, state and local governments were negotiating how to close the Red Car system in orderly fashion and replace the Red Cars with freeways. Today, Los Angeles wants more transit. Angelenos have raised their own taxes multiple times to get it. One can see the outlines of a paradigm shift un-

derway, as decarbonization has begun in earnest.

A sea change is also happening in land use policy, as post–World War II suburbia has reached the natural limit of its expansion in many cities. My home state of California made the suburban tract home an integral part of the California Dream in the 1950s. In the early 2020s, the California State Legislature abolished suburban-style single-family zoning statewide.[1] After 75 years of continuous horizontal expansion, California is making major reforms to its land use laws to encourage urban infill near transit, rather than sprawl at the urban periphery. The end results of these reforms are yet to be seen. After all, Rome wasn't built in a day. However, it appears likely that these land use reforms, coupled with large simultaneous investments in local rapid transit, will lead to a resurgence of traditional, walkable cities.

The more tricky question is how to make this transition go faster. Improving transit and building urban infill is often a painful slog in the early 21st century. Examples are everywhere in this book. New York approved the Second Avenue Subway in 2000. It took 17 years to complete its 1.8 mile first phase. The predecessor to Toronto's Ontario Line has been a political football since the 1980s, and the line won't be complete until 2031. Low-cost reforms to improve transit service often fall by the wayside, as with Philadelphia's Regional Rail. Infill, especially in wealthy cities, commonly faces prolonged timelines and persistent delays, as in Los Angeles and San Francisco.

But things don't *have* to be like this.

Today's slow, bureaucratic approach to urban growth was not always normal. After all, New York City's first subway line was approved in 1899, and it opened to the public five years later. The Montreal Metro was approved in 1960, and the initial system was fully operational six years later. Other wealthy countries are also still capable of these types of quick builds. Between 2000 and 2003, Madrid built 25 miles of subway in its southern sections, at a per-mile cost comparable to North American surface light rail.[2] This construction speed and cost is unthinkable in early 21st-century North America. Similarly, European cities stuck to transit-oriented development in a way that isn't the norm in North America anymore. The vast majority of new residential development in Madrid, for instance, is apartment buildings, rowhouses, and duplexes. The North American-style detached single-family home is comparatively uncommon.[3]

It's taken for granted in North America that better transit and infill development will face potentially high costs, slow bureaucracy, and painfully long timelines. But this system of transport and land use wasn't handed down on Mount Sinai—it was the product of decisions made by fallible people of generations past. North America's mass transit and land use problems can be tackled, but it will entail emulating international best practices and learning from past mistakes. It's going to require a long, hard look in the mirror, and the will to change our ways.

Acknowledgments

No book can be written without the help of others, and *The Lost Subways of North America* is no exception. All the same, if there are errors in this book, they are mine alone.

This book was written and illustrated with heavy reliance on primary sources. Many of the primary source documents I've used have been sitting in local government archives for decades, gathering dust, and were retrieved by filing public records requests. The rest have been sourced from contemporary newspapers and guidebooks, with support from the scores of local transit histories. Where good secondary sources exist, I've used them. I've also adapted some of the writing from my blog.

I've made extensive use of official statistics. If a source is not specifically noted, American demographic data come from the United States Census Bureau's decennial census, while American employment, gross domestic product, and other economic data come from the Federal Reserve Bank of St. Louis's FRED database. Canadian demographic and economic data are from Statistics Canada. Transit ridership figures are from the American Public Transportation Association's ridership reports.

Additionally, this book could not have come to fruition without the aid of the people who've assisted me along the way. Jarett Walker, Yonah Freemark, and Vukan Vuchic all generously provided me with their expertise in the nuts and bolts of transport planning and operations. Emily Hamilton, Jacob Anbinder, Chris Elmendorf, and Stephen Smith shaped my approach to the land use elements of the book. Eric Goldwyn, Alon Levy, and Elif Ensari's database of transit construction costs was invaluable. Steve Estes, Jason Hung, Eric Chan, Amanda Bakale, Shounak Bagchi, Mat Ahn, and Florence Gin served as local guides in my travels. Larry Bennett, Dennis Judd, Mark Ovenden, Sara Bershtel, Ellen Scordato, and Noah Brick helped me navigate the publishing world. Karlee Finch, Rob Grofe, and Katy Kassing served as sounding boards through this entire process and reviewed my early drafts.

Susannah Engstrom, my editor, was invaluable in making sure that this book truly is the best it could be. My parents, Dan and Lorna, and my brother David have unconditionally supported me as this project has grown into what it is today. I couldn't have done it without you.

Notes

A Brief Primer on Transit and Urban Development

1. Charles Lockwood, *Manhattan Moves Uptown: An Illustrated History* (New York: Barnes & Noble, 1995), 94–96.

2. *Village of Euclid v. Ambler Realty Co.*, 272 U.S. 365 (1926).

3. Andrew H. Whittemore, "Zoning Los Angeles: A Brief History of Four Regimes," *Planning Perspectives* 27, no. 3 (July 2012): 397.

4. Unless specifically cited otherwise, data used in this book comes from official sources, like the U.S. Census Bureau and Statistics Canada.

5. Conor Dougherty, "Build Build Build Build Build Build Build Build Build Build Build Build Build Build," *New York Times*, Feb. 13, 2020, https://www.nytimes.com/2020/02/13/business/economy/housing-crisis-conor-dougherty-golden-gates.html.

Chapter 1

1. W. E. B. Du Bois, *The Souls of Black Folk* (Garden City, NY: Dover Publications, 1994), 47.

2. Rebecca Burns, *Rage in the Gate City: The Story of the 1906 Atlanta Race Riot* (Athens: University of Georgia Press, 2009), 27.

3. Arthur Saltzman and Richard J. Solomon, *Historical Overview of the Decline of the Transit Industry* (Washington, DC: Transportation Research Board, 1972), https://onlinepubs.trb.org/Onlinepubs/hrr/1972/417/417-001.pdf.

4. Metropolitan Atlanta Study Commission, *A Plan and Program of Rapid Transit for the Atlanta Metropolitan Region* (Atlanta: Metropolitan Atlanta Study Commission, 1962).

5. Larry Keating, *Atlanta: Race, Class and Urban Expansion* (Philadelphia: Temple University Press, 2001), 118.

6. Kevin M. Kruse, *Atlanta and the Making of Modern Conservatism* (Princeton, NJ: Princeton University Press, 2013), 116–17.

7. Keating, *Atlanta*, 118.

8. Maddox's racism was his ultimate legacy; at his death in 2003, the headline on his obituary in the *New York Times* read "Lester Maddox, Whites-Only Restaurateur and Georgia Governor, Dies at 87."

9. Urban Mass Transit Administration, U.S. Department of Transportation, *Mass Transit Management: Case Studies of the Metropolitan Atlanta Rapid Transit Authority* (Washington, DC: U.S. Department of Transportation, 1981), I-2.

10. Jason Henderson, "Secessionist Automobility: Racism, Anti-Urbanism, and the Spatial Politics of Automobility in Atlanta, Georgia," *International Journal of Urban and Regional Research* 30, no. 2 (June 2006): 298.

11. Urban Mass Transit Administration, I-2.

12. Urban Mass Transit Administration, I-8.

13. Keating, *Atlanta*, 126.

14. Doug Monroe, "Where It All Went Wrong," *Atlanta Magazine*, August 1, 2012, https://www.atlantamagazine.com/great-reads/marta-tsplost-transportation/.

15. Monroe.

16. Nate Silver, "The Most Diverse Cities Are Often the Most Segregated," *FiveThirtyEight* (blog), May 1, 2015, https://fivethirtyeight.com/features/the-most-diverse-cities-are-often-the-most-segregated/. Silver draws his data from Brown University's American Communities Project and the 2010 U.S. Census.

17. Josh Green, "Could Georgia's Financial Support for MARTA (Gasp) Signal a New Dawn?" *Urbanize*, April 6,

2021, https://atlanta.urbanize.city/post/could-georgias
-financial-support-marta-gasp-signal-new-dawn.

18. These funds were allocated to renovate the Bankhead station, to serve a new Microsoft office campus.

19. For comparison, the San Francisco Municipal Railway received about 10 percent of its budget from the State of California that year.

Chapter 2

1. Between 1920 and 1950, the United States population grew by 42 percent. Over the same period, Boston grew by only 7 percent.

2. Massachusetts Legislative Commission on Rapid Transit, *Report of the Metropolitan Transit Recess Commission* (Boston: Massachusetts Legislative Commission on Rapid Transit, 1945).

3. Emily Sweeney, "Looking Back at Boston's West End, and the Forces That Led to Its Destruction," *Boston Globe*, February 14, 2020, https://www.bostonglobe.com/2020/02 /14/metro/looking-back-bostons-west-end-forces-that-led-its -destruction/.

4. American Institute of Architects, "For Whom the Polls Toll," *AIArchitect This Week*, April 13, 2007, https://info .aia.org/aiarchitect/thisweek07/0413/0413n_polls.htm.

5. Leon Neyfakh, "How Boston City Hall Was Born," *Boston Globe*, February 12, 2012, https://www.bostonglobe .com/ideas/2012/02/12/how-boston-city-hall-was-born /DtfspyXVbKBIKi8iSXHX6J/story.html.

6. Neyfakh.

7. Edward Mason, "Boston City Hall Named World's Ugliest Building," *Boston Herald*, November 15, 2008, https://www.bostonherald.com/2008/11/15/boston-city-hall -named-worlds-ugliest-building/.

8. Boston Transportation Planning Review, *Circumferential Transit Report* (Boston: Boston Transportation Planning Review, 1972).

9. The case that phased in Boston's school desegregation program was *Morgan v. Hennigan*, 379 F. Supp. 410 (D. Mass, 1974).

10. Jack Tager, *Boston Riots: Three Centuries of Social Violence* (Boston: Northeastern University Press, 2001), 197.

11. Bruce Gellerman, "How the Boston Busing Decision Still Affects City Schools 40 Years Later," WBUR, December 19, 2014, https://www.wbur.org/news/2014/06/20/boston-busing-ruling-anniversary.

12. Laura White, Jean-Louis Rochet, Pete Mathias, Kate O'Gorman, and Linda Bilmes, "Connecting the Northeast: A Cost Estimate for the North South Rail Link" (Working Paper No. RWP17-032, Harvard Kennedy School, Cambridge, MA, 2017), 6, http://dx.doi.org/10.2139/ssrn.3028863.

13. Virginia Greiman and Roger David Hand Warburton, "Deconstructing the Big Dig: Best Practices for Mega-Project Cost Estimating" (paper presented at PMI Global Congress 2009—North America, Orlando, FL, October 13, 2009), https://www.pmi.org/learning/library/practices-mega -project-cost-estimating-6668.

14. White et al., "Connecting the Northeast," 6.

15. Annie Weinstock, Walter Hook, Michael Replogle, and Ramon Cruz, *Recapturing Global Leadership in Bus Rapid Transit: A Survey of Select U.S. Cities* (New York: Institute for Transportation and Development Policy, 2011), 46, https://nbrti.org/wp-content/uploads/2017/05/ITDP _report_BRT_rating_2011.pdf.

16. Kristopher Carter, "Equal or Better: The Story of the Silver Line" (master's thesis, Tufts University, 2011), 13, http://hdl.handle.net/10427/011289.

17. Eric Goldwyn, Alon Levy, and Elif Ensari, *The Boston Case: The Story of the Green Line Extension* (New York: New York University Marron Institute of Urban Management, 2020), 8–9, https://marroninstitute.nyu.edu/uploads/content /Boston-Case_The-Story-of-the-Green-Line-Extension.pdf.

18. Goldwyn, Levy, and Ensari, 20–21.

19. Goldwyn, Levy, and Ensari, 23–24.

20. Goldwyn, Levy, and Ensari, 27.

Chapter 3

1. The *Chicago Examiner* editorialized in 1911: "Chicago has set its face toward real transportation reform. It will demolish the loop and substitute subways, with or without the consent and assistance of those who have bought the elevated roads [i.e., railroads] for future profit." "'High Finance' and the Doomed Loop," *Chicago Examiner*, December 5, 1911.

2. Stephen E. Schlickman and Laura Klabunde, *The History of the City of Chicago Central Area Transit Circulation Efforts* (Chicago: National University Rail Center, 2018), 11, https://rosap.ntl.bts.gov/view/dot/61456.

3. The Philadelphia & Reading Railroad (*Reading* rhymes with *bedding*) is the ancestor of the Reading Railroad, which appears in the chapter on Philadelphia.

4. Mark Carlson, *Causes of Bank Suspensions in the Panic of 1893* (Washington, DC: Federal Reserve Board, 2002), 1, https://www.federalreserve.gov/pubs/feds/2002/200211/200211pap.pdf.

5. Christina Romer, "Spurious Volatility in Historical Unemployment Data," *Journal of Political Economy* 94, no. 1 (1986): 31, https://doi.org/10.1086/261361. There are some disputes as to the exact unemployment rate, as Romer notes. Such statistical variability is unavoidable when attempting to reconstruct the late 19th century.

6. Sidney I. Roberts, "Portrait of a Robber Baron: Charles T. Yerkes," *Business History Review* 35, no. 3 (1961): 344, https://doi.org/10.2307/3111475.

7. Schlickman and Klabunde, *Central Area Transit Circulation Efforts*, 11.

8. Schlickman and Klabunde, 12.

9. Roberts, "Portrait of a Robber Baron," 357 and 371.

10. R. F. Kelker Jr., *Report on a Physical Plan for a Unified Transportation System for the City of Chicago* (Chicago: Committee on Transportation of the City Council of the City of Chicago, 1923), 35.

11. Chicago Urban Transportation District, *Annual Report* (Chicago: Chicago Urban Transportation District, 1976).

12. Sandy Riemer, "More Road Work Planned," *Arlington Heights Daily Herald*, September 24, 1979.

13. Schlickman and Klabunde, *Central Area Transit Circulation Efforts*, 11.

14. Schlickman and Klabunde, 51–52.

15. Robert C. Herguth, "CTA Floats Circle Line Plan," *Chicago Sun-Times*, March 11, 2002.

16. "Circle Line," Chicago Transit Authority, https://www.transitchicago.com/planning/circle/.

17. I have drawn L ridership figures from the City of Chicago's open data portal, at https://data.cityofchicago.org /Transportation/CTA-Ridership-Annual-Boarding-Totals/ w8km-9pzd.

18. Chicago Transit Authority, "CTA Charts Progress of the Brown Line Expansion with a New Web Site Link," news release, December 31, 2002, https://www.transitchicago.com/cta-charts-progress-of-the-brown-line-expansion-with-a-new-web-site-link/.

19. Daniel Hautzinger, "When the Green Line Shut Down for More Than Two Years," WTTW, February 26, 2020, https://interactive.wttw.com/playlist/2020/02/26/green-line-rehabilitation.

Chapter 4

1. Allen J. Singer, *The Cincinnati Subway: History of Rapid Transit* (Mount Pleasant, SC: Arcadia Publishing, 2003), 35.

2. Andrew J. Hawkins, "Train to Nowhere," *The Verge*, August 10, 2016, https://www.theverge.com/2016/8/10/12411632/public-transportation-failures-america-cincinnati-subway.

3. "Estimated Cost of Loop System: Expert Comparison of Relative Expenditure," *Cincinnati Commercial Tribune*, April 16, 1916.

4. "Traction Line Equipment Is Sold for Junk Heap," *Cincinnati Commercial Tribune*, February 1, 1920.

5. "Rapid Transit Body Dissolves after 14 Years," *Cincinnati Commercial Tribune*, January 1, 1929.

6. Jake Mecklenborg, "MetroMoves: A Decade Later," *UrbanCincy*, November 26, 2012, https://www.urbancincy.com/2012/11/metromoves-a-decade-later/.

7. Reed Albergotti and Cameron McWhirter, "A Stadium's Costly Legacy Throws Taxpayers for a Loss," *Wall Street Journal*, July 12, 2011, https://www.wsj.com/articles/SB10001424052748704461304576216330349497852.

8. James Pitcher, "Rail Side Takes No for an Answer," *Cincinnati Enquirer*, November 7, 2002.

9. Pat LaFleur, "Cincinnati Streetcar: The City Now Runs the 'Connector,' but What Does That Actually Mean?," WCPO 9 Cincinnati, January 4, 2020, https://www.wcpo.com/news/transportation-development/move-up-cincinnati/cincinnati-streetcar-the-city-now-runs-the-connector-but-what-does-that-actually-mean.

10. Jay Hanselman, "Issue 7 Transit SORTA Levy Passes," 91.7 WVXU, May 14, 2020, https://www.wvxu.org/local-news/2020-05-14/issue-7-transit-sorta-levy-passes.

Chapter 5

1. In figure 19.6, these are the aboveground sections of the F and N in the northeastern corner of San Francisco.

2. San Francisco Municipal Transportation Agency, *Short Range Transit Plan: Fiscal Year 2019–Fiscal Year 2030* (San Francisco: San Francisco Municipal Transportation Agency, December 3, 2019), 41, https://www.sfmta.com/sites/default/files/reports-and-documents/2019/12/sfmta_shortrange2019_1205_sglpg_0.pdf.

3. Tom Meyer, "Ridership in Question on Waterfront 'Ghost Train,'" WKYC 3, June 11, 2014, https://www.wkyc.com/article/news/local/cuyahoga-county/investigator-ridership-in-question-on-waterfront-ghost-train/95-315850177.

4. Mark Naymik, "RTA's Waterfront Line, aka Ghost Train, Should Be the First Service Trimmed to Help Close Budget Shortfall," *Cleveland.com*, March 11, 2016, https://www.cleveland.com/naymik/2016/03/rtas_waterfront_line_aka_ghost.html. It's hard to make a direct ridership comparison between San Francisco and Cleveland. The San Francisco data doesn't isolate the Embarcadero segments, but Cleveland isolates the Waterfront Line data. That said, even the much-maligned Detroit People Mover, discussed in chapter 7, carried 4,600 passengers per day in 2019—11.5 times the ridership of the Waterfront Line.

5. Brian Duffy, "RTA'S Waterfront Line Will Remain Shut Down through the Brown's Season," WOIO 19 News, August 17, 2022, https://www.cleveland19.com/2022/08/18/rtas-waterfront-line-will-remain-shut-down-through-browns-season/.

6. David G. Molyneux and Sue Sackman, eds., *An Informal History of Shaker Heights* (Shaker Heights, OH: Shaker Heights Public Library, 1987), 49.

7. J. Mark Souther, "A $35 Million 'Hole in the Ground': Metropolitan Fragmentation and Cleveland's Unbuilt Downtown Subway," *Journal of Planning History* 14, no. 3 (August 2015): 184–85, http://journals.sagepub.com/doi/abs/10.1177/1538513214545849.

8. Ernest Holsendolph, "4 Cities Will Get $220 Million for 'People Movers,'" *New York Times*, December 23, 1976.

9. Ernest Holsendolph, "U.S. Gift Finds a Reluctant Mayor," *New York Times*, December 16, 1977.

10. Jensen Werley, "How the Skyway's Counterpart Is a Big Success in a Nearby City," *Jacksonville Business Journal*, September 21, 2015, https://www.bizjournals.com/jacksonville/news/2015/09/21/how-the-skyways-counterpart-is-a-big-success-in-a.html.

11. Larry Hannan, "After 20 Years, Jacksonville Skyway Remains a Punchline," *Florida Times-Union*, September 5, 2010, https://www.jacksonville.com/story/business/transportation/2010/09/05/after-20-years-jacksonville-skyway-remains-punchline/15932028007/; Laura Bliss, "The Joys of Detroit's Ridiculous People Mover," *Bloomberg*, November 2, 2018, https://www.bloomberg.com/news/articles/2018-11-02/detroit-people-mover-bad-transit-but-great-attraction.

12. David Stradling and Richard Stradling, "Perceptions of the Burning River: Deindustrialization and Cleveland's Cuyahoga River," *Environmental History* 13, no. 3 (July 2008): 515–35, https://www.jstor.org/stable/25473265.

13. Kevin Smith, "Flats on Fire: As Downtown Cleveland Grows, So Will the Flats. What Does That Mean for the City?," *Cleveland Business Journal*, April 4, 2022, https://www.bizjournals.com/cleveland/news/2022/04/04/development-downtown-cleveland-flats.html.

14. Through the 2022–2023 NFL season, the Browns have not hosted a single playoff game at Browns Stadium, due to the team's poor on-field performance.

15. Patrick Cooley, "Why Aren't More Non-Browns Events Held at FirstEnergy Stadium?," *Cleveland.com*, January 19, 2017, https://www.cleveland.com/entertainment/2017/01/why_arent_there_more_non-brown.html.

16. "Full Calendar," Nationwide Arena, https://www.nationwidearena.com/calendar. To derive this figure, I assembled a list of all non-hockey ticketed events on Nationwide Arena's event calendar from October 2022 to March 2023.

17. Dillon Stewart and Henry Palattella, "Cleveland Browns Tailgate Guide: A Fan's Guide to Tailgating Downtown," *Cleveland Magazine*, September 16, 2022,

https://clevelandmagazine.com/in-the-cle/sports/articles /cleveland-browns-tailgate-guide-a-seasoned-fans-guide-to -the-muni-lot-and-tailgating-downtown.

18. Ken Prendergast, "Will Cleveland Ever Develop Its Lakefront? New Plans Are a Step Closer," *Cleveland Magazine*, July 5, 2022, https://clevelandmagazine.com/in-the-cle /the-read/articles/will-cleveland-ever-develop-its-lakefront -new-plans-are-a-step-closer.

19. Alison Grant, "What Can Be Done with Cleveland Hopkins' Vacant Concourse D? A Look to Other Airports Might Offer Clues," *Cleveland Plain Dealer*, April 18, 2014, https://www.cleveland.com/metro/2014/04/cleveland _hopkins_joins_airpor.html.

20. Prendergast, "Will Cleveland Ever Develop Its Lakefront?"

21. Prendergast.

Chapter 6

1. My original source, Weichsel, treats the Junius Heights and Tyler lines as two separate lines run as one continuous service. *Weichsel's Map and Guide of Dallas* (Dallas: C. Weichsel Co., 1919), 31–32 and 39.

2. I've calculated this by using 2020 American Community Survey data from Census Block Group 47 of Dallas County, which covers Bishop Arts and a slight bit of the surrounding area, taking the population of 2551, and dividing it by the land area of 0.36 square miles.

3. Most of the area around Parker Road Station is zoned either corridor commercial (Plano Municipal Code sec. 10.600) or light commercial (Plano Municipal Code sec. 10.500), both of which bar residential use.

4. Teri Webster, "Affordable Housing Complex to Be Built Near Parker Road DART Station after Plano Sells Land," *Dallas Morning News*, November 2, 2020, https://www.dallasnews.com/news/2020/11/02/plano-sells -land-for-affordable-housing-complex-to-be-built-near -parker-road-dart-station/.

5. For example, per commercial real estate site LoopNet, the modest two-story brick office building at 408 West Eighth Street, Bishop Arts, puts 8,900 square feet of building onto a 5,445 square foot lot.

6. See, e.g., Plano Municipal Code sec. 10.600.3.

7. Plano Municipal Code sec. 10.600.3.

8. Plano Municipal Code sec. 16.700 et seq.

9. Donald C. Shoup and Don H. Pickrell, "Problems with Parking Requirements in Zoning Ordinances," *Traffic Quarterly* 32, no. 4 (October 1978): 547–49, http://shoup.bol.ucla.edu/ ProblemsWithParkingRequirementsInZoningOrdinances .pdf. The total average space needed per parking space varies, depending on parking lot layout and type; efficiently designed surface lots can get this number down to 330 square feet per space, while a multilevel garage can use as much as 500 square feet per space.

Chapter 7

1. Karen McVeigh, "Detroit Mired in Fresh Controversy over Sale of 60,000-Piece Art Collection," *The Guardian*, August 14, 2013, https://www.theguardian.com/world/2013/ aug/14/detroit-controversy-art-collection-sale-suburbs.

2. Chrysler Group, "Imported from Detroit," YouTube video, 2:18, from a commercial played during the February 6, 2011, Super Bowl, posted by "Cause Marketing," October 30, 2016, https://www.youtube.com/watch?v=mYsFUFgOEmM.

3. Seth Schindler, "Detroit after Bankruptcy: A Case of Degrowth Machine Politics," *Urban Studies* 53, no. 4 (2014): 818–36, https://doi.org/10.1177/0042098014563485.

4. Carlos Waters, "How Detroit Moved On from Its Legendary Bankruptcy," *CNBC*, August 10, 2022, https://www.cnbc.com/2022/08/10/detroit-real-estate -developers-rebuild-city-amid-budget-shortfalls.html.

5. Paige Williams, "Drop Dead, Detroit!," *New Yorker*, January 27, 2014, https://www.newyorker.com/magazine /2014/01/27/drop-dead-detroit.

6. A few downtown express buses were exempted from this policy. Leonard N. Fleming, "SMART Urged to Change Boarding Policy in Detroit," *Detroit News*, January 11, 2015, https://www.detroitnews.com/story/news/local/wayne-county /2015/01/10/smart-urged-change-boarding-policy-detroit /21570061/. Data on worker commuting patterns is drawn from the U.S. Census Bureau's Longitudinal Employer- Household Dynamics data set covering 2015–2019, specifically, the LEHD Origin-Destination Employment Statistics.

7. Ross Schram, "Detroit Takes Over Its Street Railways," *National Municipal Review* 11, no. 7 (July 1922): 190.

8. Interurban service remained privately controlled.

9. Laurence Michelmore, "Detroit—A Tale of Two Cities," *National Municipal Review* 29, no. 11 (November 1940): 720.

10. Thomas J. Sugrue, *The Origins of the Urban Crisis: Race and Inequality in Postwar Detroit*, rev. ed. (Princeton, NJ: Princeton University Press, 2015), 64.

11. An excellent summary of how this process happened elsewhere in the United States is contained in Richard Rothstein, *The Color of Law* (New York: W. W. Norton, 2017).

12. Sugrue, *Origins of the Urban Crisis*, 47–48.

13. The Detroit monorail plan of 1958 is quite similar to the monorail plan discussed in the chapter on Los Angeles. City of Detroit Rapid Transit Commission, *Rapid Transit System and Plan Recommended for Detroit and the Metropolitan Area* (Detroit, 1958).

14. Metro Detroit's main suburban bus agency is SEMTA's legal successor, but there is little practical continuity between the two.

15. Southeastern Michigan Transportation Authority, *A Preliminary Proposal for High and Intermediate Level Transit in the Detroit Metropolitan Area* (Detroit, 1974).

16. Metropolitan Transportation Authorities Act, Public Act 204 (Mich. 1967).

17. Russell Glynn, Kevin X. Shen, and Mario Goetz, "Avenues of Institutional Change: Technology and Urban Mobility in Southeast Michigan" (working paper, MIT Industrial Performance Center, 2020), 19–20, https://workofthefuture.mit.edu/wp-content/uploads/2020/12/2020-Working-Paper-Glynn-Shen-Goetz2.pdf.

18. This was normal for the period; the federal government made a similar offer to Seattle for Seattle's Forward Thrust 1970.

19. For a good overview of the Detroit riots of 1967, please see Sidney Fine, *Violence in the Model City: The Cavanagh Administration, Race Relations, and the Detroit Riot of 1967* (East Lansing, MI: Michigan State University Press, 1987).

20. Sources differ on the number of civilian deaths attributed to the STRESS unit. Matthew Lassiter of the University of Michigan attributes 22 civilian deaths to STRESS.

The *Detroit Free Press,* in contrast, counted 20. I consider Lassiter's numbers more reliable and better-sourced. Matthew D. Lassiter, "Remembering STRESS Victims," *Detroit Under Fire: Police Violence, Crime Politics and the Struggle for Racial Justice in the Civil Rights Era,* March 2021, https://policing.umhistorylabs.lsa.umich.edu/s/detroitunderfire/page/rememberingstressvictims; Bill Laitner and Matt Helms, "Former Detroit Mayor Roman Gribbs Has Died," *Detroit Free Press*, April 5, 2016, https://www.freep.com/story/news/obituary/2016/04/05/former-detroit-mayor-roman-gribbs-has-died/82666540/.

21. This fact became a major part of Mayor Gribbs's legacy, and even made it into his *Detroit Free Press* obituary at his death in 2016. Laitner and Helms, "Former Detroit Mayor Roman Gribbs Has Died."

22. Dennis A. Deslippe, "'Do Whites Have Rights?': White Detroit Policemen and 'Reverse Discrimination' Protests in the 1970s," *Journal of American History* 91, no. 3 (2004): 932–960, https://doi.org/10.2307/3662861.

23. Deslippe.

24. Peter Benjamin, "Young Warns Criminals: 'Hit the Road,'" *Detroit Free Press*, January 3, 1974.

25. Deslippe, "Do Whites Have Rights?"

26. Deslippe.

27. Bill McGraw, "Coleman Young: The 10 Greatest Myths," *Detroit Free Press*, May 26, 2018, https://www.freep.com/story/opinion/2018/05/26/coleman-young-myths/638105002/.

28. See, e.g., "Region's Transit Future Was Bleak Until We Provided Local Leadership," *Clarkston* (MI) *News*, September 29, 1977. Every other newspaper in Oakland County published the same supplement at various points during that week.

29. Glynn, Shen, and Goetz, "Avenues of Institutional Change," 20–21.

30. Isabel Wilkerson, "Years Late, Detroit's Monorail Opens," *New York Times*, August 1, 1987. The *Times* correspondent's headline is wrong—it is a people mover, not a monorail.

31. Wilkerson.

32. Leanna First-Arai, "Failed Linkages: Detroit's People Mover in the Context of a Southeast Michigan

Regional Transportation System," *Agora Journal of Urban Planning and Design* (2010): 65–68, https://hdl.handle.net/2027.42/120366.

33. Aaron Mondry, "I Spent a Day on the World's Most Pointless Transit System," *Gizmodo*, August 14, 2015, https://gizmodo.com/i-spent-a-day-on-the-world-s-most-pointless-transit-sys-1723996796.

34. Sarah Cwiek, "It's Official: No Federal Money for Detroit Light Rail—Again," *Michigan Radio*, June 19, 2012, https://www.michiganradio.org/transportation/2012-06-19/its-official-no-federal-money-for-detroit-light-rail-again. In its early stages the QLine was named "M-1 Rail," hence the name.

35. Mark Brush, "M-1 Rail Project to get $25 million in federal support," *Michigan Radio*, January 18, 2013, https://www.michiganradio.org/transportation/2013-01-18/m-1-rail-project-to-get-25-million-in-federal-support.

Chapter 8

1. As always, there are exceptions. Some historical neighborhoods do have more restrictive land use regimes, but this isn't the norm.

2. Trip Gabriel, "Who Is LaToya Cantrell, New Orleans's First Female Mayor?," *New York Times*, November 20, 2017, https://www.nytimes.com/2017/11/20/us/latoya-cantrell-new-orleans-mayor.html.

3. "First Word from Moon Was 'Houston,'" *Blytheville* (AR) *Courier News*, April 1, 1975.

4. Another type of restrictive covenant, the racial covenant, was used to enforce segregation. Racial covenants in property deeds would say, for example, "no part of these properties should ever be used or occupied by, or sold, leased or given to, any person of the negro race or blood." *Corrigan v. Buckley*, 271 U.S. 323 (1926). The U.S. Supreme Court banned racial covenants with its decision in *Shelley v. Kraemer*, 334 U.S. 1 (1948).

5. Houston Municipal Code, sec. 26-492.

6. I've assumed, for the sake of argument, that each parking space requires 400 square feet. This is necessarily an approximation, as parking lots vary in efficiency. Shoup and Pickrell give a range of 330–500 square feet per space.

Shoup and Pickrell, "Problems with Parking Requirements," 547–49.

7. Houston Ordinance No. 2013-208 (2013).

8. Jasper Scherer, "Houston Lifts Minimum Parking Requirements in EaDo, Midtown," *Houston Chronicle*, July 17, 2019, https://www.houstonchronicle.com/news/houston-texas/houston/article/Houston-lifts-minimum-parking-requirements-in-14103862.php.

9. John Park, Luis Guajardo, Kyle Shelton, Steve Sherman, and William Fulton, *Re-Taking Stock: Understanding How Trends in the Housing Stock and Gentrification Are Connected in Houston and Harris County* (Houston: Rice University Kinder Institute for Urban Research, 2021), https://doi.org/10.25611/MHBK-WQ06.

10. Data comes from the American Community Survey.

11. It helps that Houston's homeless services division is a national model, but the single largest factor that explains the variation in homelessness rates between metropolitan areas is housing supply, and not mental health, weather, addiction rates, or even poverty. Gregg Colburn and Clayton Page Aldern, *Homelessness Is a Housing Problem* (Oakland: University of California Press, 2022): 10.

12. U.S. Department of Housing and Urban Development, *2021 AHAR: Part 1—PIT Estimates of Homelessness in the U.S.* (Washington DC, February 2022), https://www.huduser.gov/portal/sites/default/files/xls/2007-2021-PIT-Counts-by-CoC.xlsx I've used the 2019 data, as the 2020 and 2021 figures are distorted by the effects of the COVID-19 pandemic.

13. Park et al., *Re-Taking Stock*, 19.

14. Park et al., 19; see also "Home Prices in the 100 Largest Metro Areas," *Kiplinger's*, January 23, 2021, https://www.kiplinger.com/article/real-estate/t010-c000-s002-home-price-changes-in-the-100-largest-metro-areas.html.

Chapter 9

1. Dorothy Parker apocryphally said, "Los Angeles is seventy-two suburbs in search of a city."

2. William L. Kahrl, *Water and Power: The Conflict over Los Angeles' Water Supply in the Owens Valley* (Berkeley: University of California Press, 1982), 97–99.

251

3. Kahrl, 188.

4. Los Angeles City Council and Los Angeles County Board of Supervisors, *Report and Recommendations on a Comprehensive Rapid Transit Plan for the City and County of Los Angeles* (Los Angeles, 1925), 18.

5. Los Angeles City Council, 6.

6. Randolph Karr, *Rail Passenger Service History of Pacific Electric Railway Company* (Los Angeles: Southern Pacific Historical & Technical Society, 1973), 15, https://www.pacificelectric.org/wp-content/uploads/2018/06/railpassengerservice-historypacificelectricrailwayco-karr.pdf.

7. Los Angeles City Council, 35.

8. Jeremiah B. C. Axelrod, "'Keep the 'L' out of Los Angeles': Race, Discourse, and Urban Modernity in 1920s Southern California," *Journal of Urban History* 34, no. 1 (2007), 3–37, https://doi.org/10.1177/0096144207306614.

9. Karr, *Rail Passenger Service History*, 19.

10. Karr, 19.

11. Rapid Transit Action Group, Los Angeles Chamber of Commerce, *Rail Rapid Transit Now!* (Los Angeles: Angelus Press, 1948), 8, http://libraryarchives.metro.net/DPGTL/trafficplans/1948_rail_rapid_transit_now.pdf.

12. Martin Wachs, Michelle Chung, Mark Garrett, Hannah King, Ryland Lu, Michael Manville, Melissa Sather, and Bryan D. Taylor, *A Taxing Proposition: A Century of Ballot Box Transportation Planning in Los Angeles* (Los Angeles: UCLA Institute of Transportation Studies, 2018), 44, https://escholarship.org/uc/item/94c8b1rr.

13. Sy Adler, *Understanding the Dynamics of Innovation in Urban Transit* (Washington, DC: Urban Mass Transportation Administration, 1986), 13–15, https://pdxscholar.library.pdx.edu/cgi/viewcontent.cgi?article=1087&context=cus_pubs.

14. Brian D. Taylor, "Rethinking Traffic Congestion," *Access Magazine* 1, no. 21 (2002): 8–16, https://www.accessmagazine.org/wp-content/uploads/sites/7/2016/07/access21-02-rethinking-traffic-congestion.pdf.

15. Richard Simon, "Hollywood Freeway Spans Magic and Might of LA," *Los Angeles Times*, December 19, 1994.

16. Coverdale & Colpitts, *Report to the Los Angeles Metropolitan Transit Authority on a Monorail Rapid Transit Line for Los Angeles* (Los Angeles: Los Angeles Metropolitan Transit Authority, 1954).

17. Patrick Miller, S. C. Wirasinghe, Lina Kattan, and Alexandre De Barros, "Monorails for Sustainable Transportation: A Review" (paper presented at the CSCE 2014 General Conference, Halifax, NS, May 28–31, 2014), 5–6, https://www.researchgate.net/profile/Patrick-Miller-6/publication/301302321_Monorails_for_sustainable_transportation_-_a_review/links/57114aed08aeebe07c0242e2/Monorails-for-sustainable-transportation-a-review.pdf.

18. Hermann Steffen Dickemann Botzow Jr., "The Feasibility of Monorails" (master's thesis, Massachusetts Institute of Technology, 1958), 30–31.

19. See, e.g., "Anyone for Monorail?," *Fortune*, July 1954.

20. "MTA Reorganization Wins Wide Support," *Pasadena Independent*, February 6, 1964.

21. Adrian Glick Kudler, "The Secret Cowboy/Cleopatra/Tinfoil Origins of Century City," *Curbed Los Angeles*, September 26, 2013, https://la.curbed.com/2013/9/26/10193620/the-secret-cowboycleopatratinfoil-origins-of-century-city.

22. Brian Stipak, "An Analysis of the 1968 Rapid Transit Vote in Los Angeles," *Transportation* 2 (April 1973): 72.

23. "Two New Mass Transportation Plans Proposed," *Long Beach Press-Telegram*, November 8, 1968.

24. Southern California Rapid Transit District, *The Sunset Coast Line: Route of the New Red Cars* (Los Angeles, 1976), 59.

25. Southern California Rapid Transit District, 59..

26. Steve Hemmerick, "Area Cities Varied in Vote Pattern," *Star-News* (Pasadena, CA), June 13, 1976.

27. Michael Manville, Paavo Monkkonen, and Michael Lens, "It's Time to End Single-Family Zoning," *Journal of the American Planning Association* 86, no. 1 (2020): 106–112, https://doi.org/10.1080/01944363.2019.1651216.

28. Greg Morrow, "The Homeowner Revolution: Democracy, Land Use and the Los Angeles Slow-Growth Movement, 1965–1992" (PhD diss., UCLA, 2013), 3.

29. In practice, Proposition 13 means that identical properties often pay wildly different tax bills. A personal example: a good friend's condo is valued at $260,000 for tax purposes because it was purchased in 2011 and the valuation is based on the purchase price. His upstairs neighbor bought an identical apartment in 2020. The neighbor's unit is valued at $850,000 for tax purposes.

30. Proposition 13 does allow for partial reassessment if a property is renovated.

31. "California Voters Pamphlet: Primary Election," California Secretary of State, June 6, 1978, 58–59. At every election, California distributes a ballot information guide with arguments for and against the ballot propositions put before the voters. In the pro–Proposition 13 column were Howard Jarvis, conservative activist Paul Gann, and five state senators; against Proposition 13 were the mayor of Los Angeles, the former state controller, the League of Women Voters, the League of California Cities, and a half dozen other major political lobbies.

32. Evelyn Danforth, "Proposition 13 Revisited," *Stanford Law Review* 73 (2021): 528.

33. As of the 2020 U.S. Census, the Los Angeles Combined Statistical Area covers five counties: Los Angeles, Ventura, San Bernardino, Riverside, and Orange. These five counties occupy 33,954 square miles of land area; the entire nation of Switzerland is 15,940 square miles.

34. Martha Groves, "In a Reversal, Waxman Backs Westside Subway," *Los Angeles Times*, December 17, 2005.

35. Neal Broverman, "State Could Be about to Repeal Ban on Light Rail in the Valley," *Curbed Los Angeles*, February 4, 2014, https://la.curbed.com/2014/2/4/10147368/state-could-be-about-to-repeal-ban-on-light-rail-in-the-valley.

36. AB 2097 (Cal. 2022).

37. Los Angeles Municipal Code 12.21(4)(C)(4) requires four parking spaces—that is, 1,600 square feet of parking for each 1,000 square feet of retail space. This has been partially superseded by AB 2097 (Cal. 2022), the state parking reform law.

38. Dowell Myers, JungHo Park, and Eduardo Mendoza, *Housing Research Brief 2: How Much Added Housing Is Really Needed in Los Angeles?* (Los Angeles: University of Southern California, 2018), 3, https://cpb-us-e1.wpmucdn.com/sites.usc.edu/dist/6/210/files/2017/02/HRB-2-How-Much-Added-Housing-is-Really-Needed-in-Los-Angeles-1zumxf6.pdf.

39. For example, at the 1970 census, Beverly Hills had 33,416 inhabitants; in 2020, it had 32,701, or –2.7 percent population growth over 50 years. For comparison, Los Angeles County as a whole grew 42.8 percent between 1970 and 2020, from 7 million to 10 million.

40. "More Than Half of LA homes Sold above Asking Price," *The Real Deal*, April 14, 2021, https://therealdeal.com/la/2021/04/14/more-than-half-of-la-homes-sold-above-asking-price/.

41. For example, Sacramento enacted all these reforms between 2020 and 2021. See, e.g., Sacramento City Code sec. 17.860.

42. The median home sale price in Los Angeles County in August 2022 was $854,960. Assuming a 30-year mortgage, a 6.6 percent interest rate and 20 percent down, that means about $4,350 per month in mortgage payments, plus $850 per month in insurance and taxes. Personal finance experts advise spending no more than 30 percent of one's gross monthly income on housing. This implies that a person looking to buy should have a monthly income of $17,933—that is, an annual income of $208,000. California Association of Realtors, "August Home Sales and Price Report," September 16, 2022, https://www.car.org/en/aboutus/mediacenter/newsreleases/2022releases/aug2022sales.

43. "Editorial: SB 50 Is Dead. Again. LA Lawmakers Need to Stop Stonewalling and Come Up with a Housing Solution," *Los Angeles Times*, January 31, 2020, https://www.latimes.com/opinion/story/2020-01-31/sb50-dead-los-angeles-housing-crisis.

Chapter 10

1. The densest large American cities proper are, in order, New York, with a density of 29,302.66 per square mile; San Francisco, at 18,634.65 per square mile; Boston, at 13,976.98 per square mile; Miami, at 12,284.47 per square mile; Chicago, at 12,059.84 per square mile; and Philadelphia, at 11,936.92 per square mile. Miami is the only city with this low rate of mass transit usage.

2. P. H. Rolfs, "Probable Results of Draining the Everglades Extent, Altitude, Character of the Soil, Drainage-Effect on Height of Water Table, Effect on Temperature, Addition to Population," *Proceedings of the Florida State Horticultural Society* 16 (1903): 47.

3. Paul S. George, "Passage to the New Eden: Tourism in Miami from Flagler through Everest G. Sewell," *Florida Historical Quarterly* 59, no. 4 (1981): 443–45.

253

4. George, 447.

5. George, 453.

6. Edward Ridolph, *Biscayne Bay Trolleys: Street Railways of the Miami Area* (Forty Fort, PA: Harold E. Cox, 1981), 41.

7. Somerset Maugham once used the phrase "a sunny place for shady people" to describe the French Riviera, but the sentiment is equally applicable to Miami.

8. Frederick Lewis Allen, *Only Yesterday: An Informal History of the 1920s* (New York: Harper & Row, 1931), https://gutenberg.net.au/ebooks05/0500831h.html.

9. Homer B. Vanderblue, "The Florida Land Boom," *Journal of Land & Public Utility Economics* 3, no. 3 (1927): 256.

10. The *Prinz Valdemar* remained docked in Miami, where it ultimately ended its life as a floating youth center before being scrapped in 1952. "Once-Proud Ship Scrap Heap Now," *Spokesman-Review* (Spokane, WA), December 14, 1952.

11. Vanderblue, "Florida Land Boom," 259.

12. Vanderblue, 262.

13. Ridolph, *Biscayne Bay Trolleys*, 45.

14. Nearly a century later, the *Miami New Times* called Isola di Lolando "a watery testament to Miami developers' tendency to overpromise and underdeliver." Joshua Ceballos, "Isola di Lolando 'Marine Biologist: Turn Old Ruins into 'Mangrove Maze Visible from Space'" *Miami New Times*, January 21, 2022, https://www.miaminewtimes.com/news/mangrove-maze-isola-di-lolando-ruins-13755261.

15. Cited in F. L. Allen, *Only Yesterday*.

16. Greg Allen, "Paying a Local Price for I-95's Global Promise," National Public Radio, August 28, 2010, https://www.npr.org/templates/story/story.php?storyId=129475747.

17. Metropolitan Dade County (FL), *Final Project Report: Preliminary Engineering Rapid Transit System for The Dade County Transportation Improvement Program* (Miami, 1979), 97. Although the date on the cover of the report is March 1976, the Miami-Dade Metropolitan Planning Organization gives the report's date as March 1979.

18. Art Harris, "Bedecked in Gold, Miami Drug Lords Buy Luxuries for Cash," *Washington Post*, August 6, 1981.

19. Metropolitan Dade County, *Final Project Report*, 99–120.

20. Associated Press, "Reagan: Treasury a Trust, Not a Gift Shop," *Walla Walla* (WA) *Union-Bulletin*, March 4, 1985.

21. In the first quarter of 2019, Miami's Metromover carried 28,300 riders per average weekday, the Jacksonville Skyway carried 2,800, and the Detroit People Mover carried 4,600.

22. "Best of Miami 2004: Best Bait and Switch," *Miami New Times*, May 13, 2014, https://www.miaminewtimes.com/best-of/2004/people-and-places/best-bait-and-switch-6344376.

23. Data was drawn by public records request from the Los Angeles Metro.

24. Zach Patton, "The Miami Method for Zoning: Consistency Over Chaos," *Governing*, April 28, 2016, https://www.governing.com/archive/gov-miami-zoning-laws.html.

25. Miami-Dade County Department of Transportation and Public Works, *Ridership Technical Report: Division of Performance and Materials Management: January 2022* (Miami, 2022), 29–30, https://www.miamidade.gov/transit/library/rtr/2022-01-Monthly-Ridership-Report.pdf.

26. Monica Correa, "County Readies Pact for Miami–Miami Beach Monorail," *Miami Today*, August 31, 2021, https://www.miamitodaynews.com/2021/08/31/county-readies-pact-for-miami-miami-beach-monorail/.

27. Miami-Dade Department of Transportation, *Ridership Technical Report*, 2022, 47–50.

Chapter 11

1. Stephen A. Kieffer, *Transit and the Twins* (Minneapolis: Twin City Rapid Transit Company, 1958), 44.

2. Marda Woodbury, *Stopping the Presses: The Murder of Walter W. Liggett* (Minneapolis: University of Minnesota Press, 1998), 47.

3. Kieffer, *Transit and the Twins*, 45.

4. Gordon Schendel, "How Mobsters Grabbed a City's Transit Line," *Collier's*, September 29, 1951, 74.

5. Kieffer, *Transit and the Twins*, 53.

6. Peter Derrick, *Tunnelling to the Future: The Story of the Great Subway Expansion That Saved New York* (New York: NYU Press, 2002), 44.

7. Schendel, *How Mobsters Grabbed*, 75–76.

8. *Isaacs v. United States*, 301 F.2d 706 (8th Cir. 1962).

9. *In re Ossanna*, 266 Minn. 569 (1963).

10. *Blumenfield v. United States*, 284 F.2d 46 (8th Cir. 1960).

Chapter 12

1. Denis Veilleux, "Buses, Tramways, and Monopolies: The Introduction of Motor Vehicles into Montreal's Public Transport Network," *Michigan Historical Review* 22, no. 2 (1996): 103–26, https://doi.org/10.2307/20173588.

2. William Fong, *J. W. McConnell: Financier, Philanthropist, Patriot* (Montreal: McGill-Queens University Press, 2008), 168.

3. Jason Dean and Vincent Geloso, "The Linguistic Wage Gap in Quebec, 1901 to 1921, GMU Working Paper in Economics No. 21-34" (working paper, George Mason University, 2020), 3, http://dx.doi.org/10.2139/ssrn.3641844.

4. Francois Vaillancourt, Dominique Lemay, and Luc Vaillancourt, "Laggards No More: The Changed Socioeconomic Status of Francophones in Quebec," *C. D. Howe Institute Backgrounder* 103 (August 2007), 1, https://www.cdhowe .org/sites/default/files/attachments/research_papers/mixed //backgrounder_103_english.pdf.

5. With seven decades of hindsight, Duplessis's legacy is more complicated than the simplistic Great Darkness approach that prevailed in the decades after his death. The Duplessis era really did bring material improvements to the lives of the average Quebecer, even if social policy was retrograde, essential services were run by the Catholic Church, and the government was corrupt. The *Canadian Encyclopedia* pithily calls Duplessis "a modernizer, except in political methodology." Conrad M. Black, "Maurice Duplessis," *Canadian Encyclopedia*, December 17, 2020, https://www. thecanadianencyclopedia.ca/en/article/maurice-le-noblet -duplessis.

6. These major social changes were all parts of Quebec's "Quiet Revolution."

7. Ian Adams, "The Busy Little Man Who's Building Big Town," *MacLean's*, December 3, 1966, 17.

8. Douglas Martin, "After 29 Flamboyant Years, the Mayor of Montreal Is Retiring," *New York Times*, June 28, 1986.

9. Dale Gilbert, "Penser la mobilité, penser Montréal: La planification du tracé du réseau initial de métro, 1960–1966," *Revue d'Histoire de l'Amérique Française* 68, nos. 1–2 (2014): 61–64, https://doi.org/10.7202/1032019ar.

10. Anthony Perl, Matt Hern, and Jeffrey Kenworthy, "Streets Paved with Gold: Urban Expressway Building and Global City Formation in Montreal, Toronto and Vancouver," *Canadian Journal of Urban Research* 24, no. 2 (2015): 91–116, http://www.jstor.org/stable/26195293.

11. Frank Hamilton, "The Subway Nobody Wanted," *MacLean's*, August 15, 1950, 8.

12. Kenneth B. Smith, "When You Call Montreal's Metro a Subway, Smile," *Globe and Mail*, October 12, 1966.

13. American observers, in particular, were astounded by the Metro. See, e.g., Patricia Conway George, "Mass Transit: Problem and Promise," Design Quarterly, no. 71 (1968): 3–39, https://doi.org/10.2307/4047322.

14. Jack Todd, "1976 Montreal Olympics: Drapeau's Baby from Bid to Billion-Dollar Bill," *Montreal Gazette*, July 29, 2016.

15. $1.5 billion in 1976 Canadian dollars, due to the 1:1 exchange rate at the time.

Chapter 13

1. Louis C. Hennick and E. Harper Charlton, *The Streetcars of New Orleans* (Gretna, LA: Pelican Publishing Co., 1975), 14–16 and 23–27.

2. Phyllis Hutton Raabe, "Status and Its Impact: New Orleans' Carnival, The Social Upper Class and Upper-Class Power" (PhD diss., Penn State University, 1973), 16–18.

3. Calvin Trillin, "On the Possibility of Houstonization," *New Yorker*, February 17, 1975, 95.

4. A good overview of this dynamic is contained in James Gill, *Lords of Misrule: Mardi Gras and the Politics of Race in New Orleans* (Jackson: University Press of Mississippi, 1997). See also Michael L. Kurtz, "deLesseps S. Morrison: Political Reformer," *Louisiana History* 17, no. 1 (1976): 27, https://www.jstor.org/stable/4231554. Notably, Morrison had significant support among New Orleans's gentry, but this was because of his strong anticorruption stance rather than his modernization plans.

5. Hennick and Charlton, *Streetcars*, 43–45.

6. Edward J. Branley, *The Canal Streetcar Line* (Mount Pleasant, SC: Arcadia Publishing, 2004), 8.

7. "Second Progress Report on the Canal Street Transit Improvement Program," *Louisiana Weekly*, Jun. 6, 1964; see also "Street Transit Improvement Program: A Progress

Report by the President of New Orleans Public Service," *Louisiana Weekly*, May 2, 1964.

8. For example, Philadelphia's Subway-Surface trolley lines use a tunnel to get downtown.

9. Hennick and Charlton, *Streetcars*, 42; see also Robert C. Post, "The Machine in the Garden District," *Technology and Culture* 47, no. 1 (2006): 91–94, http://www.jstor.org /stable/40061009.

10. James K. Glassman, "New Orleans: I Have Seen the Future, and It's Houston," *Atlantic*, July 1978, 14, https: //www.theatlantic.com/magazine/archive/1978/07/new-orleans-i-have-seen-the-future-and-its-houston/304292/.

11. Sidney H. Bingham, *Report on the Feasibility of a Monorail System for the New Orleans Metropolitan Area* (New Orleans: New Orleans City Council, 1959).

12. Michael Lewis, "Wading toward Home," *New York Times*, October 9, 2005, https://www.nytimes.com/2005 /10/09/magazine/wading-toward-home.html.

13. Travers Mackel, "WDSU Investigates: Why Plaza Tower isn't developed," *WDSU 6 News*, May 20, 2021, https: //www.wdsu.com/article/wdsu-investigates-why-plaza-tower -isnt-developed/36279994.

14. New Orleans Regional Transit Authority (RTA), *Transit Improvement Pilots: Canal Streetcar and On-Demand Transit*, presentation to the New Orleans City Council Transportation Committee, February 27, 2019.

15. For example, Houston and Seattle use Siemens S70 trains on their light rail lines. Charlotte and Salt Lake City use the S70 as a streetcar.

16. See, e.g., Santa Clara Valley Transportation Authority, *LRT: Light Rail Transit Service Guidelines, Sustainability Policy 1-32* (San Jose, CA, 2007), https://nacto.org/docs/usdg /lrtserviceguidelines_vta.pdf.

17. Jeff Adelson, "RTA Plan Would Eliminate Many Stops, Close Intersections along Canal Street Streetcar Line," NOLA.com, February 14, 2019, https://www.nola.com /article_7bc6319f-50b9-5204-abda-24083168ae10.html.

18. I have not been able to find RTA data on mobile ticketing adoption. In my own experience, few passengers use the app to pay the fare.

19. RTA, *Transit Improvement Pilots*.

20. Jeff Adelson, "New Orleans Hits Brakes on Faster Streetcar Routes; Longer Walks to Stops Drove Outrage," NOLA.com, March 26, 2019, https://www.nola.com/article _fe081713-869e-569e-8b6b-724921dfed54.html.

Chapter 14

1. James Blaine Walker, *Fifty Years of Rapid Transit, 1864– 1917* (New York: Law Printing Co., 1917), 33–35.

2. Walker, 35.

3. Walker, 50.

4. "Our Subway Open: 150,000 Try It," *New York Times*, October 28, 1904.

5. The decades-long fights over fare hikes spilled over into the court system, making national news and even reaching the U.S. Supreme Court in 1928. "Series of Sensational Events Marks the Chronology of 1928; Election Held Press Spotlight," *Cincinnati Commercial Tribune*, January 1, 1929.

6. "100 Miles of Subway in New City Project; 52 of Them in Queens," *New York Times*, September 16, 1929.

7. "100 Miles."

8. Andrew J. Sparberg, *From a Nickel to a Token: The Journey from Board of Transportation to MTA* (New York: Fordham University Press, 2015), 59.

9. Stanley Levey, "A 2d Ave. Subway Called Unlikely," *New York Times*, March 9, 1957.

10. Richard Witkin, "2.5-Billion Transit Bonds Voted in Major Victory for Governor," *New York Times*, November 8, 1967.

11. Metropolitan Transportation Authority, *Metropolitan Transportation: A Program for Action* (New York, 1968).

12. Metropolitan Transportation Authority, *1968–1973, The Ten-Year Program at the Halfway Mark* (New York, 1973), 37.

13. Jeff Nussbaum, "The Night New York Saved Itself from Bankruptcy," *New Yorker*, October 16, 2015, https: //www.newyorker.com/news/news-desk/the-night-new-york -saved-itself-from-bankruptcy.

14. Frank Van Riper, "Ford to City: Drop Dead," *New York Daily News*, October 30, 1975.

15. Owen Fitzgerald, "Closing Mulled for 79 Stations," *New York Daily News*, April 29, 1986.

16. Andy Newman, "New Subway Line in Transit

Budget," *New York Times*, April 20, 2000, https://www.nytimes.com/2000/04/20/nyregion/new-subway-line-in-transit-budget.html.

17. Brian M. Rosenthal, "The Most Expensive Mile of Subway Track on Earth," *New York Times*, December 28, 2017, https://www.nytimes.com/2017/12/28/nyregion/new-york-subway-construction-costs.html.

18. Ben Bradford, "Why Are Subways in the U.S. So Expensive?" *Marketplace*, April 11, 2019, https://www.marketplace.org/2019/04/11/subways-us-expensive-cost-comparison/.

19. The MTA has a 23-member board, but not all members have full voting authority. Board members are appointed by the following: (a) the governor of New York (four members, four votes); (b) the mayor of New York City (four members, four votes); (c) the inner-ring suburban counties of Nassau, Suffolk, and Westchester (three members, three votes); (d) the outer-ring suburban counties of Rockland, Putnam, Dutchess, and Orange (four members, one collective vote); and (e) unions and rider advocates (six members, no voting power).

20. On-time performance data is drawn from the MTA.

21. Allana Akhtar, "New York MTA Gave Millions to Bail Out Upstate Ski Resorts as Subway Crumbles," *Jalopnik*, July 10, 2017, https://jalopnik.com/new-york-mta-gave-millions-to-bail-out-upstate-ski-reso-1796789185.

22. Ben Yakas, "Report: The Best & Worst of NYC Transit over the 2010s," *Gothamist*, December 17, 2019, https://gothamist.com/news/report-best-worst-nyc-transit-over-2010s.

23. Jim Dwyer, "How a Clash of Egos Became Bigger Than Fixing the Subway," *New York Times*, February 3, 2020, https://www.nytimes.com/2020/02/03/nyregion/cuomo-andy-byford-mta.html.

24. See, e.g., William C. Vantuono, "You Blew It, Andrew Cuomo," *Railway Age*, January 24, 2020, https://www.railwayage.com/passenger/you-blew-it-andrew-cuomo/; Ben Yakas, "Andy Byford Unchained: Train Daddy Reveals How Cuomo Made His Job 'Intolerable,'" *Gothamist*, March 9, 2020, https://gothamist.com/news/andy-byford-unchained-train-daddy-reveals-how-cuomo-made-his-job-intolerable; Nicole Gelinas, "Why Would

Anyone Take Over after Andy Byford Fled the MTA?," *New York Post*, January 23, 2020, https://nypost.com/2020/01/23/why-would-anyone-take-over-after-andy-byford-fled-the-mta/.

25. Rosenthal, "Most Expensive Mile."

26. Josh Barro, "Why New York Can't Have Nice Things," *New York Magazine*, May 30, 2019, https://nymag.com/intelligencer/2019/05/new-york-infrastructure-costs.html.

27. Bradford, "Why Are Subways So Expensive?"

28. David W. Dunlap, "Clearing the Tracks for Penn Station III," *New York Times*, January 3, 1999.

29. James S. Russell, "Penn Station's Revival Gets a $1.6 Billion Down Payment," *Bloomberg*, January 15, 2021, https://www.bloomberg.com/news/features/2021-01-15/can-the-moynihan-train-hall-redeem-penn-station.

30. Stephen Nessen, "New York's Long-Awaited Second Avenue Subway Finally Leaves the Station," NPR, January 2, 2017, https://www.npr.org/2017/01/02/507898727/new-yorks-long-awaited-second-avenue-subway-finally-leaves-the-station.

Chapter 15

1. Thomas Fitzgerald, "SEPTA Workers Might Strike Soon. It Wouldn't Be the First Time," *Philadelphia Inquirer*, October 27, 2022, https://www.inquirer.com/transportation/septa-strike-history-latest-longest-20211026.html; see also Transport Workers Union Local 234, *50 Years Today: A Vision of Tomorrow* (Philadelphia, 1993).

2. James Wolfinger, *Running the Rails: Capital and Labor in the Philadelphia Transit Industry* (Ithaca, NY: Cornell University Press, 2016), 224.

3. Wolfinger, 222.

4. Cassie Owens, "Throwback Photos: What SEPTA Strikes Looked Like from 1971–1981," *Billy Penn by WHYY*, November 2, 2016, https://billypenn.com/2016/11/02/throwback-photos-what-septa-strikes-looked-like-from-1971-1981/.

5. Local 234, *50 Years Today*, 4.

6. Vukan R. Vuchic, *General Operations Plan for the SEPTA Regional High Speed System* (Philadelphia: Southeastern Pennsylvania Transportation Authority, 1984). Notably,

257

Vuchic mentions that comparable German systems use subway fares within the city but does not directly recommend such reforms for Philadelphia.

7. Richard Voith, "Fares, Service Levels, and Demographics: What Determines Commuter Rail Ridership in the Long Run?" *Journal of Urban Economics* 41, no. 2 (1997): 176–97.

8. Pennsylvania, Act of July 23, 1970, P.L. 563, No. 195, as amended.

9. Wolfinger, *Running the Rails*, 227.

10. Local 234, *50 Years Today*, 6–7.

11. Wolfinger, *Running the Rails*, 195–97. There is a possibility that the number of wildcat strikes is higher. Nine is the absolute minimum.

12. Although SEPTA unions' willingness to strike has made it harder to run good transit service, it isn't clear whether this militancy has produced better outcomes for its workers. For example, the starting hourly wage for a SEPTA bus driver in 2022 is $40,148 per year, or $19.30 per hour. For comparison, the New York MTA pays $25.49, the Washington Metropolitan Area Transit Authority pays $28.19, and Pittsburgh's PAT pays $21.35.

13. Ryan Briggs, "SEPTA Union President Willie Brown —the 'Most Hated Man in Philadelphia'—Has a New Job," *WHYY*, October 29, 2021, https://whyy.org/articles /septa-union-president-willie-brown-the-most-hated-man -in-philadelphia-has-a-new-job/.

14. City of Philadelphia, *The Philadelphia Transit Plan: A Vision for 2045* (Philadelphia, 2021).

15. City of Philadelphia, *Philadelphia Transit Plan*, 1.

Chapter 16

1. "Pittsburgh Railways Review," *Timepoints: The Southern California Traction Review* 17, no. 4 (1959): http://www. erha.org/timepoints/v17n4.htm.

2. Port Authority of Allegheny County, *Allegheny County Rapid Transit Study* (Pittsburgh, 1967), I-8.

3. Port Authority, II-7.

4. Joe Cortright, *Surging City Center Job Growth* (Portland, OR: City Observatory, 2015), https://cityobservatory.org /wp-content/uploads/2015/02/Surging-City-Center-Jobs.pdf.

5. Port Authority, *Allegheny County Rapid Transit Study*, II-2.

6. Paul V. Didrikson and Kathryn Nickerson, "The Supply Side of the Automated People Mover Market: A Spectrum of Choices," *Journal of Advanced Transportation* 33, no. 1 (1999): 17–33, https://doi.org/10.1002/atr.5670330104.

7. U.S. Congress Office of Technology Assessment, *Automated Guideway Transit: An Assessment of PRT and Other New Systems, Including Supporting Panel Reports* (Washington DC, 1975), 155.

8. Edward K. Muller, Morton Coleman, and David Houston, "Skybus: Pittsburgh's Failed Industry Targeting Strategy of the 1960s" (working paper, University of Pittsburgh Center for Industry Studies, 2012), 25–26, https://www.engineering.pitt.edu/subsites/centers/cis /publications2/skybus-pittsburghs-failed-industry-targeting -strategy-of-the-1960s/.

9. Jonathan Williams, "Democrat Rips PAT Board for 'We Know Best' Stance," *Pittsburgh Post-Gazette*, August 21, 1969.

10. "Skybus—From the Beginning," *Pittsburgh Post-Gazette*, October 16, 1974.

11. "PAT to Speed Busway Work in South Pending Bridge Bids," *Pittsburgh Press*, April 9, 1975.

12. Urban Mass Transit Administration, *Pittsburgh Light Rail Transit Reconstruction, UMTA Project PA-03-0012: Draft Environmental Impact Report* (Washington DC, 1978), III-1 to III-3. The draft environmental impact report for the light rail conversion has a good, if bureaucratically tactful, summary of the entire saga.

13. Ridership data is pre-pandemic and has been drawn from PAT.

Chapter 17

1. Eric Morris, "From Horse Power to Horsepower," *ACCESS Magazine* 30 (2007): 7, https://escholarship.org/uc /item/6sm968t2.

2. August Meier and Elliott Rudwick, "Negro Boycotts of Segregated Streetcars in Virginia, 1904–1907," *Virginia Magazine of History and Biography* 81, no. 4 (1973): 479–87, http://www.jstor.org/stable/4247829.

3. Meier and Rudwick.

Chapter 18

1. Andrew David Lippman, "The Rochester Subway: Experiment in Municipal Rapid Transit," *Rochester History* 36, no. 2 (1974): 2.

2. Lippman, 6–7.

3. "Subway Debate Rages; Losses Big, Says RTC," *Rochester Times-Union*, November 29, 1949.

4. Lippman, "Rochester Subway," 18.

5. "Subway Will Serve It," *Rochester Democrat & Chronicle*, September 3, 1948.

6. "Subway Use Nearing Decision," *Rochester Times-Union*, December 6, 1949.

Chapter 19

1. The Bay Area's per capita metropolitan gross domestic product of $131,082 (in 2015 dollars) is the highest in North America, far exceeding that of the next-highest, Seattle ($97,987). This data is pulled from the Organisation for Economic Co-operation and Development's data set on metropolitan GDP per capita for 2020.

2. California Association of Realtors, "August Home Sales and Price Report," September 16, 2022, https://www .car.org/en/aboutus/mediacenter/newsreleases/2022releases /aug2022sales; Sabrina Speianu and Danielle Hale, "August 2022 Monthly Housing Market Trends Report," Realtor. com, September 1, 2022, https://www.realtor.com/research /august-2022-data. In August 2022, the average home in the United States sold for $435,000.

3. Over its nine-decade existence, San Francisco's principal private transit operator went by various corporate names as ownership changed hands: the Market Street Railroad (1860–1882), the Market Street Cable Railway (1882–1893), the Market Street Railway (1893–1902), the United Railroads of San Francisco (1902–1921), and the Market Street Railway (1921–1944). I use *Market Street Railway* for all these entities.

4. East of Van Ness Avenue, the road is called Geary Street. West of Van Ness, the road is called Geary Boulevard. I use simply *Geary* for both.

5. Robert Caldwell, *Transit in San Francisco: A Selected Chronology, 1850–1995* (San Francisco: San Francisco Municipal Railway, 1999), 24–25, https://archives.sfmta .com/cms/rhome/documents/TransitinSanFrancisco -CallwellChronologyweb.pdf.

6. Caldwell, 34.

7. *San Francisco Municipal Railway Merger of 1944* (San Mateo, CA: Western Railroader, 1965), 20–21. Muni's new steel streetcars were dubbed the "Magic Carpets," because of the quiet, smooth ride they provided.

8. E. G. Cahill, *Report on Rapid Transit System: Market, Mission, Geary, and Montgomery Subways* (San Francisco: San Francisco Public Utilities Commission, 1935).

9. Tom Matloff, "The Muni Paradox—A Brief Social History of the Municipal Railway," *Urbanist*, June 1, 1999, https://www.spur.org/publications/urbanist-article/1999-06 -01/muni-paradox.

10. Caldwell, *Transit in San Francisco*, 47.

11. "Savior of the Cable Cars Dies in San Francisco at Age of 90," *New York Times*, October 24, 1986.

12. Richard Grefe and Richard Smart, *A History of the Key Decisions in the Development of Bay Area Rapid Transit*, rev. ed. (San Francisco: Metropolitan Transportation Commission, 1976), 37.

13. San Mateo County was 96 percent white at the time.

14. Grefe and Smart, *History of the Key Decisions*, 40.

15. The Embarcadero Freeway and half of the Central Freeway were replaced with grand, tree-lined boulevards.

16. California Public Resources Code sec. 21000 et seq.

17. CEQA lawsuits filed against new housing developments prevail in court only 7 percent of the time. Moira O'Neill, Eric Biber, Giulia Gualco-Nelson, and Nicholas Marantz, *Examining Entitlement in California to Inform Policy and Process: Advancing Social Equity in Housing Development Patterns* (Berkeley: California Air Resources Board, 2021), 86, https://ssrn.com/abstract=3956250.

18. Jennifer L. Hernandez, *Anti-Housing CEQA Lawsuits Filed in 2020 Challenge Nearly 50% of California's Annual Housing Production* (Sacramento: California Center for Jobs and the Economy, 2022), https://centerforjobs.org/wp -content/uploads/Full-CEQA-Guest-Report.pdf.

19. Georgina McNee and Dorina Pojani, "NIMBYism

as a Barrier to Housing and Social Mix in San Francisco," *Journal of Housing and the Built Environment* 37 (2022): 564–67, https://doi.org/10.1007/s10901-021-09857-6.

20. McNee and Pojani note that "NIMBYism is sometimes the province of 'educated older hippie types'" (563).

21. California Legislative Analyst's Office, *California's High Housing Costs: Causes and Consequences* (Sacramento, 2015), 7.

22. California Legislative Analyst's Office, 22. San Francisco built less than a quarter of "estimated new housing construction needed to prevent home prices from growing faster than the rest of the country."

23. At the time, observers criticized the BART project for coming in late and over budget. By today's standards, BART was quick and cheap to build.

24. In one memorable incident, the University of California, Berkeley's decision to cut down campus trees to build a gymnasium triggered a tree-sit that lasted from 2006 to 2008. The tree-sit was led by a hard-left activist using the nom de guerre "Zachary RunningWolf." The residents of posh Panoramic Hill financed the litigation to keep RunningWolf and the other tree sitters in place.

25. San Francisco County Transportation Authority, *The Four Corridor Plan: Long Range Fixed Guideway Plan* (San Francisco, 1995).

26. Rachel Swan, "Judge Tosses Lawsuit Aimed at Slowing Geary Blvd. Bus Rapid Transit System," *San Francisco Chronicle*, October 16, 2018, https://www.sfchronicle.com/bayarea/article/Judge-tosses-lawsuit-aimed-at-slowing-Geary-Blvd-13312249.php.

27. Benjamin Schneider, "Geary BRT Changes Lanes," *SF Weekly*, May 27, 2021, https://www.sfweekly.com/news/geary-brt-changes-lanes/.

28. SB 288 (Cal. 2020) and SB 922 (Cal. 2022).

29. Jeffrey Steele, "Force Proves to Be with Developers of 'The Coronet,'" *Multi-Housing News*, July 1, 2011, https://www.multihousingnews.com/force-proves-to-be-with-developers-of-the-coronet/.

30. O'Neill et al., *Examining Entitlement*, 79.

31. Jerry Weitz, *Jobs-Housing Balance* (Chicago: American Planning Association Advisory Service, 2003), 10.

32. San Francisco Planning Department, *2020 Jobs-Housing Fit Report* (San Francisco, 2021), 15–16, https://sfplanning.org/sites/default/files/resources/2021-11/Jobs-Housing_Fit_Report_2020.pdf.

33. Colburn and Aldern, 10.

34. Christopher S. Elmendorf, *A Primer on California's "Builder's Remedy" for Housing-Element Noncompliance* (Los Angeles: UCLA Lewis Center for Regional Policy Studies, 2022), https://escholarship.org/uc/item/38x5760j. The State of California has recently passed laws allowing developers to bypass local zoning and planning rules wholesale, if cities don't plan to build enough housing to meet a state-defined quota.

Chapter 20

1. Sy Adler, *Understanding the Dynamics of Innovation in Urban Transit* (Washington, DC: Urban Mass Transportation Administration, 1986), 29–31, https://pdxscholar.library.pdx.edu/cgi/viewcontent.cgi?article=1087&context=cus_pubsAdler.

2. Fred Moody, *Seattle and the Demons of Ambition: A Love Story* (New York: St. Martin's, 2004), 66.

3. Raymond G. Deardorf, Robert J. Berg, and Chyi Kang Lu, "Measuring Station Capacity for Seattle's Bus Tunnel," *Transit Terminals* 1 (1985): 39.

4. David Schaefer, "Voters Back Transit Plan on Fourth Try," *Seattle Times*, November 6, 1996.

5. Dick Falkenbury, *Rise above It All* (Seattle: Falkenbury Enterprises, 2013), 5.

6. "Monorail Timeline," *Seattle Post-Intelligencer*, July 3, 2005, https://www.seattlepi.com/local/transportation/article/Monorail-timeline-1178148.php.

7. Jim Brunner, "Initiative 53," *Seattle Times*, November 8, 2000, https://archive.seattletimes.com/archive/?date=20001108&slug=4052054.

8. "The Recent Tumult in Seattle Sets Stage for Mayoral Race," *Everett* (WA) *Herald*, September 12, 2001, https://www.heraldnet.com/news/the-recent-tumult-in-seattle-sets-stage-for-mayoral-race/.

9. Jane Hadley, "Monorail's Building, Debt Costs Balloon to $11 billion," *Seattle Post-Intelligencer*, June 21, 2005, https://www.seattlepi.com/local/transportation/article

/Monorail-s-building-debt-costs-balloon-to-11-1176529.php.

10. Kery Murakami, "Monorail Agency Officially Dissolves; Cost Taxpayers $125 Million," *Seattle Post-Intelligencer*, January 17, 2008, https://www.seattlepi.com/local /transportation/article/Monorail-agency-officially-dissolves -cost-1262011.php.

11. Sound Transit, *Central Link: Initial Segment & Airport Link Before & After Study* (Seattle, 2014).

12. U.S. Congress Office of Technology Assessment, *An Assessment of Community Planning for Mass Transit Volume 9: Seattle Case Study* (Washington, DC, 1976), 18.

13. Sound Transit, *System Expansion Project Timelines July 2019* (Seattle, 2019), https://www.soundtransit.org/sites /default/files/documents/system-expansion-project-timelines -201907.pdf.

14. Mike Lindblom and Michelle Baruchman, "Sound Transit Faces a $6.5 Billion Shortfall: Here's What It Might Do," *Seattle Times*, August 16, 2021, https://www.seattletimes .com/seattle-news/transportation/sound-transit-faces-a-6-5 -billion-shortfall-heres-what-it-might-do/.

Chapter 21

1. Charles M. Jacobs and J. Vipond Davies, *Report of Messrs. Jacobs & Davies on Street Railway Transportation in the City of Toronto* (Toronto: City of Toronto, 1910).

2. Municipality of Toronto and Toronto Transit Commission, *Network 2011: A Rapid Transit Plan for Toronto* (Toronto, 1986). The final report is dated 1986, as it was published after the Tories' 1985 election loss.

3. Charles Trueheart, "Lonely Party on the Canadian Left," *Washington Post*, October 22, 1995.

4. Daniel Silver, Zack Taylor, and Fernando Calderon-Figueroa, "Populism in the City: The Case of Ford Nation," *International Journal of Politics, Culture, and Society* 33 (2020): 2, https://doi.org/10.1007/s10767-018-9310-1.

5. Yonah Freemark, "Hazy Future for Transit City as Toronto Gears Up for Mayoral Election," *Transport Politic*, May 10, 2010, https://www.thetransportpolitic.com/2010 /05/10/hazy-future-for-transit-city-as-toronto-gears-up-for -mayoral-election/.

6. Yonah Freemark, "When Voting for the Lesser of Two

Evils Could Save a Transit System," *Transport Politic*, October 23, 2010, https://www.thetransportpolitic.com/2010 /10/23/when-voting-for-the-lesser-of-two-evils-could-save -a-transit-system/.

7. "Rob Ford on Drug Use: 'You Name It, I Pretty Well Covered It,'" *CBC News*, July 2, 2014, https://www.cbc.ca /news/canada/toronto/rob-ford-on-drug-use-you-name-it -i-pretty-well-covered-it-1.2693774.

8. "Why Is Toronto Mayor Rob Ford Still Popular?," *BBC News*, November 5, 2013, https://www.bbc.com/news /world-us-canada-24824648.

9. Silver, Taylor, and Calderon-Figueroa, "Populism in the City," 5.

10. Don Peat, "Subways in Suburbs First, Then Downtown Relief Line: Mayor Rob Ford," *Toronto Sun*, October 11, 2013, https://torontosun.com/2013/10/11/subways-in -suburbs-first-then-downtown-relief-line-mayor-rob-ford.

11. "Rob Ford's War on Public Transit," *Toronto Star*, December 30, 2011, https://www.thestar.com/opinion /editorials/2011/12/30/rob_fords_war_on_public_transit.html.

12. Metrolinx, *Approved Changes to the Big Move* (Toronto, 2013), https://www.metrolinx.com/en/regionalplanning /bigmove/The_Big_Move_Approved_Changes_EN.pdf.

13. "Critics Slam Reports of Doug Ford's Changes to Toronto Relief Line Plan," *CBC News*, July 23, 2019, https: //www.cbc.ca/news/canada/toronto/backlash-relief-line -ford-1.5221685.

14. Ben Spurr, "TTC Hits Pause on Relief Line Work amid Provincial Transit Plans," *Toronto Star*, https: //www.thestar.com/news/gta/2019/06/18/ttc-hits-pause-on -relief-line-work-amid-provincial-transit-plans.html; Nicholas Hune-Brown, "The $11-Billion Subway War," *Toronto Life*, January 19, 2022, https://torontolife.com/city /ontario-line-toronto-ttc-11-billion-subway-war/.

Chapter 22

1. UITP (International Association of Public Transport), *World Report on Metro Automation 2018* (Brussels, Belgium: 2018), 2, https://cms.uitp.org/wp/wp-content/uploads/2020 /06/Statistics-Brief-Metro-automation_final_web03.pdf.

2. Elizabeth Arens, "The Elevated Railroad Cases:

261

Private Property and Mass Transit in Gilded Age New York," *NYU Annual Survey of American Law* 61 (2006): 629.

3. Frank H. Mackintosh, "Elevated Railroad Land-Damage Litigation," *Yale Law Journal* 2, no. 3 (February 1893): 106.

4. Mackintosh, "Elevated Railroad Land-Damage Litigation," 115.

5. Tim Stelloh, "Judge Approves $626 Million Settlement in Flint Water Crisis," *NBC News,* November 10, 2021, https://www.nbcnews.com/news/us-news/judge-approves-626-million-settlement-flint-water-crisis-rcna5183.

6. Kenneth Chan, "With Much of 'SkyTrain' Running Underground, Should It Be Renamed?," *Daily Hive,* February 6, 2021, https://dailyhive.com/vancouver/translink-ceo-kevin-desmond-vision.

7. Harry Rankin, *Beat the Traffic Rush: The Case for Rapid Transit* (Vancouver: Committee of Progressive Electors, 1970).

8. ICTS was its original name. Through the years, it has been rebranded multiple times. Its various names have included Advanced Light Rapid Transit, Advanced Rapid Transit, and now Innovia Metro.

9. Cited in Jessica Stutt, "Planning the Expo Line: Understanding the Technology Choice Behind Vancouver's First Rail Rapid Transit Line" (master's thesis, Simon Fraser University, 2011), 47.

10. De Leuw, Cather & Co. of Canada, *Greater Vancouver Area Rapid Transit Study* (Vancouver: Joint Transportation Committee, Greater Vancouver Regional District and British Columbia Power & Hydro Authority, 1970), iii.

11. C$1.5 billion in 1980.

12. John Kirkwood, "Is Vancouver Ready for Snafu 86?" *Vancouver Sun,* May 7, 1980.

13. Stutt, "Planning the Expo Line," 55.

14. Some sources suggest that Bennett conditioned the purchase of ICTS on Ontario having a working line using the ICTS technology, but I have not been able to find supporting primary sources for this.

15. Stutt, 71.

Chapter 23

1. "Why Not a Real Subway System for Washington?," *Washington Post,* December 5, 1909.

2. James C. McKay Jr., "Separation of Powers in the District of Columbia under Home Rule," *Catholic University Law Review* 27, no. 3 (1977): 515.

3. Freeway plans necessarily varied over time; the version I've drawn is derived from De Leuw, Cather Associates, *District of Columbia Interstate System 1971* (Washington, DC: District of Columbia Dept of Highways and Traffic, 1971).

4. Richard F. Weingroff, *The D.C. Freeway Revolt and the Coming of Metro* (Washington, DC: U.S. Department of Transportation, 2019), https://www.fhwa.dot.gov/highwayhistory/dcrevolt/.

5. National Capital Transportation Agency, *Transportation in the National Capital Region: Finance and Organization* (Washington, DC, 1962).

6. Howard Gillette, "A National Workshop for Urban Policy: The Metropolitanization of Washington, 1946–1968," *Public Historian* 7, no. 1 (1985): 21, https://doi.org/10.2307/3377297.

7. Henry Jaffe, "The Insane Highway Plan That Would Have Bulldozed DC's Most Charming Neighborhoods," *Washingtonian,* October 21, 2015, https://www.washingtonian.com/2015/10/21/the-insane-highway-plan-that-would-have-bulldozed-washington-dcs-most-charming-neighborhoods/.

8. Walterene Swanson and Ronald Sarro, "District of Columbia Council Vote Follows Wild Melee," *Washington Star,* August 10, 1969.

9. Swanson and Sarro.

10. George Pipkin, "Georgetown Students Jailed in Three Sisters Incidents," *The Hoya,* October 23, 1969.

11. "Arson Probed in Bridge Site Blaze," *Washington Star,* November 3, 1969.

12. Jack Eisen, "House Releases District Subway Funds: Leadership Rebuffed, 195 to 174," *Washington Post,* December 3, 1971.

13. Stephen Repetski, "While Still Working on Train Automation, Metro Makes Moves toward a New Signaling System," *Greater Washington,* September 9, 2021, https://ggwash.org/view/82467/while-still-working-on-train-automation-metro-makes-moves-toward-a-new-signaling-system.

Conclusion

1. SB 9 (Cal. 2022).

2. David Lepeska, "Why $1 Billion Doesn't Buy Much Transit Infrastructure Anymore," *Bloomberg*, November 9, 2011, https://www.bloomberg.com/news/articles/2011-11-09/why-1-billion-doesn-t-buy-much-transit-infrastructure-anymore. Madrid Metro Line 12 cost $93 million per mile ($58 million per kilometer). This is a lower per-mile cost than the Green and Purple Lines in Houston, the Southwest LRT in Minneapolis, and the Orange Line in Portland.

3. For example, on October 28, 2022, *Idealista*, a large Spanish real estate listing site, showed 699 ads for new-construction residential units in the province of Madrid. Of those, 71 were for North American-style single-family homes (*chalets independientes*), a little over 10 percent of the total. A Redfin search on the same date in Los Angeles County shows 350 listings for new construction homes; 230 are advertised as detached single-family, or 66 percent of the total.

Further Reading

Baron, Steven M. *Houston Electric: The Street Railways of Houston, Texas.* Self-published, 1996.

Baumbach, Richard, and William Borah. *The Second Battle of New Orleans: A History of the Vieux Carré Riverfront-Expressway Controversy.* Lafayette: University of Louisiana at Lafayette Press, 2019.

Brechin, Gray. *Imperial San Francisco: Urban Power, Earthly Ruin.* Berkeley: University of California Press, 1999.

Diers, John W., and Aaron Isaacs. *Twin Cities by Trolley: The Streetcar Era in Minneapolis and St. Paul.* Minneapolis: University of Minnesota Press, 2007.

DiMento, Joseph F. C., and Cliff Ellis. *Changing Lanes: Visions and Histories of Urban Freeways.* Cambridge, MA: MIT Press, 2012.

Dougherty, Conor. *Golden Gates: Fighting for Housing in America.* New York: Penguin Random House, 2020.

Elkind, Ethan. *Railtown: The Fight for the Los Angeles Metro Rail and the Future of the City.* Berkeley: University of California Press, 2014.

Hinton, George. *The Electric Interurban Railways in America.* Stanford, CA: Stanford University Press, 2000.

Jacobs, Jane. *The Death and Life of Great American Cities.* New York: Random House, 1961.

Karcher, Alan J. *New Jersey's Multiple Municipal Madness.* New Brunswick, NJ: Rutgers University Press, 1998.

Keating, Larry. *Atlanta: Race, Class and Urban Expansion.* Philadelphia: Temple University Press, 2001.

Kieffer, Stephen A. *Transit and the Twins: A Survey of the History of the Transportation Company in Minneapolis and Saint Paul.* Minneapolis: Twin City Rapid Transit Co., 1958.

Lewyn, Michael. *Government Intervention and Suburban Sprawl: The Case for Market Urbanism.* London: Palgrave Macmillan, 2017.

Lockwood, Charles. *Manhattan Moves Uptown: An Illustrated History.* New York: Barnes & Noble, 1995.

Marohn, Charles L., Jr. *Strong Towns: A Bottom-Up Revolution to Rebuild American Prosperity.* Hoboken, NJ: John Wiley & Sons, 2020.

Mecklenborg, Jacob. *Cincinnati's Incomplete Subway: The Complete History.* Charleston, SC: History Press, 2010.

Morris, John E. *Subway: The Curiosities, Secrets, and Unofficial History of the New York City Transit System.* New York: Black Dog & Leventhal, 2020.

Ovenden, Mark. *Transit Maps of the World: The World's First Collection of Every Urban Train Map on Earth.* New York: Penguin, 2015.

Parolek, Daniel, and Arthur C. Nelson. *Missing Middle Housing: Thinking Big and Building Small to Respond to Today's Housing Crisis.* Washington, DC: Island Press, 2020.

Reardon, Patrick T. *The Loop: The 'L' Tracks That Shaped and Saved Chicago.* Carbondale: Southern Illinois University Press, 2020.

Rothstein, Richard. *The Color of Law: A Forgotten History of How Our Government Segregated America.* New York: W. W. Norton, 2017.

Safford, Sean. *Why the Garden Club Couldn't Save Youngstown: The Transformation of the Rust Belt.* Cambridge, MA: Harvard University Press, 2009.

Schrag, Zachary M. *The Great Society Subway: A History of the Washington Metro.* Baltimore: Johns Hopkins University Press, 2014.

Shoup, Donald. *The High Cost of Free Parking*. Milton Park, UK: Routledge, 2011.

Spieler, Christof. *Trains, Buses, People: An Opinionated Atlas of U.S. Transit*. Washington, DC: Island Press, 2018.

Sugrue, Thomas. *The Origins of the Urban Crisis: Race and Inequality in Postwar Detroit*. Princeton, NJ: Princeton University Press, 2014.

Violette, Zachary J. *The Decorated Tenement: How Immigrant Builders and Architects Transformed the Slum in the Gilded Age*. Minneapolis: University of Minnesota Press, 2019.

Walker, Jarrett. *Human Transit: How Clearer Thinking about Public Transit Can Enrich Our Communities and Our Lives*. Washington, DC: Island Press, 2011.

Weaver, Russell, Sharmistha Bagchi-Sen, Jason Knight, and Amy E. Frazier. *Shrinking Cities: Understanding Urban Decline in the United States*. Milton Park, UK: Routledge, 2016.

Weingroff, Richard F. *The D.C. Freeway Revolt and the Coming of Metro*. Washington, DC: Federal Highway Administration, 2019.

Archives Used

Boston Public Library

City Archives & Special Collections, New Orleans Public Library

City of Portland Archives and Records Center

David Rumsey Historical Map Collection

Georgia Department of Transportation

HathiTrust Digital Library

Houston Metropolitan Transit Authority

Internet Archive

Los Angeles Metro, Dorothy Peyton Grey Transportation Library and Archive

King County, Washington Archives

New York Public Library

New York University Library

Metro Vancouver Library

Michigan State University Library

Philadelphia Free Library

San Francisco Municipal Transportation Agency Photo Department and Archive

Seattle Municipal Archives

Southwest Ohio Regional Transit Authority

University of Alabama Library

University of California, Berkeley Library

University of Chicago Library

Texas General Land Office

Texas State Library and Archives

Index

Page numbers in italics refer to maps.

Allen, Fredrick Lewis, 105
Allen, Ivan, 10, 13
American Institute of Architects, 21
Amtrak, 55–56
Atlanta, Georgia, 6–15; business culture and
  population growth compared to New
  Orleans, 137, 139; early MARTA system
  plans, *9*; map of, *6*; proposed MARTA
  rapid transit system (1968), *12*; proposed
  rapid transit system (1962), *11*; subway
  system (2022), *15*
Atlanta Transit System, 13

Baltimore, Maryland, 136, 232
BART (Bay Area Rapid Transit), San Fran-
  cisco Bay Area, 2, 53, 94, 185, 188, *190*, 191,
  192, 193, *195*, 223, 225
Beame, Abraham, 149
Beebe, James, 93
Bennett, Bill, 227
Bingham, S. H., 137
Blumenfeld, Isadore, alias Kid Cann, 116,
  118–19
BMT (Brooklyn-Manhattan Transit), 147,
  149
Bohannon, David, 191
Boston, Massachusetts, 16–26; the Big Dig,
  23; Boston Elevated Railway (1925), *19*;
  Coolidge Commission transit expansion
  proposal (1945), *20*; distinct types of light
  rail cars used on the Green Line, 225; electric streetcars placed underground, 179;
  map of, *16*; MBTA rapid transit service
  (2023), *26*; MBTA subways and light rail
  (1967), *22*; MBTA subways and light rail
  (2009), *25*; shift from elevated to subway
  construction in, 222–23; streetcars updated
  to light rail standards in, 223; waterfront
  development in, 49
Boston Elevated Railway, 18, *19*
Bowron, Fletcher, 93–94
British Columbia Electric Railway, 223
Brooklyn-Manhattan Transit. *See* BMT
Brooklyn Rapid Transit Company, 144, 147
Brown, Willie, 163
Browns Stadium (FirstEnergy Stadium),
  Cleveland, 55
Bryan, William Jennings, 105
Buffalo, New York, 225
Burke Lakefront Airport, Cleveland, 56–57
Burton, Richard, 96
Bush, George H. W., 80, 137
bus rapid transit. *See* busways
busways: in Atlanta, 13; in Boston, 18, 23–24,
  25–26; in Chicago, 35; in Cleveland, 53;
  components of well-designed, 23–24;
  in Dallas, 63; in Detroit, 68, 72, 75, 77;
  in Houston, 80; in Los Angeles, 87,
  89, 91, 96, 99, *100*; in Miami, 111; in
  Minneapolis–St. Paul, 116, 119, 121; in
  Philadelphia, 157, 159; in Pittsburgh, 169,
  171–72, *174*, *175*; in Richmond, 179; in San
Francisco, 186, 188, 191, 193–94; in Seattle,
  200, 203; as a technology for mass transit,
  4; in Toronto, 215–16
Byford, Andy, 152–53, 216
Byrne, Jane, 35

cable cars: in San Francisco, 177, 185–86,
  188, 191; in Seattle, 199–200, *201*; as a
  technology for mass transit, 4
Calgary, Alberta, Canada, 225
California Environmental Quality Act. *See*
  CEQA
Capital Transit Company, Washington, DC,
  232
Capitol Subway, Washington, DC, 232
Castro, Fidel, 107–8
CEQA (California Environmental Quality
  Act), 99, 191, 193–94
Chesebro, Ray, 93
Chicago, Illinois, 28–40; Chicago Transit
  Authority Circle Line proposal (2002),
  37; Chicago Transit Authority L system
  (2022), *39*; elevated Loop, survival of, 223;
  elevated railroads, 1898 service guide, *32*;
  elevated system (1921), *33*; expressways
  routed through black neighborhoods
  in, 10; map of, *28*; proposed central area
  circulator (1994), *36*; proposed subways
  (1976), *34*; skyscrapers built in, 89; transit
  union strikes in, 158
Chicago Transit Authority. *See* CTA

Cincinnati, Ohio, 42–47; electric interurban railways, 1912, *44*; job density of central, 171; map of, *42*; unfinished subway (1929), *46*

Cincinnati Bengals, 47

cities: remaking with urban freeways, 235; types of, 2

Cleveland, Ohio, 48–57; interurban electric railways (1898), *51*; job density of central, 171; map of, *48*; RTA Rapid Transit (2022), *54*; subway proposal (1955), *52*

Cleveland Browns, FirstEnergy Stadium, 55

Committee for Sensible Rapid Transit, Atlanta, 13

commuter rail. *See* regional rail

Coolidge, Arthur, 18

Coolidge Commission, Boston, 21

cost of rapid transit systems. *See* financing rapid transit systems

Couzens, James, 71

Cox, George "Boss," 45

CTA (Chicago Transit Authority), 35, *37*, 38, 40

Cuomo, Andrew, 152–53

Daley, Richard M., 35

Dallas, Texas, 58–65; Bishop Arts District, 60, 63; Dallas Railway streetcars and connecting interurban service (1919), *61*; DART light rail system (2022), *62*; Houston and, comparison of light rail in, 83; map of, *58*

Dallas Area Rapid Transit. *See* DART

Dallas Railway, 60

DART (Dallas Area Rapid Transit), 59, *62*, 63

Davies, J. Vipond, 211

Davis, Bill, 211

Detroit, Michigan, 66–77; Detroit People Mover and QLine streetcar (2022), *76*; Detroit United Railway streetcar and interurban lines, 1905, *69*; expressways routed through black neighborhoods in, 10; map of, *66*; people mover in, 53; population loss in, 21; subway proposal (1918), *70*; subway proposal (1974), *73*

Detroit Institute of Arts, 67

Detroit People Mover, 68, 72, 75, *76*, 108

Detroit United Railway (DUR), 68, *69*, 71

Doheny, Edward, 88

Drapeau, Jean, 123–24, 127–28, 131

Du Bois, W. E. B., 7

Duplessis, Maurice, 124, 127

DUR. *See* Detroit United Railway

Eastman Kodak, 182

Eaton, Fred, 88

economic development: in Atlanta, 7–8, 10, 137, 139; in Boston, 17–18, 21, 23; in Chicago, 30–31, 35, 38; in Cleveland, 49–50; in Dallas, 63; in Detroit, 67–68, 71–72; in Houston, 80, 82, 137, 139; in Los Angeles, 88–89; in Miami, 104–5, 107; in Montreal, 127; in New Orleans, 133–34, 136–37, 139; in New York, 144, 149, 152; in Pittsburgh, 171. *See also* real estate development

Edmonton, Alberta, Canada, 225

Electric Railroaders Association of Southern California, 1

elevateds. *See* subway/elevated/metro

Emeryville, California, 5

Erie Canal, 181

Ernst & Young (accounting firm), 53

ESPN, 53

Expo 86, 225, 227

expressways. *See* freeways/expressways/turnpikes

Falkenbury, Dick, 203, 205

Faulkner, William, 14

Federal-Aid Highway Act of 1956, 71, 235

Federal Housing Administration (FHA), 71

Fenton, Reuben, 144

Fimia, Maggi, 203

financing rapid transit systems: in Atlanta, 13–14, 200; in Boston, 23–24; in Chicago, 35; in Cincinnati, 45, 47; in Cleveland, 53; in Detroit, 72, 75; in Los Angeles, 91, 94; in Miami, 108, 111; in Montreal, 123; in New Orleans, 137; in New York, 143, 147, 149, 152–53; in Philadelphia, 159; in Pittsburgh, 172; in Rochester, 182; in San Francisco, 186, 193–94; in Seattle, 200, 203, 205, 207; in Washington, DC, 239

Fisher, Carl, 104–5, 107

Flagler, Henry, 103–4

Flaherty, Pete, 172

Flint, Michigan, 222

Florida East Coast Railway, 104–5

Ford, Doug, 216, 219

Ford, Gerald, 75, 149

Ford, Rob, 216

Ford, Tirey, 186

Ford Motor Company, 67

Forward Thrust Committee, Seattle, 200, 207

freeways/expressways/turnpikes: in American cities, 231; in Atlanta, 8, 10, 14; in Boston, 18, 21, 23; in Chicago, 35; in Cleveland, 50; in Detroit, 71; in Houston, 79–80, 82; induced demand and, 93; in Los Angeles, 1–2, 87–88, 91, 93–94, 241; in Miami, 103–4, 107, 111; nationwide, 235; in New Orleans, 136; in Rochester, 182; in San Francisco, 188, 191; in San Francisco, the Freeway Revolt, 185, 188, 191, 196; as a technology for mass transit, 4; in Toronto, 215; in Vancouver, 223; in Washington, DC, 231–32, 235, 237, *238*, 239

Frito-Lay, 63

Galveston, Texas, 80

Georgia Power, 7–8, 13

Goetz, Bernie, 152

Goodrich, Calvin, 116

Grady, Henry, 8
Green, Charles, 116, 118–19
Gribbs, Roman, 74

Harcourt, Mike, 227
Harris, Mike, 215
Haslam, Jimmy, 57
Hébert, Napoléon, 124
Hocken, Horatio, 211
Hofheinz, Fred, 80, 82
Hopkins International Airport, Cleveland, 56
horsecars: in Boston, 18; last of the, 179; in
    New Orleans, 134; in Philadelphia, 158–59,
    160; in Richmond, 178; as a technology for
    mass transit, 4, 177
Houston, Texas, 78–85; Harris County
    METRORail (2022), 84; Houston City
    Street Railway (1895), 81; map of, 78; New
    Orleans, competition/comparison with,
    137, 139; social culture and population
    growth in, 136–37
Huntington, Henry, 87–89

ICTS. See Intermediate Capacity Transit
    System
Independent Subway (IND), New York, 147,
    149
induced demand, 93
Interborough Rapid Transit (IRT), New York,
    147, 149, 153
Intermediate Capacity Transit System (ICTS),
    225, 227, 229
Interurban Railway & Terminal Company, 45
interurbans: in Cincinnati, 43, 44, 45; in
    Cleveland, 50, 51; in Dallas, 60, 63; in De-
    troit, 68, 69; in Los Angeles, 87–89, 90, 91,
    92, 93; in Rochester, 182; in San Francisco,
    188; as a technology for mass transit, 3; in
    Vancouver, 223
IRT. See Interborough Rapid Transit

Jacksonville, Florida, 53, 108
Jacobs, Charles M., 211
Jarvis, Howard, 98
JC Penney, 63
Johnson, Lyndon, 72, 235

Kucinich, Dennis, 53

Lafayette, California, 5
land use law: changes in, 241; in Los Angeles,
    88–89, 96, 98–99; in Miami, 111; transit
    policy and, 4–5, 79. See also real estate
    development
Levittown, New York, 5
Lewis, Michael, 137
light rail: best practices for, 139; in Boston,
    23–24, 25–26, 223, 225; in Chicago, 35, 36;
    in Cincinnati, 47; in Cleveland, 49–50,
    53, 54, 55–57; in Dallas, 59, 62, 63–65,
    83; in Houston, 80, 82–83, 84, 85, 139; in
    Los Angeles, 93, 99; in Minneapolis–St.
    Paul, 119, 120, 121; in Pittsburgh, 172, 174,
    175; in San Francisco, 194, 195; in Seattle,
    203, 205, 206, 207; as a technology for
    mass transit, 3, 225; in Toronto, 215–16; in
    Vancouver, 225
London, England, 3, 31, 153, 235
London Underground, 31, 91, 144, 153
Los Angeles, California, 86–101; Baxter Ward
    subway proposal (1976), 97; car ownership
    in, 89; elevated system rejected in, 222;
    freeway system and traffic in, 1–2, 87, 241;
    Hollywood Park, 57; housing availability
    and cost in, 83, 85, 88, 98–99, 101; Los An-
    geles Metro (2023), 100; map of, 86; Rail
    Rapid Transit Now proposal (1948), 92;
    Red Cars, 1–3, 87–89, 90, 91, 93–94, 241;
    RTD subway proposal (1968), 95; Seventh
    Street station, daily passenger usage at, 111;
    spacing of stations on the B Line subway,

38; subways entirely underground in, 223
Lowry, Horace, 116
Lowry, Thomas, 115–16

Mackintosh, Frank, 222
Maddox, Lester, 10, 13–14
Madrid, Spain, 235, 242
maglev, 211
maps, drawing of, 2
Market Street Railway, San Francisco, 2,
    185–86, 188
MARTA (Metropolitan Atlanta Rapid Transit
    Authority), 7–8, 9, 10, 12, 13–14, 15, 200
Massachusetts Bay Transportation Authority.
    See MBTA
mass transit/public transportation systems:
    past and future of, 241–42; real estate
    development and (see real estate develop-
    ment); types of, 3–4
MBTA (Massachusetts Bay Transportation
    Authority), 18, 21, 22, 24, 25–26, 153
McKeithen, John J., 137
McKenzie, Troy, 193
McKinsey (consulting firm), 53
Merrick, George, 105, 107
metro. See subway/elevated/metro
Metrolinx, Toronto, 215–16
Metropolitan Atlanta Rapid Transit Authority.
    See MARTA
Metropolitan Transit Authority, Los Angeles,
    94
Metropolitan Transportation Authority, New
    York. See MTA
Miami, Florida, 102–13; early streetcar system
    (1926), 106; elevated system required in,
    223; map of, 102; Metromover in, 53, 108;
    Metrorail and connecting services (2022),
    112; proposed Metrorail system (1979),
    109; proposed People's Transportation Plan
    (2002), 110

Miller, David, 215
Minneapolis–St. Paul, Minnesota, 114–21; map of, *114*; Metro Transit (2022), *120*; streetcar network (1948), *117*
Monorail Inc., 137
monorails: demonstration track at the Texas State Fair, 137; proposed in Detroit, 72; proposed in Los Angeles, 94; proposed in Miami, 111, 113; proposed in Montreal, 127; proposed in New Orleans, 137, *138*; proposed in Seattle, 203, *204*, 205; proposed in Washington, DC, 232; in Seattle, 200; as a technology for mass transit, 4
Montreal, Quebec, Canada, 122–31; map of, *122*; Metro (2023), *130*; metro system proposal (1944), *126*; metro system proposal (1953), *129*; Montreal Metro, construction of, 242; Montreal Metro and Toronto Subway, comparisons of, 128; Montreal Metro entirely underground, 223; Montreal Tramways Company streetcar network, *125*; rubber-tired metro introduced in, 223
Montreal Summer Olympics, 131
Montreal Tramways Company, 124, *125*
Montreal Transportation Commission, 124
Morrison, deLesseps "Chep," 134, 137
MTA (Metropolitan Transportation Authority), New York, 24, 144, 149, 152–53
Mulholland, William, 88
Muni. *See* San Francisco Municipal Railway

Natcher, William, 235, 237, 239
National Capital Transportation Agency (NCTA), Washington, DC, 235, *236*
National Cordage Company, 30
NCTA. *See* National Capital Transportation Agency
Neiman Marcus, Dallas, 60
Newark, New Jersey, 45
New Orleans, Louisiana, 132–41; Atlanta,

comparison with, 137, 139; economy of, 80; Houston, competition/comparison with, 137, 139; map of, *132*; monorail proposed routes, *138*; as the South's largest city, 8; streetcar network (1945), *135*; streetcar network (2022), *140*
New Orleans Public Service, Inc. (NOPSI), 134, 136, 139
New Orleans Regional Transit Authority (RTA), 139, 141
New York City, New York, 142–55; Brooklyn Rapid Transit lines (1912), *148*; elevated lines of Manhattan with connecting streetcars and cable cars (1899), *146*; elevateds in Manhattan, 222; first subway line in, 242; map of, *142*; Metropolitan Transportation Authority Program for Action (1968), *151*; shift from elevated to subway construction in, 222–23; skyscrapers built in, 89; subway (2022), *154*; subway and elevated lines (1939), *150*; subway fares in, 91; subway ridership decrease (1946–50), 118; transit union strikes in, 158; waterfront development in, 49; Willson's Metropolitan Railroad proposal (1865), *145*; zoning policy in, 5
New York State Barge Canal, 181
Nichols, John, 74
Nickels, Greg, 205
NIMBYs, San Francisco, 193
Nixon, Richard, 237, 239
NOPSI. *See* New Orleans Public Service, Inc.

Ohio Electric Railway, 45
Only Yesterday (Allen), 105
Ossana, Fred, 116, 118–19

Pacific Electric Railway, Los Angeles, 1–2, 87–89, *90*, 91, 93, 241
Pacific Railway Society, 1

Pantalone, Joe, 216
Paris, France, 235
PAT (Port Authority Transit), Pittsburgh, 171–72, *174*, 175
Patterson, L. Brooks, 68, 75
Pennsylvania Railroad, 159
people movers: in Detroit, 53, 68, 72, 75, *76*; in Jacksonville, 53; in Miami, 53, 108, 111; proposed in Cleveland, 53; proposed in Los Angeles, 57; proposed in Pittsburgh, 171–72, *173*; as a technology for mass transit, 4
People's Transportation Plan, Miami, 108, *110*
Peterson, David, 215
Philadelphia, Pennsylvania, 156–66; horsecars and urban steam railroads (1875), *160*; labor issues/transit union strikes in, 158–59, 163, 166; map of, *156*; Philadelphia Transportation Co. and connecting services (1940), *161*; rowhouses in, 83; SEPTA rapid transit (1974), *162*; SEPTA rapid transit and regional rail (2022), *165*; shift from elevated to subway construction in, 222–23; Vuchic plan for commuter tunnel service (1984), *164*
Philadelphia & Reading Railroad. *See* Reading Railroad
Philadelphia Transportation Company (PTC), 159, 163
Pittsburgh, Pennsylvania, 168–75; job density of central, 171; map of, *168*; Pittsburgh Railways trolleys (1954), *170*; Port Authority Transit light rail and busways (2022), *174*; Skybus proposal (1967), *173*
Pittsburgh Railways, 169, *170*, 171
pod cars, 4, 21, 119
Port Authority Transit, Pittsburgh. *See* PAT
Porter, Albert, 50
Portland, Oregon, 225
Proposition 13, California, 98, 193

PTC. *See* Philadelphia Transportation Company

race and race relations: in Atlanta, 7–8, 10, 13–14; in Boston, 21, 23; in Chicago, 30, 38; in Detroit, 68, 71–72, 74–75, 159; in Miami, 107; in New Orleans, 136; in New York, 152; in Philadelphia, 159; in Richmond, 179; in Washington, DC, 237

Rae, Bob, 215

Rapid Transit Action Group (RTAG), Los Angeles, 93

Rapid Transit Commission, Cincinnati, 45

Reading Railroad, 30, 159

Reagan, Ronald, 108, 191

real estate development: in Atlanta, 10; in Cleveland, 49–50, 53, 55–57; in Dallas, 59–60, 63–65; in Detroit, redlining and, 71; in Houston, 79–80, 82–83, 85; land use regulation and zoning laws, 4–5; in Los Angeles, 1, 57, 87–89, 93–94; in Miami, 103, 105, 107, 111; in Montreal, 128; in North America compared to Europe, 242; in Rochester, 182; in San Francisco, 55, 191, 193–94; zoning laws and, 64–65. *See also* economic development; land use law

Red Arrow Lines, Philadelphia, 159

regional rail: best practices for a modern system, 158, 163; in Boston, 22, 23; in Detroit, 75; in Philadelphia, 23, 157–59, *160*, *162*, 163, *164–65*, 166; as a technology for mass transit, 4. *See also* interurbans; light rail

Regional Transit Authority, Detroit, 75

Richmond, Virginia, 176–79; map of, *176*; street railway lines (1891), *178*

Richmond Union Passenger Railway, 177, 179

Rochester, New York, 180–83; canal-to-subway conversion in, 45; map of, *180*; subway and connecting streetcars (1929), *183*

Rochester Subway, 2, 182

Roosevelt, Franklin D., 163

RTA. *See* New Orleans Regional Transit Authority

RTAG. *See* Rapid Transit Action Group

RTD (Southern California Rapid Transit District), 94, 96

Sacramento, California, 194

San Diego, California, 225

San Fernando Mission Land Company, 88–89

San Francisco Bay Area, California, 184–96; BART and connecting services (2022), *195*; BART system crash, 53; cable car system (1892), *187*; early BART system plans (1960), *190*; elevated sections of BART, 223; expressways routed through black neighborhoods in, 10; gentrification in, 60; housing/homelessness and rowhouses in, 83, 185–86, 193–94; map of, *184*; Muni Rapid proposal (1966), *192*; public transit in, 2; streetcar lines (1929), *189*; waterfront development in, 49–50

San Francisco Giants, Oracle Park, 55

San Francisco Municipal Railway (Muni), 2, 185–86, 188, 191, 194, 196

San Juan, Puerto Rico, 182

Santa Fe Railroad, 88

Sargent, Francis, 21, 23

Saulnier, Lucien, 127–28

Savannah, Georgia, 8

Schmitz, Eugene, 186

Seasongood, Murray, 45

Seattle, Washington, 198–207; early streetcar and cable car system (1896), *201*; Forward Thrust subway proposal (1970), *202*; Link light rail (2022), *206*; map of, *198*; monorail proposal (1997), *204*

SEMTA (Southeastern Michigan Transportation Authority), 72, 75

SEPTA (Southeastern Pennsylvania Transportation Authority), 157–59, *162*, 163, *164–65*, 166

Sherman, William Tecumseh, 8

Skybus, Pittsburgh, 171–72, *173*

Smitherman, George, 216

Société de Transport de Montréal, 124

Soo Line Railroad, 116

SORTA (Southeast Ohio Regional Transit Authority), 47

Sound Transit, Seattle, 203, 205, 207

Southeastern Michigan Transportation Authority. *See* SEMTA

Southeastern Pennsylvania Transportation Authority. *See* SEPTA

Southeast Ohio Regional Transit Authority. *See* SORTA

Southern California Rapid Transit District. *See* RTD

Southern Methodist University (SMU), 63

Sprague, Frank J., 177, 179

Stern, Robert A. M., 21

St. Louis, Missouri, 171

Stolzenbach, Darwin, 235, 239

St. Paul, Minnesota. *See* Minneapolis–St. Paul, Minnesota

streetcars/trolleys: in Atlanta, 7–8, *9*, *15*; in Boston, 18, 179, 223; in Chicago, 30; in Cincinnati, 43, 45, 47; in Cleveland, 50; conspiracy against, 115; in Dallas, 59–60, *62*, 63; in Detroit, 69, 71, 75, 76; in Houston, 80, *81*; in Los Angeles, 91; in Miami, 103, 105, *106*, 107; in Minneapolis–St. Paul, 115–16, *117*, 118–19, 121; in Montreal, 124, *125*; in New Orleans, 133–34, *135*, 136, 139, *140*, 141; in New York, *146*; in Philadelphia, 157, 159, *161–62*; in Pittsburgh, 169, *170*, 171; in Richmond, 177, *178*, 179; in Rochester, *183*; in San Francisco, 119, 185–86, 188, *189*, 191, 196; in Seattle, 199–200,

271

201; as a technology for mass transit, 3; in Toronto, 209, 211, *212*, *213*; in Vancouver, 223, *224*; in Washington, DC, 232, *233*

STRESS (Stop the Robberies, Enjoy Safe Streets), Detroit, 74

Strouse, D. J., 116, 118

subway/elevated/metro: in Atlanta, 8, 10, *12*, 13–14, *15*, 200; in Boston, 18, 21, 23, *25–26*, 222–23; in Chicago, 29–31, *32–34*, 35, *37*, 38, *39*, 40, 223; in Cincinnati, 43, 45, *46*, 47; in Cleveland, 50, *52*, 53; in Detroit, 68, *70*, 71–72, *73*, *75*; elevateds, advantages and disadvantages of, 221; elevateds, history of, 222–23; Intermediate Capacity Transit System (ICTS), 225, 227; in Los Angeles, 94, *95*, 96, *97*, 99, *100*, 222–23; in Miami, 104, 107–8, *109*, *110*, 111, *112*, 113, 223; in Minneapolis–St. Paul (proposed), 119; in Montreal, 123–24, *126*, 127–28, *129–30*, 131, 223; in New York, 91, 118, 143–44, *145–46*, 147, *148*, 149, *150–51*, 152–53, *154*, 155, 222–23, 241–42; in Philadelphia, 159, *161–62*, 222–23; in Rochester, 181–82, *183*; in San Francisco, 188, 191, *192*, *195*, 196; in Seattle, *202*; as a technology for mass transit, 3; in Toronto, 128, 209, *210*, 211, *213–14*, 215–16, *217–18*, 219; in Vancouver, 221, 223, 227, *228*, 229; in Washington, DC, 231–32, *234*, *236*, *238*, 239

Taylor, Elizabeth, 96

Tebbetts, Merle, 107

Thompson, James "Big Jim," 35

Toronto, Ontario, Canada, 208–19; comparison of proposed Ontario Line and scuttled Relief Line subways (2019), *217*; ICTS forced upon, 227; map of, *208*; metro opened in 1954, 124; Montreal Metro and Toronto Subway, comparisons of, 128;

Network 2011 proposed subways (1985), *214*; proposed subways and underground streetcar routes (1945), *213*; subway (2022), *218*; subway proposal (1910), *210*; TTC streetcars (1932), *212*

Toronto Railway Company, 211

Toronto Transit Commission. *See* TTC

TrafficRelief.com, 108

Transit Oriented Communities program, Los Angeles, 101

Triborough Bridge and Tunnel Authority, New York, 149

trolleybus, 3, 8, 13, 159, 200, 203

trolleys. *See* streetcars/trolleys

TTC (Toronto Transit Commission), 209, 211, 216, 219, 227

turnpikes. *See* freeways/expressways/turnpikes

Tuttle, Julia, 103–4

20th Century Fox, 96

Twin City Rapid Transit Company, Minneapolis–St. Paul, 115–16, *117*, 118–19, 121

Union Labor Party, San Francisco, 186

Union Nationale (National Union Party), Montreal, 124, 127

Urban Transportation Development Corporation (Ontario), 225

Vancouver, British Columbia, Canada, 220–29; map of, *220*; proposed rapid transit system (1970), *226*; Skytrain and connecting services (2022), *228*; spacing of stations on the Millennium Line, 38; streetcar and interurban network (1945), *224*

Villard, Henry S., 107

Virginia Passenger and Power Company, 179

Vuchic, Vukan, 163, *164*, 166

WABCO, 171

Wallace, George, 10

Ward, Baxter, 96

Washington, DC, 230–39; Capital Transit streetcars, *233*; map of, *230*; Metro and unbuilt downtown freeways (2022), *238*; percentage of the Metro underground, 223; proposed downtown subways (1944), *234*; proposed subway and commuter rail lines (1962), *236*

Washington Metropolitan Area Transit Authority. *See* WMATA

Waxman, Henry, 99

Western and Atlantic Railroad, 8

Westinghouse Electric, 171–72

Wheeler-Rayburn Act of 1935, 8

Who Framed Roger Rabbit, v, 88, 115

Wilde, Oscar, 8

Willson, Hugh, 144

WMATA (Washington Metropolitan Area Transit Authority), 235, 239

Wolf, Frank, 35

Yerkes, Charles Tyson, 31

Young, Coleman, 68, 74–75

zoning laws: in California, 241; in Houston, 82–83; impact on urban development of, 4–5; in Los Angeles, 98, 101; in Miami, 111; in Plano/Dallas, 64–65. *See also* real estate development